ADVANCES IN FUNCTIONAL CONTROL
Principles and Industrial Applications
Handbook

ADVANCES IN FUNCTIONAL CONTROL
Principles and Industrial Applications
Handbook

Contributors :
Zhifu Pei, Xingfa Ma,
Pengfei Ding, *et al.*

AURIS REFERENCE LTD.
London, UK

Advances in Functional Control : *Principles and Industrial Applications Handbook*
Contributors : Zhifu Pei, Xingfa Ma, *and* Pengfei Ding, *et al.*

Auris Reference Ltd., UK

www.aurisreference.com
United Kingdom

Advances in Functional Control : *Principles and Industrial Applications Handbook*

ISBN: 978-1-78154-483-9

British Library Cataloguing in Publication Data
A CIP record for this book is available from the British Library

Exclusively distributed by CBS Publishers & Distributors Pvt. Ltd.

Sales & Distribution Rights only for India, Pakistan, Bangladesh, Sri Lanka, Nepal and Bhutan.This book is not to be sold outside these territories.

PREFACE

The demands of the modern economic climate have led to a dramatic increase in the industrial application of model-based predictive control techniques. In fact, apart from PID, predictive control is probably the most popular control approach in use today. The functional control technique was first used to develop a model-based predictive controller that was easy to understand, implement and tune from an instrumentation engineer's perspective. In the forty years since, there have been thousands of successful applications of PFC controllers in a large and diverse group of industries.

This book is intended for technical staff in the process industries, familiar with classical control techniques, who need to take up the challenges posed by today's economic environment; engineering graduate students requiring a background in modern control techniques; and industrial managers who require an overview of the PFC technique with a view to assessing its suitability for use in future projects.

This page left intentionally blank.

CONTENTS

LIST OF CONTRIBUTORS

Zhifu Pei

National Key Laboratory of Industrial Control Technology, Department of Control Science and Engineering, Zhejiang University, Hangzhou 310027, China

Xingfa Ma

State Key Laboratory of Silicon Materials, Zhejiang University, Hangzhou 310027, China

and

School of Environmental and Material Engineering, Center of Advanced Functional Materials, Yantai University, Yantai 264005, China

Pengfei Ding

National Key Laboratory of Industrial Control Technology, Department of Control Science and Engineering, Zhejiang University, Hangzhou 310027, China

Wuming Zhang

National Key Laboratory of Industrial Control Technology, Department of Control Science and Engineering, Zhejiang University, Hangzhou 310027, China

Zhiyuan Luo

Computer Learning Research Centre, Royal Holloway, University of London, Egham, Surrey TW20 0EX, UK

Guang Li

National Key Laboratory of Industrial Control Technology, Department of Control Science and Engineering, Zhejiang University, Hangzhou 310027, China

This page left intentionally blank.

Chapter 1

CONTROL ENGINEERING

Control engineering or control systems engineering is the *engineering* discipline that applies *control theory* to design systems with desired behaviors. The practice uses *sensors* to measure the output performance of the device being controlled and those measurements can be used to give *feedback* to the input *actuators* that can make corrections toward desired performance. When a device is designed to perform without the need of human inputs for correction it is called *automatic control* (such as *cruise control* for regulating a car's speed). *Multi–disciplinary* in nature, control systems engineering activities focus on implementation of control systems mainly derived by *mathematical modeling* of *systems* of a diverse range.

OVERVIEW

Modern day control engineering (also called control systems engineering) is a relatively new field of study that gained a significant attention during 20th century with the advancement in technology. It can be broadly defined or classified as practical application of *control theory*. Control engineering has an essential role in a wide range of control systems, from simple household washing machines to high–performance *F–16* fighter aircraft. It seeks to understand physical systems, using mathematical modeling, in terms of inputs, outputs and various components with different behaviors; use control systems design tools to develop *controllers* for those systems; and implement controllers in physical systems employing available technology. A *system* can be *mechanical, electrical, fluid, chemical, financial* and even *biological*, and the mathematical modeling, analysis and controller design uses *control theory* in one or many of the *time, frequency* and *complex–s* domains, depending on the nature of the design problem.

HISTORY

Automatic control systems were first developed over two thousand years ago. The first feedback control device on record is thought to be the ancient *Ktesibios's water clock* in *Alexandria*, Egypt around the third century B.C. It kept time by regulating the water level in a vessel and, therefore, the water flow from that vessel. This certainly was a successful device as water clocks of similar design were still being made in Baghdad when the Mongols *captured* the city in 1258 A.D. A variety of automatic devices have been used over the centuries to accomplish useful tasks or simply to just entertain. The latter includes the automata, popular in Europe in the 17th and 18th centuries, featuring dancing figures that would repeat the same task over and over again; these automata are examples of open–loop control. Milestones among feedback, or "closed–loop" automatic control devices, include the temperature regulator of a furnace attributed to Drebbel, circa 1620, and the centrifugal flyball governor used for regulating the speed of steam engines by James Watt in 1788.

In his 1868 paper "On Governors", J. C. Maxwell (who discovered the Maxwell electromagnetic field equations) was able to explain instabilities exhibited by the flyball governor using differential equations to describe the control system. This demonstrated the importance and usefulness of mathematical models and methods in understanding complex phenomena, and signaled the beginning of mathematical control and systems theory. Elements of control theory had appeared earlier but not as dramatically and convincingly as in Maxwell's analysis.

Control theory made significant strides in the next 100 years. New mathematical techniques made it possible to control, more accurately, significantly more complex dynamical systems than the original flyball governor. These techniques include developments in optimal control in the 1950s and 1960s, followed by progress in stochastic, robust, adaptive and optimal control methods in the 1970s and 1980s. Applications of control methodology have helped make possible space travel and communication satellites, safer and more efficient aircraft, cleaner auto engines, cleaner and more efficient chemical processes, to mention but a few.

Before it emerged as a unique discipline, control engineering was practiced as a part of mechanical engineering and control theory was studied as a part of *electrical engineering*, since *electrical circuits* can often be easily described using control theory techniques. In the very first control relationships, a current output was represented with a voltage control input. However, not having proper technology to implement electrical control systems, designers left with the option of less efficient and slow responding mechanical systems. A very effective mechanical controller that is still widely used in some hydro plants is the *governor*. Later on, previous to modern *power electronics*, process control systems for industrial applications were devised by mechanical engineers using *pneumatic* and *hydraulic* control devices, many of which are still in use today.

CONTROL SYSTEMS

Control engineering is the engineering *discipline* that focuses on the *modeling* of a diverse range of *dynamic systems* (e.g. *mechanical systems*) and the design of *controllers* that will cause these systems to behave in the desired manner. Although such controllers need not be electrical many are and hence control engineering is often viewed as a subfield of electrical engineering. However, the falling price of microprocessors is making the actual implementation of a control system essentially trivial. As a result, focus is shifting back to the mechanical and process engineering discipline, as intimate knowledge of the physical system being controlled is often desired.

Electrical circuits, digital signal processors and *microcontrollers* can all be used to implement *Control systems*. Control engineering has a wide range of applications from the flight and propulsion systems of *commercial airliners* to the *cruise control* present in many modern *automobiles*.

In most of the cases, control engineers utilize *feedback* when designing *control systems*. This is often accomplished using a *PID controller* system. For example, in an *automobile* with *cruise control* the vehicle's *speed* is continuously monitored and fed back to the system, which adjusts the *motor's torque* accordingly. Where there is regular feedback, *control theory* can be used to determine how the system responds to such feedback. In practically all such systems *stability* is important and control theory can help ensure stability is achieved.

Although feedback is an important aspect of control engineering, control engineers may also work on the control of systems without feedback. This is known as *open loop control*. A classic example of *open loop control* is a *washing machine* that runs through a pre–determined cycle without the use of *sensors*.

CONTROL ENGINEERING EDUCATION

At many universities, control engineering courses are taught in *Electrical* and *Electronic Engineering, Mechatronics Engineering, Mechanical engineering,* and *Aerospace engineering;* in others it is connected to *computer science,* as most control techniques today are implemented through computers, often as *embedded systems* (as in the automotive field). The field of control within *chemical engineering* is often known as *process control*. It deals primarily with the control of variables in a chemical process in a plant. It is taught as part of the undergraduate curriculum of any chemical engineering program, and employs many of the same principles in control engineering. Other engineering disciplines also overlap with control engineering, as it can be applied to any system for which a suitable model can be derived. However, specialised control engineering departments do exist, for example the Department of Automatic Control and Systems Engineering at the University of Sheffield.

Control engineering has diversified applications that include science, finance management, and even human behavior. Students of control engineering may start with a linear control system course dealing with the time and complex–s domain, which requires a thorough background in elementary mathematics and *Laplace transform* (called classical control theory). In linear control, the student does frequency and time domain analysis. *Digital control* and *nonlinear control* courses require *z transformation* and algebra respectively, and could be said to complete a basic control education. From here onwards there are several sub branches.

RECENT ADVANCEMENT

Originally, control engineering was all about continuous systems. Development of computer control tools posed a requirement of discrete control system engineering because the communications between the computer–based digital controller and the physical system are governed by a *computer clock*. The equivalent to *Laplace transform* in the discrete domain is the *z–transform*. Today many of the control systems are computer controlled and they consist of both digital and analog components.

Therefore, at the design stage either digital components are mapped into the continuous domain and the design is carried out in the continuous domain, or analog components are mapped into discrete domain and design is carried out there. The first of these two methods is more commonly encountered in practice because many industrial systems have many continuous systems components, including mechanical, fluid, biological and analog electrical components, with a few digital controllers.

Similarly, the design technique has progressed from paper–and–ruler based manual design to *computer–aided design*, and now to *computer–automated design* (CAutoD), which has been made possible by *evolutionary computation*. CAutoD can be applied not just to tuning a predefined control scheme, but also to controller structure optimisation, system identification and invention of novel control systems, based purely upon a performance requirement, independent of any specific control scheme.

CONTROL THEORY

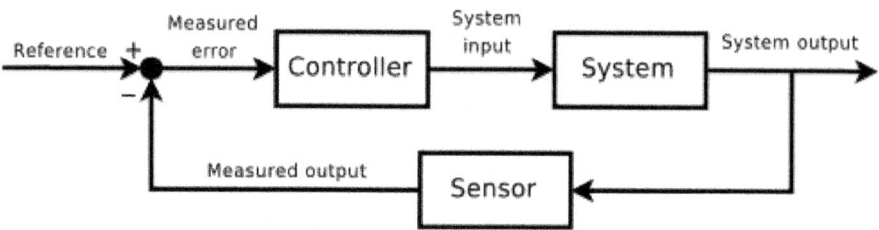

The concept of the feedback loop to control the dynamic behavior of the system : this is negative feedback, because the sensed value is subtracted from the desired value to create the error signal, which is amplified by the controller.

Control theory is an interdisciplinary branch of engineering and *mathematics* that deals with the behavior of *dynamical systems* with inputs. The external input of a system is called the *reference*. When one or more output variables of a system need to follow a certain reference over time, a *controller* manipulates the inputs to a system to obtain the desired effect on the output of the system.

The usual objective of a control theory is to calculate solutions for the proper corrective action from the controller that result in system stability, that is, the system will hold the set point and not oscillate around it.

The inputs and outputs of a continuous control system are generally related by differential equations. If these are linear with constant coefficients, then a transfer function relating the input and output can be obtained by taking their *Laplace transform*. If the differential equations are nonlinear and have a known solution, then it may be possible to linearize the nonlinear differential equations at that solution. If the resulting linear differential equations have constant coefficients, then one can take their Laplace transform to obtain a transfer function.

The *transfer function* is also known as the system function or network function. The transfer function is a mathematical representation, in terms of spatial or temporal frequency, of the relation between the input and output of a linear time-invariant solution of the nonlinear differential equations describing the system.

Extensive use is usually made of a diagrammatic style known as the *block diagram*.

Overview

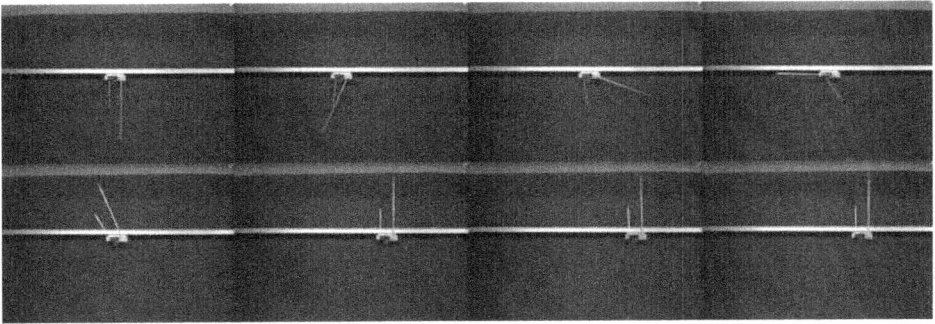

Smooth nonlinear trajectory planning with linear quadratic Gaussian feedback (LQR) control on a dual pendula system.

Control theory is :

- A theory that deals with influencing the behavior of *dynamical systems*
- An interdisciplinary subfield of science, which originated in *engineering* and *mathematics*, and evolved into use by the social sciences, such as *psychology, sociology, criminology* and in the *financial system*.

Control systems may be thought of as having four functions : Measure, Compare, Compute, and Correct. These four functions are completed by five elements : *Detector, Transducer, Transmitter, Controller*, and Final Control Element. The measuring function is completed by the detector, transducer and transmitter. In practical applications these three elements are typically contained in one unit. A standard example of a measuring unit is a *resistance thermometer*. The compare and compute functions are completed within the controller, which may be implemented electronically by *proportional control*, a *PI controller*, *PID controller*, bistable, hysteretic control or *programmable logic controller*. Older controller units have been mechanical, as in a *centrifugal governor* or a *carburetor*. The correct function is completed with a final control element. The final control element changes an input or output in the control system that affects the manipulated or controlled variable.

An Example

Consider a car's *cruise control*, which is a device designed to maintain vehicle speed at a constant *desired* or *reference* speed provided by the driver. The *controller* is the cruise control, the *plant* is the car, and the *system* is the car and the cruise control. The system output is the car's speed, and the control itself is the engine's *throttle* position which determines how much power the engine generates.

A primitive way to implement cruise control is simply to lock the throttle position when the driver engages cruise control. However, if the cruise control is engaged on a stretch of flat road, then the car will travel slower going uphill and faster when going downhill. This type of controller is called an *open–loop controller* because no measurement of the system output (the car's speed) is used to alter the control (the throttle position.) As a result, the controller cannot compensate for changes acting on the car, like a change in the slope of the road.

In a **closed–loop control system**, a sensor monitors the system output (the car's speed) and feeds the data to a controller which adjusts the control (the throttle position) as necessary to maintain the desired system output (match the car's speed to the reference speed.) Now, when the car goes uphill, the decrease in speed is measured, and the throttle position changed to increase engine power, speeding up the vehicle. Feedback from measuring the car's speed has allowed the controller to dynamically compensate for changes to the car's speed. It is from this feedback that the paradigm of the control *loop* arises : the control affects the system output, which in turn is measured and looped back to alter the control.

History

Although control systems of various types date back to antiquity, a more formal analysis of the field began with a dynamics analysis of the *centrifugal governor*, conducted by the physicist *James Clerk Maxwell* in 1868, entitled *On Governors*. This described and analyzed the phenomenon of *self-oscillation*, in which lags in the system may lead to overcompensation and unstable behavior. This generated a flurry of interest in the topic, during which Maxwell's classmate, *Edward John Routh*, abstracted Maxwell's results for the general class of linear systems. Independently, *Adolf Hurwitz* analyzed system stability using differential equations in 1877, resulting in what is now known as the *Routh–Hurwitz theorem*.

A notable application of dynamic control was in the area of manned flight. The *Wright brothers* made their first successful test flights on December 17, 1903 and were distinguished by their ability to control their flights for substantial periods (more so than the ability to produce lift from an airfoil, which was known).

Continuous, reliable control of the airplane was necessary for flights lasting longer than a few seconds.

By *World War II*, control theory was an important part of *fire–control systems*, *guidance systems* and *electronics*.

Sometimes mechanical methods are used to improve the stability of systems. For example, *ship stabilizers* are fins mounted beneath the waterline and emerging laterally. In contemporary vessels, they may be gyroscopically controlled active fins, which have the capacity to change their angle of attack to counteract roll caused by wind or waves acting on the ship.

The *Sidewinder missile* uses small control surfaces placed at the rear of the missile with spinning disks on their outer surfaces; these are known as *rollerons*. Airflow over the disks spins them to a high speed. If the missile starts to roll, the gyroscopic force of the disks drives the control surface into the airflow, cancelling the motion. Thus, the Sidewinder team replaced a potentially complex control system with a simple mechanical solution.

The *Space Race* also depended on accurate spacecraft control, and control theory has also seen an increasing use in fields such as *economics*.

People in Systems and Control

Many active and historical figures made significant contribution to control theory, including, for example :

*   *Pierre–Simon Laplace* (1749–1827) invented the *Z–transform* in his work on *probability theory*, now used to solve discrete–time control theory problems. The Z–transform is a discrete–time equivalent of the *Laplace transform* which is named after him.
*   *Alexander Lyapunov* (1857–1918) in the 1890s marks the beginning of *stability theory*.
*   *Harold S. Black* (1898–1983), invented the concept of *negative feedback amplifiers* in 1927. He managed to develop stable negative feedback amplifiers in the 1930s.
*   *Harry Nyquist* (1889–1976), developed the *Nyquist stability criterion* for feedback systems in the 1930s.
*   *Richard Bellman* (1920–1984), developed *dynamic programming* since the 1940s.
*   *Andrey Kolmogorov* (1903–1987) co–developed the *Wiener–Kolmogorov filter* (1941).
*   *Norbert Wiener* (1894–1964) co–developed the Wiener–Kolmogorov filter and coined the term *cybernetics* in the 1940s.
*   *John R. Ragazzini* (1912–1988) introduced *digital control* and the use of *Z–transform* in control theory (invented by Laplace) in the 1950s.
*   *Lev Pontryagin* (1908–1988) introduced the *maximum principle* and the *bang–bang principle*.

- *Pierre-Louis Lions* (1956) who developed *viscosity solutions* into stochastic control and *optimal control* methods.

Classical Control Theory

To overcome the limitations of the open–loop controller, control theory introduces *feedback*. A closed–loop *controller* uses feedback to control *states* or *outputs* of a *dynamical system*. Its name comes from the information path in the system : process inputs (*e.g., voltage* applied to an *electric motor*) have an effect on the process outputs (*e.g.*, speed or torque of the motor), which is measured with *sensors* and processed by the controller; the result (the control signal) is "fed back" as input to the process, closing the loop.

Closed–loop controllers have the following advantages over *open–loop controllers* :

- Disturbance rejection
- Guaranteed performance even with *model* uncertainties, when the model structure does not match perfectly the real process and the model parameters are not exact
- *Unstable* processes can be stabilized
- Reduced sensitivity to parameter variations
- Improved reference tracking performance.

    In some systems, closed–loop and open–loop control are used simultaneously. In such systems, the open–loop control is termed *feedforward* and serves to further improve reference tracking performance.

    A common closed–loop controller architecture is the *PID controller*.

Closed–loop transfer function

The output of the system *y(t)* is fed back through a sensor measurement *F* to the reference value *r(t)*. The controller *C* then takes the error *e* (difference) between the reference and the output to change the inputs *u* to the system under control *P*. This is shown in the figure. This kind of controller is a closed–loop controller or feedback controller.

    This is called a single–input–single–output (*SISO*) control system; *MIMO* (*i.e.*, Multi–Input–Multi–Output) systems, with more than one input/output, are common. In such cases variables are represented through *vectors* instead of simple *scalar* values. For some *distributed parameter systems* the vectors may be infinite–*dimensional* (typically functions).

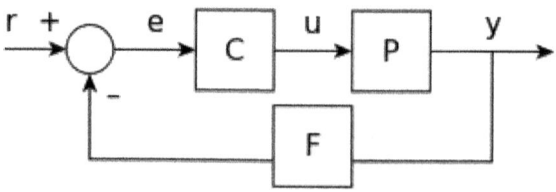

If we assume the controller C, the plant P, and the sensor F are *linear* and *time–invariant* (i.e., elements of their *transfer function* $C(s)$, $P(s)$, and $F(s)$ do not depend on time), the systems above can be analysed using the *Laplace transform* on the variables. This gives the following relations :

$$Y(s) = P(s)U(s)$$

$$U(s) = C(s)E(s)$$

$$E(s) = R(s) - F(s)Y(s).$$

Solving for $Y(s)$ in terms of $R(s)$ gives :

$$Y(s) = \left(\frac{P(s)C(s)}{1 + F(s)P(s)C(s)} \right) R(s) = H(s)R(s).$$

The expression $H(s) = \dfrac{P(s)C(s)}{1 + F(s)P(s)C(s)}$ is referred to as the *closed–loop transfer function* of the system. The numerator is the forward (open–loop) gain from r to y, and the denominator is one plus the gain in going around the feedback loop, the so-called loop gain. If $|P(s)C(s)| \gg 1$, *i.e.*, it has a large *norm* with each value of s, and if $|F(s)| \approx 1$, then $Y(s)$ is approximately equal to $R(s)$ and the output closely tracks the reference input.

Modern Control Theory

In contrast to the frequency domain analysis of the classical control theory, modern control theory utilizes the time–domain *state space* representation, a mathematical model of a physical system as a set of input, output and state variables related by first–order differential equations. To abstract from the number of inputs, outputs and states, the variables are expressed as vectors and the differential and algebraic equations are written in matrix form (the latter only being possible when the dynamical system is linear). The state space representation (also known as the "time–domain approach") provides a convenient and compact way to model and analyze systems with multiple inputs and outputs. With inputs and outputs, we would otherwise have to write down Laplace transforms to encode all the information about a system. Unlike the frequency domain approach, the use of the state space representation is not limited to systems with linear components and zero initial conditions. "State space" refers to the space whose axes are the state variables. The state of the system can be represented as a vector within that space.

Topics in Control Theory

Stability

The *stability* of a general *dynamical system* with no input can be described with *Lyapunov stability* criteria. A *linear system* that takes an input is called *bounded–input bounded–output (BIBO) stable* if its output will stay *bounded* for any bounded input.

Stability for *nonlinear systems* that take an input is *input–to–state stability* (ISS), which combines Lyapunov stability and a notion similar to BIBO stability. For simplicity, the following descriptions focus on continuous–time and discrete–time linear systems.

Mathematically, this means that for a causal linear system to be stable all of the *poles* of its *transfer function* must have negative–real values, *i.e.* the real part of all the poles are less than zero. Practically speaking, stability requires that the transfer function complex poles reside.

- In the open left half of the *complex plane* for continuous time, when the *Laplace transform* is used to obtain the transfer function.
- Inside the *unit circle* for discrete time, when the *Z–transform* is used.

The difference between the two cases is simply due to the traditional method of plotting continuous time versus discrete time transfer functions. The continuous Laplace transform is in *Cartesian coordinates* where the x-axis is the real axis and the discrete Z–transform is in *circular coordinates* where the ρ-axis is the real axis.

When the appropriate conditions above are satisfied a system is said to be *asymptotically stable* : the variables of an asymptotically stable control system always decrease from their initial value and do not show permanent oscillations. Permanent oscillations occur when a pole has a real part exactly equal to zero (in the continuous time case) or a modulus equal to one (in the discrete time case). If a simply stable system response neither decays nor grows over time, and has no oscillations, it is *marginally stable* : in this case the system transfer function has non–repeated poles at complex plane origin (*i.e.* their real and complex component is zero in the continuous time case). Oscillations are present when poles with real part equal to zero have an imaginary part not equal to zero.

If a system in question has an *impulse response* of

$x[n] = 0.5^n u[n]$

then the Z–transform, is given by

$$X(z) = \frac{1}{1 - 0.5z^{-1}}$$

which has a pole in $z = 0.5$ (zero *imaginary part*). This system is BIBO (asymptotically) stable since the pole is *inside* the unit circle.

However, if the impulse response was

$x[n] = 1.5^n u[n]$

then the Z–transform is

$$X(z) = \frac{1}{1 - 1.5z^{-1}}$$

which has a pole at $z = 1.5$ and is not BIBO stable since the pole has a modulus strictly greater than one.

Numerous tools exist for the analysis of the poles of a system. These include graphical systems like the *root locus*, *Bode plots* or the *Nyquist plots*.

Mechanical changes can make equipment (and control systems) more stable. Sailors add ballast to improve the stability of ships. Cruise ships use antiroll fins that extend transversely from the side of the ship for perhaps 30 feet (10 m) and are continuously rotated about their axes to develop forces that oppose the roll.

Control Specification

Several different control strategies have been devised in the past years. These vary from extremely general ones (*PID controller*), to others devoted to very particular classes of systems (especially *robotics* or *aircraft* cruise control).

A control problem can have several specifications. Stability, of course, is always present : the controller must ensure that the closed–loop system is stable, regardless of the open–loop stability. A poor choice of controller can even worsen the stability of the open–loop system, which must normally be avoided. Sometimes it would be desired to obtain particular dynamics in the closed loop : *i.e.* that the poles have $Re[\lambda] < -\lambda$, where λ is a fixed value strictly greater than zero, instead of simply asking that $Re[\lambda] < 0$.

Another typical specification is the rejection of a step disturbance; including an *integrator* in the open–loop chain (*i.e.* directly before the system under control) easily achieves this. Other classes of disturbances need different types of sub–systems to be included.

Other "classical" control theory specifications regard the time–response of the closed–loop system : these include the *rise time* (the time needed by the control system to reach the desired value after a perturbation), peak *overshoot* (the highest value reached by the response before reaching the desired value) and others (*settling time*, quarter–decay). Frequency domain specifications are usually related to *robustness*.

Modern performance assessments use some variation of integrated tracking error (IAE,ISA,CQI).

Model Identification and Robustness

A control system must always have some robustness property. A *robust controller* is such that its properties do not change much if applied to a system slightly different from the mathematical one used for its synthesis. This specification is important : no real physical system truly behaves like the series of differential equations used to represent it mathematically. Typically a simpler mathematical model is chosen in order to simplify calculations, otherwise the true system dynamics can be so complicated that a complete model is impossible.

System Identification

The process of determining the equations that govern the model's dynamics is called *system identification*. This can be done off-line : for example, executing a series of measures from which to calculate an approximated mathematical model, typically its *transfer function* or matrix. Such identification from the output, however, cannot take account of unobservable dynamics. Sometimes the model is built directly starting from known physical equations : for example, in the case of a *mass–spring–damper* system we know that $m\ddot{x}(t) = -Kx(t) - B\dot{x}(t)$. Even assuming that a "complete" model is used in designing the controller, all the parameters included in these equations (called "nominal parameters") are never known with absolute precision; the control system will have to behave correctly even when connected to physical system with true parameter values away from nominal.

Some advanced control techniques include an "on–line" identification process. The parameters of the model are calculated ("identified") while the controller itself is running : in this way, if a drastic variation of the parameters ensues (for example, if the robot's arm releases a weight), the controller will adjust itself consequently in order to ensure the correct performance.

Analysis

Analysis of the robustness of a SISO (single input single output) control system can be performed in the frequency domain, considering the system's transfer function and using *Nyquist* and *Bode diagrams*. Topics include *gain and phase margin* and amplitude margin. For MIMO (multi input multi output) and, in general, more complicated control systems one must consider the theoretical results devised for each control technique : *i.e.,* if particular robustness qualities are needed, the engineer must shift his attention to a control technique by including them in its properties.

Constraints

A particular robustness issue is the requirement for a control system to perform properly in the presence of input and state constraints. In the physical world every signal is limited. It could happen that a controller will send control signals that cannot be followed by the physical system : for example, trying to rotate a valve at excessive speed. This can produce undesired behavior of the closed–loop system, or even damage or break actuators or other subsystems. Specific control techniques are available to solve the problem : *model predictive control*, and *anti–wind up systems*. The latter consists of an additional control block that ensures that the control signal never exceeds a given threshold.

System Classifications

Linear Systems Control

For MIMO systems, pole placement can be performed mathematically using a *state space representation* of the open–loop system and calculating a feedback matrix assigning poles in the desired positions. In complicated systems this can require computer–assisted calculation capabilities, and cannot always ensure robustness.

Furthermore, all system states are not in general measured and so observers must be included and incorporated in pole placement design.

Nonlinear Systems Control

Processes in industries like *robotics* and the *aerospace industry* typically have strong nonlinear dynamics. In control theory it is sometimes possible to linearize such classes of systems and apply linear techniques, but in many cases it can be necessary to devise from scratch theories permitting control of nonlinear systems. These, *e.g., feedback linearization, backstepping, sliding mode control*, trajectory linearization control normally take advantage of results based on *Lyapunov's theory. Differential geometry* has been widely used as a tool for generalizing well–known linear control concepts to the non–linear case, as well as showing the subtleties that make it a more challenging problem.

Decentralized Systems

When the system is controlled by multiple controllers, the problem is one of decentralized control. Decentralization is helpful in many ways, for instance, it helps control systems operate over a larger geographical area. The agents in decentralized control systems can interact using communication channels and coordinate their actions.

Main Control Strategies

Every control system must guarantee first the stability of the closed–loop behavior. For *linear systems*, this can be obtained by directly placing the poles. Non–linear control systems use specific theories (normally based on *Aleksandr Lyapunov's* Theory) to ensure stability without regard to the inner dynamics of the system. The possibility to fulfill different specifications varies from the model considered and the control strategy chosen. Here a summary list of the main control techniques is shown :

Adaptive control uses on–line identification of the process parameters, or modification of controller gains, thereby obtaining strong robustness properties. Adaptive controls were applied for the first time in the *aerospace industry* in the 1950s, and have found particular success in that field.

A *hierarchical control system* is a type of *control system* in which a set of devices and governing software is arranged in a *hierarchical tree*. When the links in the tree are implemented by a *computer network*, then that hierarchical control system is also a form of *Networked control system*.

Intelligent control uses various AI computing approaches like *neural networks, Bayesian probability, fuzzy logic, machine learning, evolutionary computation* and *genetic algorithms* to control a *dynamic system*.

Optimal control is a particular control technique in which the control signal optimizes a certain "cost index" : for example, in the case of a satellite, the jet thrusts needed to bring it to desired trajectory that consume the least amount of fuel. Two optimal control design methods have been widely used in industrial applications, as it has been shown they can guarantee closed–loop stability. These are *Model Predictive Control* (MPC) and *linear–quadratic–Gaussian control* (LQG). The first can more explicitly take into account constraints on the signals in the system, which is an important feature in many industrial processes. However, the "optimal control" structure in MPC is only a means to achieve such a result, as it does not optimize a true performance index of the closed–loop control system. Together with PID controllers, MPC systems are the most widely used control technique in *process control.*

Robust control deals explicitly with uncertainty in its approach to controller design. Controllers designed using *robust control* methods tend to be able to cope with small differences between the true system and the nominal model used for design. The early methods of *Bode* and others were fairly robust; the state–space methods invented in the 1960s and 1970s were sometimes found to lack robustness. Examples of modern robust control techniques include *H–infinity loop–shaping* developed by *Duncan McFarlane* and *Keith Glover* of *Cambridge University, United Kingdom* and *Sliding mode control* (SMC) developed by *Vadim Utkin.* Robust methods aim to achieve robust performance and/or *stability* in the presence of small modeling errors.

Stochastic control deals with control design with uncertainty in the model. In typical stochastic control problems, it is assumed that there exist random noise and disturbances in the model and the controller, and the control design must take into account these random deviations.

Energy–shaping control view the plant and the controller as energy–transformation devices. The control strategy is formulated in terms of interconnection (in a power–preserving manner) in order to achieve a desired behavior.

Self–organized criticality control may be defined as attempts to interfere in the processes by which the *self–organized* system dissipates energy.

CONTROLLABILITY

Controllability is an important property of a *control system*, and the controllability property plays a crucial role in many control problems, such as stabilization of *unstable systems* by feedback, or optimal control.

Controllability and *observability* are *dual* aspects of the same problem.

Roughly, the concept of controllability denotes the ability to move a system around in its entire configuration space using only certain admissible manipulations. The exact definition varies slightly within the framework or the type of models applied.

The following are examples of variations of controllability notions which have been introduced in the systems and control literature :

• State controllability

- Output controllability
- Controllability in the behavioural framework.

State Controllability

The *state* of a system, which is a collection of the system's variables values, completely describes the system at any given time. In particular, no information on the past of a system will help in predicting the future, if the states at the present time are known.

Complete state controllability (or simply *controllability* if no other context is given) describes the ability of an external input to move the internal state of a system from any initial state to any other final state in a finite time interval.

Continuous Linear Systems

Consider the *continuous linear time–variant system*

$$\dot{x}(t) = A(t)x(t) + B(t)u(t)$$

$$y(t) = C(t)x(t) + D(t)u(t).$$

There exists a control u from state x_0 at time t_0 to state x_1 at time $t_1 > t_0$ if and only if $x_1 - \phi(t_0, t_1)x_0$ is in the *column space* of

$$W(t_0, t_1) = \int_{t_0}^{t_1} \phi(t_0, t)B(t)B(t)^T \phi(t_0, t)^T \, dt$$

where ϕ is the *state–transition matrix*.

In fact, if η_0 is a solution to $W(t_0, t_1)\eta = x_1 - \phi(t_0, t_1)x_0$ then a control given by $u(t) = -B(t)^T \phi(t_0, t)^T \eta_0$ would make the desired transfer.

Note that the matrix W defined as above has the following properties :

- $W(t_0, t_1)$ is *symmetric*
- $W(t_0, t_1)$ is *positive semidefinite* for $t_1 \geq t_0$
- $W(t_0, t_1)$ satisfies the linear *matrix differential equation*

$$\frac{d}{dt}W(t, t_1) = A(t)W(t, t_1) + W(t, t_1)A(t)^T - B(t)B(t)^T, W(t_1, t_1) = 0$$

- $W(t_0, t_1)$ satisfies the equation
$$W(t_0, t_1) = W(t_0, t) + \phi(t_0, t)W(t, t_1)\phi(t_0, t)^T$$

Continuous Linear Time–invariant (LTI) Systems

Consider the continuous linear *time–invariant system*

$$\dot{x}(t) = Ax(t) + Bu(t)$$

$$y(t) = Cx(t) + Du(t)$$

where

x is the $n \times 1$ "state vector",

y is the $m \times 1$ "output vector",

u is the $r \times 1$ "input (or control) vector",

A is the $n \times n$ "state matrix",

B is the $n \times r$ "input matrix",

C is the $m \times n$ "output matrix",

D is the $m \times r$ "feed through (or feedforward) matrix".

The $n \times nr$ controllability matrix is given by

$R = [B \ AB \ A^2B \ ... \ A^{n-1}B$

The system is controllable if the controllability matrix has full *rank* (*i.e.* rank $(R) = n$).

Discrete Linear Time–Invariant (LTI) Systems

For a discrete–time linear state-space system (*i.e.* time variable $k \in Z$) the state equation is

$x(k + 1) = Ax(k) + Bu(k)$

Where A is an $n \times n$ matrix and B is a $n \times r$ matrix (*i.e.* u is r inputs collected in a $r \times 1$ vector. The test for controllability is that the $n \times nr$ matrix

$c = [B \ AB \ A^2B \ ... \ A^{n-1}B]$

has full row *rank* (*i.e.*, rank(C) = n). That is, if the system is controllable, C will have n columns that are *linearly independent*; if n columns of C are *linearly independent*, each of the n states is reachable giving the system proper inputs through the variable $u(k)$.

Example

For example, consider the case when $n = 2$ and $r = 1$ (*i.e.* only one control input). Thus, B and AB are $n \times 1$ vectors. If $[B \ AB]$ has rank 2 (full rank), and so B and AB are *linearly independent* and span the entire plane. If the rank is 1, then B and AB are *collinear* and do not span the plane.

Assume that the initial state is zero.

At time $k = 0 : x(1) = Ax(0) + Bu(0) = Bu(0)$

At time $k = 1 : x(2) = Ax(1) + Bu(1) = ABu(0) + Bu(1)$

At time $k = 0$ all of the reachable states are on the line formed by the vector B. At time $k = 1$ all of the reachable states are linear combinations of AB and B. If the system is controllable then these two vectors can span the entire plane and can be done so for time $k = 2$. The assumption made that the initial state is zero is merely for convenience. Clearly if all states can be reached from the origin then any state can be reached from another state (merely a shift in coordinates).

This example holds for all positive n, but the case of $n - 2$ is easier to visualize.

Analogy for Example of n = 2

Consider an *analogy* to the previous example system. You are sitting in your car on an infinite, flat plane and facing north. The goal is to reach any point in the plane by driving a distance in a straight line, come to a full stop, turn, and driving another distance, again, in a straight line. If your car has no steering then you can only drive straight, which means you can only drive on a line (in this case the north–south line since you started facing north). The lack of steering case would be analogous to when the rank of C is 1 (the two distances you drove are on the same line).

Now, if your car did have steering then you could easily drive to any point in the plane and this would be the analogous case to when the rank of C is 2.

If you change this example to $n = 3$ then the analogy would be flying in space to reach any position in 3D space (ignoring the *orientation* of the *aircraft*). You are allowed to :

* Fly in a straight line
* Turn left or right by any amount (*Yaw*)
* Direct the plane upwards or downwards by any amount (*Pitch*)

Although the 3–dimensional case is harder to visualize, the concept of controllability is still analogous.

Nonlinear Systems

Nonlinear systems in the control–affine form

$$\dot{x} = f(x) + \sum_{i=1}^{m} g_i(x)u_i$$

is locally accessible about x_0 if the accessibility distribution R spans n space, when n equals the rank of x and R is given by :

$R = [g_1 \cdots g_m \; [ad^k_{gi}gj] \cdots [ad^k_f gi]].$

Here, $[ad^k_f g]$ is the repeated *Lie bracket* operation defined by

$[ad^k_f g] = [f \cdots j \; [f, g]].$

The controllability matrix for linear systems in the previous section can in fact be derived from this equation.

Output Controllability

Output controllability is the related notion for the output of the system; the output controllability describes the ability of an external input to move the output from any initial condition to any final condition in a finite time interval. It is not necessary that there is any relationship between state controllability and output controllability. In particular :

- A controllable system is not necessarily output controllable. For example, if matrix $D = 0$ and matrix C does not have full row rank, then some positions of the output are masked by the limiting structure of the output matrix. Moreover, even though the system can be moved to any state in finite time, there may be some outputs that are inaccessible by all states. A trivial numerical example uses $D = 0$ and a C matrix with at least one row of zeros; thus, the system is not able to produce a non–zero output along that dimension.

- An output controllable system is not necessarily state controllable. For example, if the dimension of the state space is greater than the dimension of the output, then there will be a set of possible state configurations for each individual output. That is, the system can have significant *zero dynamics*, which are trajectories of the system that are not observable from the output. Consequently, being able to drive an output to a particular position in finite time says nothing about the state configuration of the system.

For a linear continuous–time system, like the example above, described by matrices A, B, C, and D, the $m \times (n + 1)r$ *output controllability matrix*

$$[CB \quad CAB \quad CA^2B \quad \cdots \quad CA^{n-1}B \quad D]$$

must have full row rank (*i.e.* rank m) if and only if the system is output controllable. This result is known as Kalman's criteria of controllability.

Controllability Under Input Constraints

In systems with limited control authority, it is often no longer possible to move any initial state to any final state inside the controllable subspace. This phenomenon is caused by constraints on the input that could be inherent to the system (*e.g.* due to saturating actuator) or imposed on the system for other reasons (*e.g.* due to safety–related concerns). The controllability of systems with input and state constraints is studied in the context of *reachability* and *viability theory*.

Controllability in the Behavioural Framework

In the so–called *behavioral system theoretic approach* due to Willems, models considered do not directly define an input–output structure. In this framework systems are described by admissible trajectories of a collection of variables, some of which might be interpreted as inputs or outputs.

A system is then defined to be controllable in this setting, if any past part of a behavior (trajectory of the external veriables) can be concatenated with any future trajectory of the behavior in such a way that the concatenation is contained in the behavior, *i.e.* is part of the admissible system behavior.

Stabilizability

A slightly weaker notion than controllability is that of **stabilizability**. A system is determined to be stabilizable when all uncontrollable states have stable dynamics. Thus, even though some of the states cannot be controlled all the states will still remain bounded during the system's behavior.

OBSERVABILITY

In *control theory*, **observability** is a measure for how well internal states of a *system* can be inferred by *knowledge* of its external *outputs*. The observability and *controllability* of a system are mathematical *duals*. The concept of observability was introduced by American–Hungarian scientist *Rudolf E. Kalman* for linear dynamic systems.

Definition

Formally, a system is said to be **observable** if, for any possible sequence of state and control vectors, the current state can be determined in finite time using only the outputs (this definition is slanted towards the *state space* representation). Less formally, this means that from the system's outputs it is possible to determine the behaviour of the entire system. If a system is not observable, this means the current values of some of its states cannot be determined through output *sensors*. This implies that their value is unknown to the *controller* (although they can be estimated through various means).

For *time–invariant linear systems* in the state space representation, a convenient test to check if a system is observable exists. Consider a *SISO* system with n states, if the row *rank* of the following *observability matrix* :

$$O = \begin{bmatrix} C \\ CA \\ CA^2 \\ \vdots \\ CA^{n-1} \end{bmatrix}$$

is equal to n, then the system is observable. The rationale for this test is that if n rows are linearly independent, then each of the n states is viewable through linear combinations of the output variables $y(k)$.

A module designed to estimate the state of a system from measurements of the outputs is called a *state observer* or simply an observer for that system.

Observability Index

The Observability index v of a linear time–invariant discrete system is the smallest natural number for which is satisfied that rank $(O_v) = $ rank(O_{v+1}), where

$$O_v = \begin{bmatrix} C \\ CA \\ CA^2 \\ \vdots \\ CA^{v-1} \end{bmatrix}.$$

Detectability

A slightly weaker notion than observability is detectability. A system is detectable if and only if all of its unobservable modes are asymptotically stable. Thus even though not all system modes are observable, the ones that are not observable do not require stabilization.

Continuous Time–Varying System

Consider the *continuous linear time–variant system*

$$\dot{x}(t) = A(t)x(t) + B(t)u(t)$$
$$y(t) = C(t)x(t).$$

Suppose that the matrices A, B, and C are given as well as inputs and outputs u and y for all $t \in [t_0, t_1]$ then it is possible to determine $x(t_0)$ to within an additive constant vector which lies in the *null space* of $M(t_0, t_1)$ defined by

$$M(t_0, t_1) = \int_{t_0}^{t_1} \phi(t, t_0)^T C(t)^T C(t) \phi(t, t_0) dt$$

where ϕ is the *state–transition matrix*.

It is possible to determine a unique $x(t_0)$ if $M(t_0, t_1)$ is *nonsingular*. In fact, it is not possible to distinguish the initial state for x_1 from that of x_2 if $x_1 - x_2$ is in the null space of $M(t_0, t_1)$.

Note that the matrix M defined as above has the following properties :

- $M(t_0, t_1)$ is *symmetric*
- $M(t_0, t_1)$ is *positive semidefinite* for $t_1 \geq t_0$
- $M(t_0, t_1)$ satisfies the linear *matrix differential equation*

$$\frac{d}{dt} M(t, t_1) = -A(t)^T M(t, t_1) - M(t, t_1)A(t) - C(t)^T C(t), M(t_1, t_1) = 0$$

- $M(t_0, t_1)$ satisfies the equation
$$M(t_0, t_1) = M(t_0, t) + \phi(t, t_0)^T M(t, t_1)\phi(t, t_0)$$

Nonlinear Case

Given the system $\dot{x} = f(x) + \sum_{j=1}^{m} g_j(x)u_j$, $y_i = h_i(x), i \in p$. Where $x \in R^n$ the state vector, $u \in R^m$ the input vector and $y \in R^p$ the output vector f, g, h are to be smooth vectorfields.

Now define the observation space O_s to be the space containing all repeated *Lie derivatives*. Now the system is observable in x_0 if and only if $\dim(dO_s(x_0)) = n$.

Note : $dO_s(x_0) = \mathrm{span}\ (dh_1(x_0), \ldots, dh_p(x_0), dL_{vi}L_{vi-1}, \ldots, L_{v1}h_j(x_0)), j \in p, k = 1, 2, \ldots$.

Early criteria for observability in nonlinear dynamic systems were discovered by Griffith and Kumar, Kou, Elliot and Tarn, and Singh.

Static Systems and General Topological Spaces

Observability may also be characterized for steady state systems (systems typically defined in terms of algebraic equations and inequalities), or more generally, for sets in R^n,. Just as observability criteria are used to predict the behavior of *Kalman filters* or other observers in the dynamic system case, observability criteria for sets in R^n are used to predict the behavior of *data reconciliation* and other static estimators. In the nonlinear case, observability can be characterized for individual variables, and also for local estimator behavior rather than just global behavior.

ADVANCED PROCESS CONTROL

In *control theory* **Advanced process control** (APC) refers to a broad range of techniques and technologies implemented within industrial process control systems. Advanced process controls are usually deployed optionally and in addition to *basic* process controls. Basic process controls are designed and built with the process itself, to facilitate basic operation, control and automation requirements. Advanced process controls are typically added subsequently, often over the course of many years, to address particular performance or economic improvement opportunities in the process.

Process control (basic and advanced) normally implies the process industries, which includes chemicals, petrochemicals, oil and mineral refining, food processing, pharmaceuticals, power generation, etc. These industries are characterized by continuous processes and fluid processing, as opposed to discrete parts manufacturing, such as automobile and electronics manufacturing. The term process automation is essentially synonymous with process control.

Process controls (basic as well as advanced) are implemented within the process control system, which usually means a *distributed control system (DCS)*, *programmable logic controller (PLC)*, and/or a supervisory control computer. DCSs and PLCs are typically industrially hardened and fault–tolerant. Supervisory control computers are often not hardened or fault–tolerant, but they bring a higher level of computational capability to the control system, to host valuable, but not critical, advanced control applications. Advanced controls may reside in either the DCS or the supervisory computer, depending on the application. Basic controls reside in the DCS and its subsystems, including PLCs.

Types of Advanced Process Control

Following is a list of the best known types of advanced process control :

* Advanced regulatory control (ARC) refers to several proven advanced control techniques, such as feed forward, override or adaptive gain. ARC is also a catch–all term used to refer to any customized or non–simple technique that does not fall into any other category. ARCs are typically implemented using function blocks or custom programming capabilities at the DCS level. In some cases, ARCs reside at the supervisory control computer level.

- Multivariable *Model predictive control* (MPC) is a popular technology, usually deployed on a supervisory control computer, that identifies important independent and dependent process variables and the dynamic relationships (models) between them, and uses matrix–math based control and optimization algorithms, to control multiple variables simultaneously. MPC has been a prominent part of APC ever since supervisory computers first brought the necessary computational capabilities to control systems in the 1980s.

- *Inferential control :* The concept behind inferentials is to calculate a stream property from readily available process measurements, such as temperature and pressure, that otherwise would require either an expensive and complicated online analyzer or periodic laboratory analysis. Inferentials can be utilized in place of actual online analyzers, whether for operator information, cascaded to base–layer process controllers, or multivariable controller CVs.

- Sequential control refers to dis–continuous time and event based automation sequences that occur within continuous processes. These may be implemented as a collection of time and logic function blocks, a custom algorithm, or using a formalized *Sequential function chart* methodology.

- Compressor control typically includes compressor anti–surge and performance control.

Related Technologies

- *Statistical process control* (SPC), despite its name, is much more common in discrete parts manufacturing and batch process control than in continuous process control. In SPC, "process" refers to the work and quality control process, rather than continuous process control.

- Batch process control is employed in non–continuous batch processes, such as many pharmaceuticals, chemicals, and foods.

- Simulation–based optimization incorporates dynamic or steady–state computer–based process simulation models to determine more optimal operating targets in real–time, *i.e.* on a periodic basis, ranging from hourly to daily. This is sometimes considered a part of APC, but in practice it is still an emerging technology and is more often part of MPO.

- Manufacturing planning and optimization (MPO) refers to ongoing business activity to arrive at optimal operating targets that are then implemented in the operating organization, either manually or in some cases automatically communicated to the process control system.

- *Safety instrumented system* refers to a system that is independent of the process control system, both physically and administratively, whose purpose is to assure basic safety of the process.

APC Business and Professionals

Those responsible for the design, implementation and maintenance of APC applications are often referred to as APC Engineers or Control Application Engineers.

Usually their education is dependent upon the field of specialization. For example, in the process industries many APC Engineers have a chemical engineering background, combining process control and chemical processing expertise.

Most large operating facilities, such as oil refineries, employ a number of control system specialists and professionals, ranging from field instrumentation, regulatory control system (DCS and PLC), advanced process control, and control system network and security. Depending on facility size and circumstances, these personnel may have responsibilities across multiple areas, or be dedicated to each area. There are also many process control service companies that can be hired for support and services in each area.

Terminology

- *APC* : Advanced process control
- *ARC* : Advanced regulatory control, including feedforward, adaptive gain, override, logic, fuzzy logic, sequence control, device control, inferentials, and custom algorithms; usually implies DCS–based.
- Base–Layer : Includes DCS, SIS, field devices, and other DCS subsystems, such as analyzers, equipment health systems, and PLCs.
- *BPCS* : Basic process control system
- *DCS* : Distributed control system, often synonymous with BPCS
- *MPO* : Manufacturing planning optimization
- *MPC* : Multivariable *Model predictive control*
- *SIS* : *Safety instrumented system*
- *SME* : Subject matter expert.

ADAPTIVE CONTROL

Adaptive control is the control method used by a controller which must adapt to a controlled system with parameters which vary, or are initially uncertain. For example, as an aircraft flies, its mass will slowly decrease as a result of fuel consumption; a control law is needed that adapts itself to such changing conditions. Adaptive control is different from *robust control* in that it does not need *a priori* information about the bounds on these uncertain or time–varying parameters; robust control guarantees that if the changes are within given bounds the control law need not be changed, while adaptive control is concerned with control law changing themselves.

Parameter Estimation

The foundation of adaptive control is *parameter estimation*. Common methods of estimation include *recursive least squares* and *gradient descent*. Both of these methods provide update laws which are used to modify estimates in real time (*i.e.*, as the system operates). *Lyapunov stability* is used to derive these update laws and show convergence criterion (typically persistent excitation). *Projection (mathematics)* and normalization are commonly used to improve the robustness of estimation algorithms.

Classification of Adaptive Control Techniques

In general one should distinguish between :

1. Feedforward Adaptive Control
2. Feedback Adaptive Control
   as well as between
1. Direct Methods and
2. Indirect Methods

Direct methods are ones wherein the estimated parameters are those directly used in the adaptive controller. In contrast, indirect methods are those in which the estimated parameters are used to calculate required controller parameters

There are several broad categories of feedback adaptive control (classification can vary) :

- Dual Adaptive Controllers [based on *Dual control theory*]
  - Optimal Dual Controllers [difficult to design]
  - Suboptimal Dual Controllers
- Nondual Adaptive Controllers
  - Adaptive Pole Placement
  - Extremum Seeking Controllers
  - *Iterative learning control*
  - *Gain scheduling*
  - Model Reference Adaptive Controllers (MRACs) [incorporate a *reference model* defining desired closed loop performance]

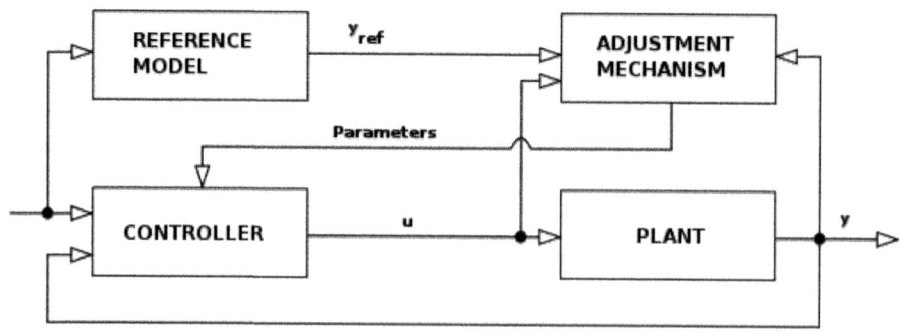

MODEL REFERENCE ADAPTIVE CONTROL (MRAC)

MRAC

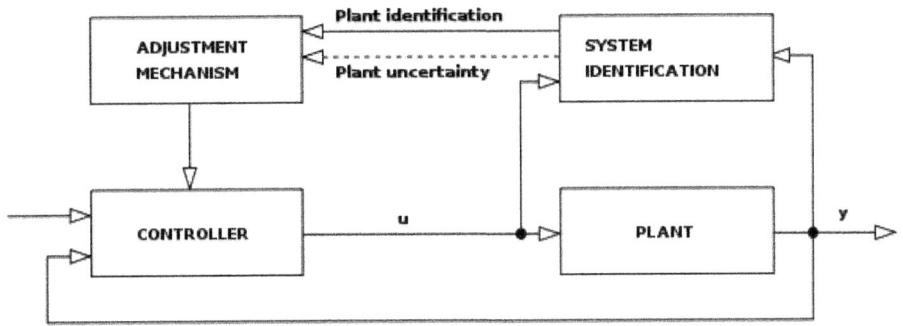

MODEL IDENTIFICATION ADAPTIVE CONTROL (MIAC)

MIAC

- Gradient Optimization MRACs [use local rule for adjusting params when performance differs from reference. Ex. : "MIT rule".]
- Stability Optimized MRACs
    - Model Identification Adaptive Controllers (MIACs) [perform *System identification* while the system is running]
- Cautious Adaptive Controllers [use current SI to modify control law, allowing for SI uncertainty]
- Certainty Equivalent Adaptive Controllers [take current SI to be the true system, assume no uncertainty]
- Nonparametric Adaptive Controllers
- Parametric Adaptive Controllers
- Explicit Parameter Adaptive Controllers
- Implicit Parameter Adaptive Controllers.

Some special topics in adaptive control can be introduced as well :

1. Adaptive Control Based on Discrete–Time Process Identification
2. Adaptive Control Based on the Model Reference Technique
3. Adaptive Control based on Continuous–Time Process Models
4. Adaptive Control of Multivariable Processes
5. Adaptive Control of Nonlinear Processes.

Applications

When designing adaptive control systems, special consideration is necessary of *convergence* and *robustness* issues. *Lyapunov stability* is typically used to derive control adaptation laws and show convergence.

Typical applications of adaptive control are (in general) :

- Self–tuning of subsequently fixed linear controllers during the implementation phase for one operating point;

- Self-tuning of subsequently fixed robust controllers during the implementation phase for whole range of operating points;
- Self-tuning of fixed controllers on request if the process behaviour changes due to ageing, drift, wear etc.;
- Adaptive control of linear controllers for nonlinear or time-varying processes;
- Adaptive control or self-tuning control of nonlinear controllers for nonlinear processes;
- Adaptive control or self-tuning control of multivariable controllers for multivariable processes (MIMO systems);

Usually these methods adapt the controllers to both the process statics and dynamics. In special cases the adaptation can be limited to the static behavior alone, leading to adaptive control based on characteristic curves for the steady-states or to extremum value control, optimizing the steady state. Hence, there are several ways to apply adaptive control algorithms.

Chapter 2

INDUSTRIAL CONTROL SYSTEM

Industrial control system (ICS) is a general term that encompasses several types of *control systems* used in industrial production, including supervisory control and data acquisition (*SCADA*) systems, *distributed control systems* (DCS), and other smaller control system configurations such as *programmable logic controllers* (PLC) often found in the industrial sectors and critical infrastructures.

ICSs are typically used in industries such as electrical, water, oil, gas and data. Based on data received from remote stations, automated or operator-driven supervisory commands can be pushed to remote station control devices, which are often referred to as field devices. Field devices control local operations such as opening and closing valves and breakers, collecting data from sensor systems, and monitoring the local environment for alarm conditions.

A HISTORICAL PERSPECTIVE

Industrial control system technology has evolved over the decades.

DCS systems generally refer to the particular functional distributed control system design that exist in industrial process plants (*e.g.*, oil and gas, refining, chemical, pharmaceutical, some food and beverage, water and wastewater, pulp and paper, utility power, mining, metals). The DCS concept came about from a need to gather data and control the systems on a large campus in real time on high-bandwidth, low-latency data networks. It is common for loop controls to extend all the way to the top level controllers in a DCS, as everything works in real time. These systems evolved from a need to extend *pneumatic control* systems beyond just a small cell area of a *refinery*.

PLC (programmable logic controller) evolved out of a need to replace racks of relays in ladder form. The latter were not particularly reliable, were difficult to rewire, and were difficult to diagnose. PLC control tends to be used in very regular, high-speed *binary* controls, such as controlling a high-speed printing press. Originally, PLC equipment did not have remote *I/O* racks, and many couldn't even perform more than rudimentary *analog* controls.

SCADA's history is rooted in distribution applications, such as power, natural gas, and water pipelines, where there is a need to gather remote data through potentially unreliable or intermittent low-bandwidth/high-latency links. SCADA systems use open-loop control with sites that are widely separated geographically. A SCADA system uses RTUs (remote terminal units, also referred to as remote telemetry units) to send supervisory data back to a control center. Most RTU systems always did have some limited capacity to handle local controls while the master station is not available. However, over the years RTU systems have grown more and more capable of handling local controls.

The boundaries between these system definitions are blurring as time goes on. The technical limits that drove the designs of these various systems are no longer as much of an issue. Many PLC platforms can now perform quite well as a small DCS, using remote I/O and are sufficiently reliable that some SCADA systems actually manage closed loop control over long distances. With the increasing speed of today's processors, many DCS products have a full line of PLC-like subsystems that weren't offered when they were initially developed.

This led to the concept of a PAC (*programmable automation controller* or process automation controller), that is an amalgamation of these three concepts. Time and the market will determine whether this can simplify some of the terminology and confusion that surrounds these concepts today.

DISTRIBUTED CONTROL SYSTEM

A **distributed control system** (DCS) refers to a *control system* usually of a manufacturing system, *process* or any kind of *dynamic system*, in which the *controller* elements are not central in location (like the brain) but are distributed throughout the system with each component sub-system controlled by one or more controllers.

DCS (Distributed Control System) is a computerized control system used to control production in various industries.

The entire system of controllers is connected by networks for communication and monitoring.

DCS is a very broad term used in a variety of industries, to monitor and control distributed equipment.

- *Electrical power grids* and electrical generation plants
- Environmental control systems
- Traffic signals
- Radio signals
- Water management systems
- Oil refining plants
- *Metallurgical* process plants
- *Chemical plants*
- Pharmaceutical manufacturing

- *Sensor networks*
- Dry cargo and bulk oil carrier ships

Elements

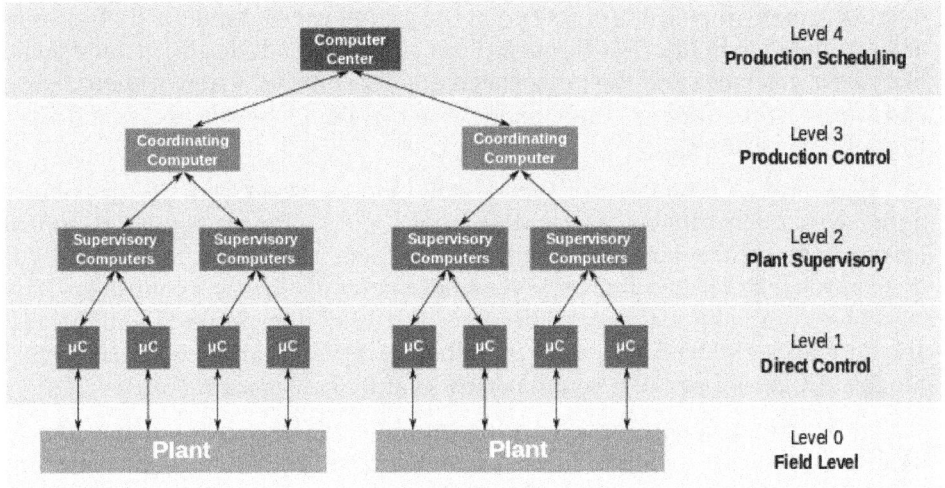

Figure : Functional levels of a typical Distributed Control System.

A DCS typically uses custom designed processors as controllers and uses both proprietary interconnections and communications protocol for communication. Input and output modules form component parts of the DCS. The processor receives information from input modules and sends information to output modules. The input modules receive information from input instruments in the process (or field) and the output modules transmit instructions to the output instruments in the field. Computer buses or electrical buses connect the processor and modules through multiplexer or demultiplexers. Buses also connect the distributed controllers with the central controller and finally to the *Human–machine interface* (HMI) or control consoles.

The elements of a DCS may connect directly to physical equipment such as switches, pumps and valves and to Human Machine Interface (HMI) via *SCADA*. The differences between a DCS and SCADA is often subtle, especially with advances in technology allowing the functionality of each to overlap.

Applications

Distributed control systems (DCSs) are dedicated systems used to control manufacturing processes that are continuous or batch-oriented, such as oil refining, petrochemicals, central station power generation, fertilizers, pharmaceuticals, food and beverage manufacturing, cement production, steelmaking, and papermaking. DCSs are connected to sensors and actuators and use setpoint control to control the flow of material through the plant. The most common example is a setpoint *control loop* consisting of a pressure sensor, controller, and *control valve*. Pressure or flow measurements are transmitted to the controller, usually through the aid of a signal conditioning input/output (I/O) device. When the measured variable reaches a certain point, the controller instructs a valve or actuation device to open or close until the fluidic flow process reaches the desired setpoint. Large oil refineries have many thousands of I/O points and employ very large DCSs. Processes are not limited to fluidic flow through pipes, however, and can also include things like paper machines and their associated quality controls, variable speed drives and motor control centers, cement kilns, mining operations, ore processing facilities, and many others.

A typical DCS consists of functionally and/or geographically distributed digital controllers capable of executing from 1 to 256 or more regulatory control loops in one control box. The input/output devices (I/O) can be integral with the controller or located remotely via a field network. Today's controllers have extensive computational capabilities and, in addition to proportional, integral, and derivative (PID) control, can generally perform logic and sequential control. Modern DCSs also support neural networks and fuzzy application.

DCSs are usually designed with redundant processors to enhance the reliability of the control system. Most systems come with canned displays and configuration software which enables the end user to set up the control system without a lot of low level programming. This allows the user to better focus on the application rather than the equipment, although a lot of system knowledge and skill is still required to support the hardware and software as well as the applications. Many plants have dedicated groups that focus on this task. These groups are in many cases augmented by vendor support personnel and/or maintenance support contracts.

DCSs may employ one or more workstations and can be configured at the workstation or by an off-line personal computer. Local communication is handled by a control network with transmission over twisted pair, coaxial, or fiber optic cable. A server and/or applications processor may be included in the system for extra computational, data collection, and reporting capability.

History

Early *minicomputers* were used in the control of industrial processes since the beginning of the 1960s. The *IBM 1800*, for example, was an early computer that had input/output hardware to gather process signals in a plant for conversion from field contact levels (for digital points) and analog signals to the digital domain.

The first industrial control computer system was built 1959 at the Texaco Port Arthur, Texas, refinery with an *RW-300* of the *Ramo-Wooldridge* Company.

The DCS was introduced in 1975. Both *Honeywell* and Japanese electrical engineering firm *Yokogawa* introduced their own independently produced DCSs at roughly the same time, with the TDC 2000 and CENTUM systems, respectively. US-based Bristol also introduced their UCS 3000 universal controller in 1975. In

1978 *Metso* (known as Valmet in 1978) introduced their own DCS system called Damatic (latest generation named Metso DNA). In 1980, Bailey (now part of ABB) introduced the NETWORK 90 system, Fisher Controls (now part of *Emerson Electric*) introduced the PROVoX system, *Fischer & Porter Company* (now also part of ABB) introduced DCI-4000 (DCI stands for Distributed Control Instrumentation).

The DCS largely came about due to the increased availability of microcomputers and the proliferation of microprocessors in the world of process control. Computers had already been applied to process automation for some time in the form of both direct digital control (DDC) and set point control. In the early 1970s *Taylor Instrument Company*, (now part of ABB) developed the 1010 system, Foxboro the FOX1 system, Fisher Controls the DC^2 system and *Bailey Controls* the 1055 systems. All of these were DDC applications implemented within minicomputers (*DEC PDP-11*, *Varian Data Machines*, *MODCOMP* etc.) and connected to proprietary Input/Output hardware. Sophisticated (for the time) continuous as well as batch control was implemented in this way. A more conservative approach was set point control, where process computers supervised clusters of analog process controllers. A CRT-based workstation provided visibility into the process using text and crude character graphics. Availability of a fully functional graphical user interface was a way away.

Central to the DCS model was the inclusion of control function blocks. Function blocks evolved from early, more primitive DDC concepts of "Table Driven" software. One of the first embodiments of object-oriented software, function blocks were self-contained "blocks" of code that emulated analog hardware control components and performed tasks that were essential to process control, such as execution of PID algorithms. Function blocks continue to endure as the predominant method of control for DCS suppliers, and are supported by key technologies such as **Foundation Fieldbus** today.

Midac Systems, of Sydney, Australia, developed an objected-oriented distributed direct digital control system in 1982. The central system ran 11 microprocessors sharing tasks and common memory and connected to a serial communication network of distributed controllers each running two Z80s. The system was installed at the University of Melbourne.

Digital communication between distributed controllers, workstations and other computing elements (peer to peer access) was one of the primary advantages of the DCS. Attention was duly focused on the networks, which provided the all-important lines of communication that, for process applications, had to incorporate specific functions such as determinism and redundancy. As a result, many suppliers embraced the IEEE 802.4 networking standard. This decision set the stage for the wave of migrations necessary when information technology moved into process automation and IEEE 802.3 rather than IEEE 802.4 prevailed as the control LAN.

The Network Centric Era of the 1980s

In the 1980s, users began to look at DCSs as more than just basic process control. A very early example of a *Direct Digital Control* DCS was completed by the Australian business *Midac* in 1981–82 using R-Tec Australian designed hardware. The system installed at the *University of Melbourne* used a serial communications network, connecting campus buildings back to a control room "front end". Each remote unit ran 2 *Z80* microprocessors whilst the front end ran 11 in a Parallel Processing configuration with paged common memory to share tasks and could run up to 20,000 concurrent controls objects.

It was believed that if openness could be achieved and greater amounts of data could be shared throughout the enterprise that even greater things could be achieved. The first attempts to increase the openness of DCSs resulted in the adoption of the predominant operating system of the day : *UNIX*. UNIX and its companion networking technology TCP-IP were developed by the US Department of Defense for openness, which was precisely the issue the process industries were looking to resolve.

As a result suppliers also began to adopt Ethernet-based networks with their own proprietary protocol layers. The full TCP/IP standard was not implemented, but the use of Ethernet made it possible to implement the first instances of object management and global data access technology. The 1980s also witnessed the first PLCs integrated into the DCS infrastructure. Plant-wide historians also emerged to capitalize on the extended reach of automation systems. The first DCS supplier to adopt UNIX and Ethernet networking technologies was Foxboro, who introduced the I/A Series system in 1987.

The Application-Centric Era of the 1990s

The drive toward openness in the 1980s gained momentum through the 1990s with the increased adoption of *commercial off-the-shelf* (COTS) components and IT standards. Probably the biggest transition undertaken during this time was the move from the UNIX operating system to the Windows environment. While the realm of the real time operating system (*RTOS*) for control applications remains dominated by real time commercial variants of UNIX or proprietary operating systems, everything above real-time control has made the transition to Windows.

The introduction of Microsoft at the desktop and server layers resulted in the development of technologies such as *OLE for process control (OPC)*, which is now a de facto industry connectivity standard. Internet technology also began to make its mark in automation and the DCS world, with most DCS HMI supporting Internet connectivity. The 1990s were also known for the "Fieldbus Wars", where rival organizations competed to define what would become the IEC fieldbus standard for digital communication with field instrumentation instead of 4–20 milliamp analog communications. The first fieldbus installations occurred in the 1990s. Towards the end of the decade, the technology began to develop significant momentum, with the market consolidated around Ethernet I/P, Foundation Fieldbus and Profibus

PA for process automation applications. Some suppliers built new systems from the ground up to maximize functionality with fieldbus, such as *Rockwell* Plant PAX System, *Honeywell* with *Experion* & Plantscape SCADA systems, *ABB* with System 800xA, Emerson Process Management with the *Emerson Process Management DeltaV* control system, *Siemens* with the SPPA-T3000 or *Simatic PCS 7* and *Azbil Corporation* with the *Harmonas-DEO* system. Fieldbus technics have been used to integrate machine, drives, quality and *condition monitoring* applications to one DCS with Metso DNA system.

The impact of COTS, however, was most pronounced at the hardware layer. For years, the primary business of DCS suppliers had been the supply of large amounts of hardware, particularly I/O and controllers. The initial proliferation of DCSs required the installation of prodigious amounts of this hardware, most of it manufactured from the bottom up by DCS suppliers. Standard computer components from manufacturers such as Intel and Motorola, however, made it cost prohibitive for DCS suppliers to continue making their own components, workstations, and networking hardware.

As the suppliers made the transition to COTS components, they also discovered that the hardware market was shrinking fast. COTS not only resulted in lower manufacturing costs for the supplier, but also steadily decreasing prices for the end users, who were also becoming increasingly vocal over what they perceived to be unduly high hardware costs. Some suppliers that were previously stronger in the *PLC* business, such as Rockwell Automation and Siemens, were able to leverage their expertise in manufacturing control hardware to enter the DCS marketplace with cost effective offerings, while the stability/scalability/reliability and functionality of these emerging systems are still improving. The traditional DCS suppliers introduced new generation DCS System based on the latest Communication and IEC Standards, which resulting in a trend of combining the traditional concepts/functionalities for PLC and DCS into a one for all solution — named "Process Automation System". The gaps among the various systems remain at the areas such as : the database integrity, pre-engineering functionality, system maturity, communication transparency and reliability. While it is expected the cost ratio is relatively the same (the more powerful the systems are, the more expensive they will be), the reality of the automation business is often operating strategically case by case. The current next evolution step is called *Collaborative Process Automation Systems*.

To compound the issue, suppliers were also realizing that the hardware market was becoming saturated. The life cycle of hardware components such as I/O and wiring is also typically in the range of 15 to over 20 years, making for a challenging replacement market. Many of the older systems that were installed in the 1970s and 1980s are still in use today, and there is a considerable installed base of systems in the market that are approaching the end of their useful life. Developed industrial economies in North America, Europe, and Japan already had many thousands of DCSs installed, and with few if any new plants being built, the market for new hardware was shifting rapidly to smaller, albeit faster growing regions such as China, Latin America, and Eastern Europe.

Because of the shrinking hardware business, suppliers began to make the challenging transition from a hardware-based business model to one based on software and value-added services. It is a transition that is still being made today. The applications portfolio offered by suppliers expanded considerably in the '90s to include areas such as production management, model-based control, real-time optimization, plant asset management (PAM), Real-time performance management (RPM) tools, *alarm management*, and many others. To obtain the true value from these applications, however, often requires a considerable service content, which the suppliers also provide.

PROGRAMMABLE LOGIC CONTROLLER

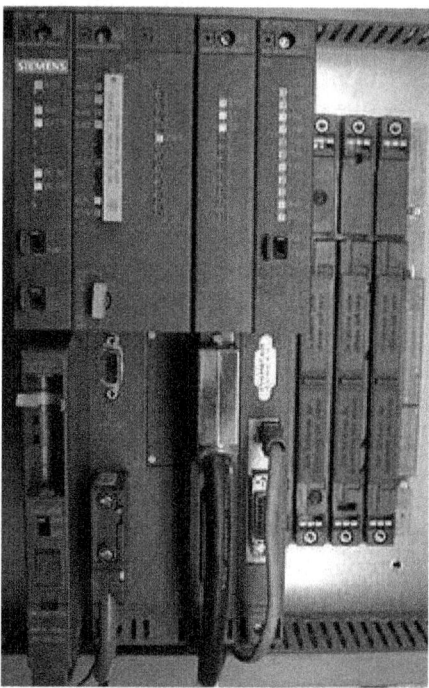

Figure : Siemens Simatic S7-400 system at rack, left-to-right : power supply unit PS407 4A,CPU 416-3, interface module IM 460-0 and communication processor CP 443-1.

A **Programmable Logic Controller, PLC** or **Programmable Controller** is a *digital computer* used for *automation* of *electromechanical* processes, such as control of machinery on factory *assembly lines, amusement rides*, or *light fixtures*. PLCs are used in many industries and machines. Unlike general-purpose computers, the PLC is designed for multiple inputs and output arrangements, extended temperature ranges, immunity to electrical noise, and resistance to vibration and impact. Programs to control machine operation are typically stored in battery-backed-up or *non-volatile memory*. A PLC is an example of a *hard real-time* system since output results must be produced in response to input conditions within a limited time, otherwise unintended operation will result.

History

Before the PLC, control, sequencing, and safety interlock logic for manufacturing automobiles was mainly composed of *relays, cam timers, drum sequencers,* and dedicated closed-loop controllers. Since these could number in the hundreds or even thousands, the process for updating such facilities for the yearly model *change-over* was very time consuming and expensive, as *electricians* needed to individually rewire relays to change the logic.

Digital computers, being general-purpose programmable devices, were soon applied to control of industrial processes. Early computers required specialist programmers, and stringent operating environmental control for temperature, cleanliness, and power quality. Using a general-purpose computer for process control required protecting the computer from the plant floor conditions. An industrial control computer would have several attributes : it would tolerate the shop-floor environment, it would support discrete (bit-form) input and output in an easily extensible manner, it would not require years of training to use, and it would permit its operation to be monitored. The response time of any computer system must be fast enough to be useful for control; the required speed varying according to the nature of the process.

In 1968 GM *Hydra-Matic* (the *automatic transmission* division of *General Motors*) issued a request for proposals for an electronic replacement for hard-wired relay systems based on a white paper written by engineer Edward R. Clark. The winning proposal came from Bedford Associates of *Bedford, Massachusetts*. The first PLC, designated the 084 because it was Bedford Associates' eighty-fourth project, was the result. Bedford Associates started a new company dedicated to developing, manufacturing, selling, and servicing this new product : Modicon, which stood for **MO**dular **DI**gital **CON**troller. One of the people who worked on that project was *Dick Morley*, who is considered to be the "father" of the PLC. The Modicon brand was sold in 1977 to *Gould Electronics*, and later acquired by German Company *AEG* and then by French *Schneider Electric*, the current owner.

One of the very first 084 models built is now on display at Modicon's headquarters in *North Andover, Massachusetts*. It was presented to Modicon by *GM*, when the unit was retired after nearly twenty years of uninterrupted service. Modicon used the 84 moniker at the end of its product range until the 984 made its appearance.

The automotive industry is still one of the largest users of PLCs.

Development

Early PLCs were designed to replace relay logic systems. These PLCs were programmed in *"ladder logic"*, which strongly resembles a schematic diagram of relay logic. This program notation was chosen to reduce training demands for the existing technicians. Other early PLCs used a form of *instruction list* programming, based on a stack-based logic solver.

Modern PLCs can be programmed in a variety of ways, from the relay-derived ladder logic to programming languages such as specially adapted dialects of *BASIC* and *C*. Another method is *State Logic, a very high-level programming language* designed to program PLCs based on *state transition diagrams*.

Many early PLCs did not have accompanying programming terminals that were capable of graphical representation of the logic, and so the logic was instead represented as a series of logic expressions in some version of *Boolean format*, similar to *Boolean algebra*. As programming terminals evolved, it became more common for ladder logic to be used, for the aforementioned reasons and because it was a familiar format used for electromechanical control panels. Newer formats such as State Logic and Function Block (which is similar to the way logic is depicted when using digital integrated logic circuits) exist, but they are still not as popular as ladder logic. A primary reason for this is that PLCs solve the logic in a predictable and repeating sequence, and ladder logic allows the programmer (the person writing the logic) to see any issues with the timing of the logic sequence more easily than would be possible in other formats.

Programming

Early PLCs, up to the mid-1990s, were programmed using proprietary programming panels or special-purpose programming *terminals,* which often had dedicated function keys representing the various logical elements of PLC programs. Some proprietary programming terminals displayed the elements of PLC programs as graphic symbols, but plain *ASCII* character representations of contacts, coils, and wires were common. Programs were stored on *cassette tape cartridges*. Facilities for printing and documentation were minimal due to lack of memory capacity. The very oldest PLCs used non-volatile *magnetic core memory*.

More recently, PLCs are programmed using application software on personal computers, which now represent the logic in graphic form instead of character symbols. The computer is connected to the PLC through *Ethernet, RS-232, RS-485* or *RS-422* cabling. The programming software allows entry and editing of the ladder-style logic. Generally the software provides functions for debugging and troubleshooting the PLC software, for example, by highlighting portions of the logic to show current status during operation or via simulation. The software will upload and download the PLC program, for backup and restoration purposes. In some models of programmable controller, the program is transferred from a personal computer to the PLC through a *programming board* which writes the program into a removable chip such as an *EEPROM* or *EPROM*.

Functionality

The functionality of the PLC has evolved over the years to include sequential relay control, motion control, *process control, distributed control systems* and *networking*. The data handling, storage, processing power and communication capabilities of some modern PLCs are approximately equivalent to *desktop computers*. PLC-like programming combined with remote I/O hardware, allow a general-purpose desktop computer to overlap some PLCs in certain applications. Regarding the practicality of these desktop computer based logic controllers, it is important to note that they have not been generally accepted in heavy industry because the desktop computers run on less stable operating systems than do PLCs, and because the desktop computer hardware is typically not designed to the same levels of tolerance to temperature, humidity, vibration, and longevity as the processors used in PLCs. In addition to the hardware limitations of desktop based logic, operating systems such as Windows do not lend themselves to deterministic logic execution, with the result that the logic may not always respond to changes in logic state or input status with the extreme consistency in timing as is expected from PLCs. Still, such desktop logic applications find use in less critical situations, such as laboratory automation and use in small facilities where the application is less demanding and critical, because they are generally much less expensive than PLCs.

Programmable Logic Relay (PLR)

In more recent years, small products called PLRs (programmable logic relays), and also by similar names, have become more common and accepted. These are very much like PLCs, and are used in light industry where only a few points of I/O (*i.e.* a few signals coming in from the real world and a few going out) are involved, and low cost is desired. These small devices are typically made in a common physical size and shape by several manufacturers, and branded by the makers of larger PLCs to fill out their low end product range. Popular names include PICO Controller, NANO PLC, and other names implying very small controllers. Most of these have between 8 and 12 discrete inputs, 4 and 8 discrete outputs, and up to 2 analog inputs. Size is usually about 4" wide, 3" high, and 3" deep. Most such devices include a tiny postage stamp sized LCD screen for viewing simplified ladder logic (only a very small portion of the program being visible at a given time) and status of I/O points, and typically these screens are accompanied by a 4-way rocker push-button plus four more separate push-buttons, similar to the key buttons on a VCR remote control, and used to navigate and edit the logic. Most have a small plug for connecting via RS-232 or RS-485 to a personal computer so that programmers can use simple Windows applications for programming instead of being forced to use the tiny LCD and push-button set for this purpose. Unlike regular PLCs that are usually modular and greatly expandable, the PLRs are usually not modular or expandable, but their price can be two *orders of magnitude* less than a PLC and they still offer robust design and deterministic execution of the logic.

PLC Topics

Features

The main difference from other computers is that PLCs are armored for severe conditions (such as dust, moisture, heat, cold) and have the facility for extensive *input/output* (I/O) arrangements. These connect the PLC to *sensors* and *actuators*. PLCs read *limit switches*, analog process variables (such as temperature and pressure), and the positions of complex positioning systems. Some uses *machine vision*. On the actuator side, PLCs operate *electric motors, pneumatic* or *hydraulic* cylinders, magnetic *relays, solenoids,* or *analog* outputs. The input/output arrangements may be built into a simple PLC, or the PLC may have external I/O modules attached to a computer network that plugs into the PLC.

Scan Time

A PLC program is generally executed repeatedly as long as the controlled system is running. The status of physical input points is copied to an area of memory accessible to the processor, sometimes called the "I/O Image Table". The program is then run from its first instruction rung down to the last rung. It takes some time for the processor of the PLC to evaluate all the rungs and update the I/O image table with the status of outputs. This scan time may be a few milliseconds for a small program or on a fast processor, but older PLCs running very large programs could take much longer (say, up to 100 ms) to execute the program. If the scan time were too long, the response of the PLC to process conditions would be too slow to be useful.

As PLCs became more advanced, methods were developed to change the sequence of ladder execution, and subroutines were implemented. This simplified programming could be used to save scan time for high-speed processes; for example, parts of the program used only for setting up the machine could be segregated from those parts required to operate at higher speed.

Special-purpose I/O modules, such as timer modules or counter modules such as encoders, can be used where the scan time of the processor is too long to reliably pick up, for example, counting pulses and interpreting quadrature from a shaft encoder. The relatively slow PLC can still interpret the counted values to control a machine, but the accumulation of pulses is done by a dedicated module that is unaffected by the speed of the program execution.

System Scale

A small PLC will have a fixed number of connections built in for inputs and outputs. Typically, expansions are available if the base model has insufficient I/O.

Modular PLCs have a chassis (also called a rack) into which are placed modules with different functions. The processor and selection of I/O modules are customized for the particular application. Several racks can be administered by a single processor, and may have thousands of inputs and outputs. A special high speed serial I/O link is used so that racks can be distributed away from the processor, reducing the wiring costs for large plants.

User Interface

PLCs may need to interact with people for the purpose of configuration, alarm reporting or everyday control. A *human-machine interface* (HMI) is employed for this purpose. HMIs are also referred to as man-machine interfaces (MMIs) and graphical user interfaces (GUIs). A simple system may use buttons and lights to interact with the user. Text displays are available as well as graphical touch screens. More complex systems use programming and monitoring software installed on a computer, with the PLC connected via a communication interface.

Communications

PLCs have built in communications ports, usually 9-pin *RS-232*, but optionally *EIA-485* or *Ethernet*. *Modbus*, *BACnet* or *DF1* is usually included as one of the *communications protocols*. Other options include various *fieldbuses* such as *Device Net* or *Profibus*. Other communications protocols that may be used are listed in the *List of automation protocols*.

Most modern PLCs can communicate over a network to some other system, such as a computer running a *SCADA* (Supervisory Control And Data Acquisition) system or web browser.

PLCs used in larger I/O systems may have *peer-to-peer* (P2P) communication between processors. This allows separate parts of a complex process to have individual control while allowing the subsystems to coordinate over the communication link. These communication links are also often used for *HMI* devices such as keypads or *PC*-type workstations.

Programming

PLC programs are typically written in a special application on a personal computer, then downloaded by a direct-connection cable or over a network to the PLC. The program is stored in the PLC either in battery-backed-up *RAM* or some other non-volatile *flash memory*. Often, a single PLC can be programmed to replace thousands of *relays*.

Under the *IEC 61131-3* standard, PLCs can be programmed using standards-based programming languages. A graphical programming notation called *Sequential Function Charts* is available on certain programmable controllers. Initially most PLCs utilized Ladder Logic Diagram Programming, a model which emulated electromechanical control panel devices (such as the contact and coils of relays) which PLCs replaced. This model remains common today.

IEC 61131-3 currently defines five programming languages for programmable control systems : *function block diagram* (FBD), *ladder diagram* (LD), *structured text* (ST; similar to the *Pascal programming language*), *instruction list* (IL; similar to *assembly language*) and *sequential function chart* (SFC). These techniques emphasize logical organization of operations.

While the fundamental concepts of PLC programming are common to all manufacturers, differences in I/O addressing, memory organization and instruction sets mean that PLC programs are never perfectly interchangeable between different makers. Even within the same product line of a single manufacturer, different models may not be directly compatible.

Security

Prior to the discovery of the *Stuxnet computer virus* in June 2010, security of PLCs received little attention. PLCs generally contain a real-time operating system such as *OS-9* or *VxWorks* and exploits for these systems exist much as they do for desktop computer operating systems such as *Microsoft Windows*. PLCs can also be attacked by gaining control of a computer they communicate with.

Simulation

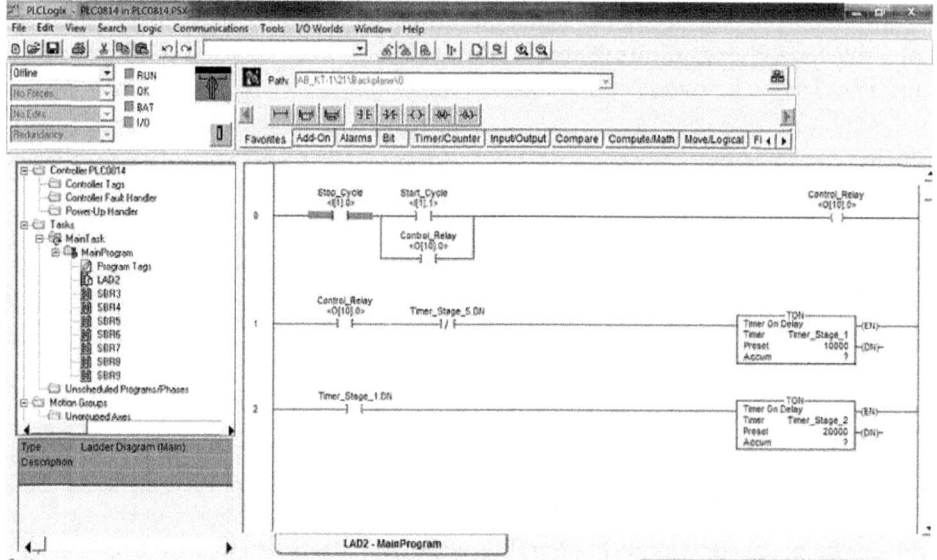

Figure : *PLCLogix* Simulation Software.

In order to properly understand the operation of a PLC, it is necessary to spend considerable time *programming*, testing, and *debugging* PLC programs. PLC systems are inherently expensive, and down-time is often very costly. In addition, if a PLC is programmed incorrectly it can result in lost productivity and dangerous conditions. PLC simulation software is a valuable tool in the understanding and learning of PLCs and to keep this knowledge refreshed and up to date. The advantages of using PLC simulation tools such as *PLCLogix* are that they save time in the design of automated control applications and they can also increase the level of safety associated with equipment since various "what if" scenarios can be tried and tested before the system is activated.

Redundancy

Some special processes need to work permanently with minimum unwanted stopping. Therefore, it is necessary to design a system which is fault tolerant and capable of handling the process with faulty modules. In such cases to increase the system availability in the event of hardware component failure, redundant

CPU or I/O modules with the same functionality can be added to hardware configuration for preventing total or partial process shutdown due to hardware failure from any kind.

PLC Compared with Other Control Systems

PLCs are well adapted to a range of *automation* tasks. These are typically industrial processes in manufacturing where the cost of developing and maintaining the automation system is high relative to the total cost of the automation, and where changes to the system would be expected during its operational life. PLCs contain input and output devices compatible with industrial pilot devices and controls; little electrical design is required, and the design problem centers on expressing the desired sequence of operations. PLC applications are typically highly customized systems, so the cost of a packaged PLC is low compared to the cost of a specific custom-built controller design. On the other hand, in the case of mass-produced goods, customized control systems are economical. This is due to the lower cost of the components, which can be optimally chosen instead of a "generic" solution, and where the non-recurring engineering charges are spread over thousands or millions of units.

For high volume or very simple fixed automation tasks, different techniques are used. For example, a consumer *dishwasher* would be controlled by an electromechanical *cam timer* costing only a few dollars in production quantities.

A *microcontroller*-based design would be appropriate where hundreds or thousands of units will be produced and so the development cost (design of power supplies, input/output hardware and necessary testing and certification) can be spread over many sales, and where the end-user would not need to alter the control. Automotive applications are an example; millions of units are built each year, and very few end-users alter the programming of these controllers. However, some specialty vehicles such as transit buses economically use PLCs instead of custom-designed controls, because the volumes are low and the development cost would be uneconomical.

Very complex process control, such as used in the chemical industry, may require algorithms and performance beyond the capability of even high-performance PLCs. Very high-speed or precision controls may also require customized solutions; for example, aircraft flight controls. *Single-board computers* using semi-customized or fully proprietary hardware may be chosen for very demanding control applications where the high development and maintenance cost can be supported. "Soft PLCs" running on desktop-type computers can interface with industrial I/O hardware while executing programs within a version of commercial operating systems adapted for process control needs.

Programmable controllers are widely used in motion control, positioning control and torque control. Some manufacturers produce motion control units to be integrated with PLC so that *G-code* (involving a *CNC* machine) can be used to instruct machine movements.

PLCs may include logic for single-variable feedback analog control loop, a "proportional, integral, derivative" or "*PID controller*". A PID loop could be used to control the temperature of a manufacturing process, for example. Historically PLCs were usually configured with only a few analog control loops; where processes required hundreds or thousands of loops, a *distributed control system* (DCS) would instead be used. As PLCs have become more powerful, the boundary between DCS and PLC applications has become less distinct.

PLCs have similar functionality as *Remote Terminal Units*. An RTU, however, usually does not support control algorithms or control loops. As hardware rapidly becomes more powerful and cheaper, *RTUs*, PLCs and *DCSs* are increasingly beginning to overlap in responsibilities, and many vendors sell RTUs with PLC-like features and *vice versa*. The industry has standardized on the *IEC 61131-3* functional block language for creating programs to run on RTUs and PLCs, although nearly all vendors also offer proprietary alternatives and associated development environments.

In recent years "Safety" PLCs have started to become popular, either as stand-alone models or as functionality and safety-rated hardware added to existing controller architectures (Allen Bradley Guardlogix, Siemens F-series etc.). These differ from conventional PLC types as being suitable for use in safety-critical applications for which PLCs have traditionally been supplemented with hard-wired safety relays. For example, a Safety PLC might be used to control access to a robot cell with *trapped-key access*, or perhaps to manage the shutdown response to an emergency stop on a conveyor production line. Such PLCs typically have a restricted regular instruction set augmented with safety-specific instructions designed to interface with emergency stops, light screens and so forth. The flexibility that such systems offer has resulted in rapid growth of demand for these controllers.

Discrete and Analog Signals

Discrete signals behave as binary switches, yielding simply an On or Off signal (1 or 0, True or False, respectively). Push buttons, *Limit switches*, and *photoelectric sensors* are examples of devices providing a discrete signal. Discrete signals are sent using either *voltage* or *current*, where a specific range is designated as *On* and another as *Off*. For example, a PLC might use 24 V DC I/O, with values above 22 V DC representing *On*, values below 2VDC representing *Off*, and intermediate values undefined. Initially, PLCs had only discrete I/O.

Analog signals are like volume controls, with a range of values between zero and full-scale. These are typically interpreted as integer values (counts) by the PLC, with various ranges of accuracy depending on the device and the number of bits available to store the data. As PLCs typically use 16-bit signed binary processors, the integer values are limited between -32,768 and +32,767. Pressure, temperature, flow, and weight are often represented by analog signals. Analog signals can use *voltage* or *current* with a magnitude proportional to the value of the process signal. For example, an analog 0-10 V input or *4-20 mA* would be *converted* into an integer value of 0–32767.

Current inputs are less sensitive to electrical noise (*i.e.* from welders or electric motor starts) than voltage inputs.

Example

As an example, say a facility needs to store water in a tank. The water is drawn from the tank by another system, as needed, and our example system must manage the water level in the tank by controlling the valve that refills the tank. Shown is a *"ladder diagram"* which shows the control system. A ladder diagram is a method of drawing control circuits which pre-dates PLCs. The ladder diagram resembles the schematic diagram of a system built with electromechanical relays. Shown are :

- Two inputs (from the low and high level switches) represented by contacts of the float switches.
- An output to the fill valve, labelled as the fill valve which it controls.
- An "internal" contact, representing the output signal to the fill valve which is created in the program.
- A logical control scheme created by the interconnection of these items in software.

In ladder diagram, the contact symbols represent the state of bits in processor memory, which corresponds to the state of physical inputs to the system. If a discrete input is energized, the memory bit is a 1, and a "normally open" contact controlled by that bit will pass a logic "true" signal on to the next element of the ladder. **Therefore, the contacts in the PLC program that "read" or look at the physical switch contacts in this case must be "opposite" or open in order to return a TRUE for the closed physical switches.** Internal status bits, corresponding to the state of discrete outputs, are also available to the program.

In the example, the physical state of the float switch contacts must be considered when choosing "normally open" or "normally closed" symbols in the ladder diagram. The PLC has two discrete inputs from *float switches* (Low Level and High Level). Both float switches (normally closed) open their contacts when the water level in the tank is above the physical location of the switch.

When the water level is below both switches, the float switch physical contacts are both closed, and a true (logic 1) value is passed to the Fill Valve output. Water begins to fill the tank. The internal "Fill Valve" contact latches the circuit so that even when the "Low Level" contact opens (as the water passes the lower switch), the fill valve remains on. Since the High Level is also normally closed, water continues to flow as the water level remains between the two switch levels. Once the water level rises enough so that the "High Level" switch is off (opened), the PLC will shut the inlet to stop the water from overflowing; this is an example of seal-in (latching) logic. The output is sealed in until a high level condition breaks the circuit. After that the fill valve remains off until the level drops so low that the Low Level switch is activated, and the process repeats again.

```
| (N.C. physical     (N.C. physical                                       |
|   Switch)            Switch)                                            |
|   Low Level          High Level                    Fill Valve          |
|------[ ]------|------[ ]---------------------(OUT)---------|
|              |                                                          |
|              |                                                          |
|              |                                                          |
|   Fill Valve |                                                          |
|------[ ]------|                                                        |
|                                                                         |
|                                                                         |
```

A complete program may contain thousands of rungs, evaluated in sequence. Typically the PLC processor will alternately scan all its inputs and update outputs, then evaluate the ladder logic; input changes during a program scan will not be effective until the next I/O update. A complete program scan may take only a few milliseconds, much faster than changes in the controlled process.

Programmable controllers vary in their capabilities for a "rung" of a ladder diagram. Some only allow a single output bit. There are typically limits to the number of series contacts in line, and the number of branches that can be used. Each element of the rung is evaluated sequentially. If elements change their state during evaluation of a rung, hard-to-diagnose faults can be generated, although sometimes the technique is useful. Some implementations forced evaluation from left-to-right as displayed and did not allow reverse flow of a logic signal (in multi-branched rungs) to affect the output.

MODEL PREDICTIVE CONTROL

Model predictive control (MPC) is an advanced method of *process control* that has been in use in the *process* industries in *chemical plants* and *oil refineries* since the 1980s. In recent years it has also been used in *power system* balancing models. Model predictive controllers rely on dynamic models of the process, most often linear *empirical* models obtained by *system identification*. The main advantage of MPC is the fact that it allows the current timeslot to be optimized, while keeping future timeslots in account. This is achieved by optimizing a finite time-horizon, but only implementing the current timeslot. MPC has the ability to anticipate future events and can take control actions accordingly. PID and LQR controllers do not have this predictive ability. MPC is a digital control.

Overview

The models used in MPC are generally intended to represent the behavior of complex *dynamical systems*. The additional complexity of the MPC control algorithm is not generally needed to provide adequate control of simple systems, which are often controlled well by generic *PID controllers*. Common dynamic characteristics that are difficult for PID controllers include large time delays and high-order dynamics.

MPC models predict the change in the *dependent variables* of the modeled system that will be caused by changes in the *independent variables*. In a chemical process, independent variables that can be adjusted by the controller are often either the setpoints of regulatory PID controllers (pressure, flow, temperature, etc.) or the final control element (valves, dampers, etc.). Independent variables that cannot be adjusted by the controller are used as disturbances. Dependent variables in these processes are other measurements that represent either control objectives or process constraints.

MPC uses the current plant measurements, the current dynamic state of the process, the MPC models, and the process variable targets and limits to calculate future changes in the dependent variables. These changes are calculated to hold the dependent variables close to target while honoring constraints on both independent and dependent variables. The MPC typically sends out only the first change in each independent variable to be implemented, and repeats the calculation when the next change is required.

While many real processes are not linear, they can often be considered to be approximately linear over a small operating range. Linear MPC approaches are used in the majority of applications with the feedback mechanism of the MPC compensating for prediction errors due to structural mismatch between the model and the process. In model predictive controllers that consist only of linear models, the *superposition principle* of *linear algebra* enables the effect of changes in multiple independent variables to be added together to predict the response of the dependent variables. This simplifies the control problem to a series of direct matrix algebra calculations that are fast and robust.

When linear models are not sufficiently accurate to represent the real process nonlinearities, several approaches can be used. In some cases, the process variables can be transformed before and/or after the linear MPC model to reduce the nonlinearity. The process can be controlled with nonlinear MPC that uses a nonlinear model directly in the control application. The nonlinear model may be in the form of an empirical data fit (*e.g.* artificial neural networks) or a high-fidelity dynamic model based on fundamental mass and energy balances. The nonlinear model may be linearized to derive a *Kalman filter* or specify a model for linear MPC.

Theory Behind MPC

MPC is based on iterative, finite horizon optimization of a plant model. At time t the current plant state is sampled and a cost minimizing control strategy is computed (via a numerical minimization algorithm) for a relatively short time horizon in the future : $[t, t + T]$. Specifically, an online or on-the-fly calculation is used to explore state trajectories that emanate from the current state and find (via the solution of *Euler–Lagrange equations*) a cost-minimizing control strategy until time $t + T$. Only the first step of the control strategy is implemented, then the plant state is sampled again and the calculations are repeated starting from the now current state, yielding a new control and new predicted state path. The prediction horizon keeps being shifted forward and for this reason MPC is also called **receding horizon control**. Although this approach is not optimal, in practice it has given very good results. Much academic research has been done to find fast methods of solution of Euler–Lagrange type equations, to understand the global stability properties of MPC's local optimization, and in general to improve the MPC method. To some extent the theoreticians have been trying to catch up with the control engineers when it comes to MPC.

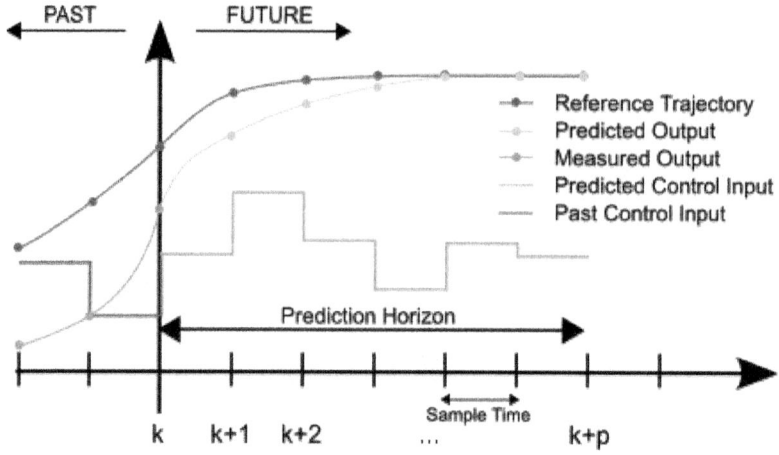

Figure : A discrete MPC scheme.

Principles of MPC

Model Predictive Control (MPC) is a multivariable control algorithm that uses :

- An internal dynamic model of the process
- A history of past control moves and
- An optimization cost function J over the receding prediction horizon, to calculate the optimum control moves.

The optimization cost function is given by :

$$J = \sum_{i=1}^{N} w_{xi}(r_i - x_i)^2 + \sum_{i=1}^{N} w_{ui}\Delta u_i^2$$

without violating constraints (low/high limits)

With :

$x_i = i$ -th controlled variable (*e.g.* measured temperature)

$r_i = i$ -th reference variable (*e.g.* required temperature)

$u_i = i$ -th manipulated variable (*e.g.* control valve)

w_{xi} = weighting coefficient reflecting the relative importance of x_i

w_{ui} = weighting coefficient penalizing relative big changes in u_i

etc.

Nonlinear MPC

Nonlinear Model Predictive Control, or NMPC, is a variant of model predictive control (MPC) that is characterized by the use of nonlinear system models in the prediction. As in linear MPC, NMPC requires the iterative solution of optimal control problems on a finite prediction horizon. While these problems are convex in linear MPC, in nonlinear MPC they are not convex anymore. This poses challenges for both, NMPC stability theory and numerical solution.

The numerical solution of the NMPC optimal control problems is typically based on direct optimal control methods using Newton-type optimization schemes, in one of the variants : *direct single shooting, direct multiple shooting methods,* or *direct collocation.* NMPC algorithms typically exploit the fact that consecutive optimal control problems are similar to each other.

This allows to initialize the Newton-type solution procedure efficiently by a suitably shifted guess from the previously computed optimal solution, saving considerable amounts of computation time. The similarity of subsequent problems is even further exploited by path following algorithms (or "real-time iterations") that never attempt to iterate any optimization problem to convergence, but instead only take one iteration towards the solution of the most current NMPC problem, before proceeding to the next one, which is suitably initialized.

While NMPC applications have in the past been mostly used in the process and chemical industries with comparatively slow sampling rates, NMPC is more and more being applied to applications with high sampling rates, *e.g.,* in the automotive industry, or even when the states are distributed in space (*Distributed parameter systems*).

Chapter 3

PID Controller

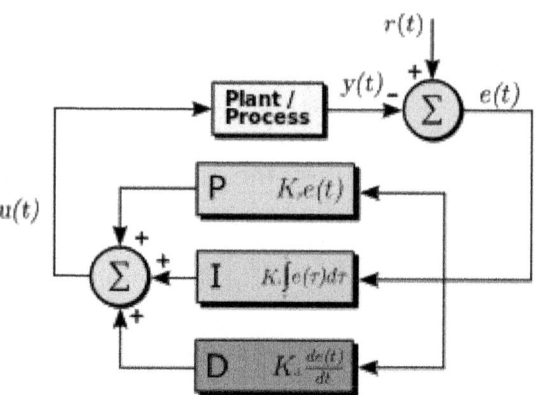

Figure : A *block diagram* of a PID controller in a feedback loop.

A **proportional-integral-derivative controller (PID controller)** is a *control loop feedback mechanism* (*controller*) widely used in *industrial control systems*. A PID controller calculates an "error" value as the difference between a measured *process variable* and a desired *setpoint*. The controller attempts to minimize the error in outputs by adjusting the process control inputs.

The PID controller *algorithm* involves three separate constant parameters, and is accordingly sometimes called **three-term control** : the *proportional*, the *integral* and *derivative* values, denoted P, I, and D. Simply put, these values can be interpreted in terms of time : P depends on the *present* error, I on the accumulation of *past* errors, and D is a prediction of *future* errors, based on current rate of change. The weighted sum of these three actions is used to adjust the process via a control element such as the position of a *control valve*, a *damper*, or the power supplied to a heating element.

In the absence of knowledge of the underlying process, a PID controller has historically been considered to be the best controller. By tuning the three parameters in the PID controller algorithm, the controller can provide control action designed for specific process requirements. The response of the controller can be described in terms of the responsiveness of the controller to an error, the degree to which the controller *overshoots* the setpoint, and the degree of system oscillation. Note that the use of the PID algorithm for control does not guarantee *optimal control* of the system or system stability.

Some applications may require using only one or two actions to provide the appropriate system control. This is achieved by setting the other parameters to zero. A PID controller will be called a PI, PD, P or I controller in the absence of the respective control actions. PI controllers are fairly common, since derivative action is sensitive to measurement noise, whereas the absence of an integral term may prevent the system from reaching its target value due to the control action.

HISTORY AND APPLICATIONS

PID controllers date to 1890s *governor* design. PID controllers were subsequently developed in automatic ship steering. One of the earliest examples of a PID-type controller was developed by *Elmer Sperry* in 1911, while the first published theoretical analysis of a PID controller was by *Russian American* engineer *Nicolas Minorsky*, in (*Minorsky 1922*). Minorsky was designing automatic steering systems for the US Navy, and based his analysis on observations of a *helmsman*, observing that the helmsman controlled the ship not only based on the current error, but also on past error and current rate of change; this was then made mathematical by Minorsky. His goal was stability, not general control, which significantly simplified the problem. While proportional control provides stability against small disturbances, it was insufficient for dealing with a steady disturbance, notably a stiff gale (due to *droop*), which required adding the integral term. Finally, the derivative term was added to improve control.

Trials were carried out on the *USS New Mexico*, with the controller controlling the *angular velocity* (not angle) of the rudder. PI control yielded sustained yaw (angular error) of ±2°, while adding D yielded yaw of ±1/6°, better than most helmsmen could achieve.

The Navy ultimately did not adopt the system, due to resistance by personnel. Similar work was carried out and published by several others in the 1930s.

In the early history of automatic process control the PID controller was implemented as a mechanical device. These mechanical controllers used a *lever*, *spring* and a *mass* and were often energized by compressed air. These *pneumatic* controllers were once the industry standard.

Electronic *analog* controllers can be made from a *solid-state* or *tube amplifier*, a *capacitor* and a *resistor*. Electronic analog PID control loops were often found within more complex electronic systems, for example, the head positioning of a *disk drive*, the power conditioning of a *power supply*, or even the movement-detection circuit of a modern *seismometer*. Nowadays, electronic controllers have largely been replaced by digital controllers implemented with *microcontrollers* or *FPGAs*.

Most modern PID controllers in industry are implemented in *programmable logic controllers* (PLCs) or as a panel-mounted digital controller. Software implementations have the advantages that they are relatively cheap and are flexible with respect to the implementation of the PID algorithm. PID temperature controllers are applied in industrial ovens, plastics injection machinery, hot stamping machines and packing industry.

Variable voltages may be applied by the *time proportioning* form of *pulse-width modulation* (PWM) — a *cycle time* is fixed, and variation is achieved by varying the proportion of the time during this cycle that the controller outputs +1 (or −1) instead of 0. On a digital system the possible proportions are discrete — *e.g.*, increments of 0.1 second within a 2 second cycle time yields 20 possible steps : percentage increments of 5%; so there is a *discretization error*, but for high enough time resolution this yields satisfactory performance.

CONTROL LOOP BASICS

A familiar example of a control loop is the action taken when adjusting hot and cold *faucets* to fill a container with water at a desired temperature by mixing hot and cold water. The person touches the water in the container as it fills to sense its temperature. Based on this feedback they perform a control action by adjusting the hot and cold faucets until the temperature stabilizes as desired.

The sensed water temperature is the *process variable* (PV). The desired temperature is called the setpoint (SP). The input to the process (the water valve position), and the output of the PID controller, is called the manipulated variable (MV) or the control variable (CV). The difference between the temperature measurement and the setpoint is the error (e) and quantifies whether the water in the container is too hot or too cold and by how much.

After measuring the temperature (PV), and then calculating the error, the controller decides how much to change the tap position (MV). Because the taps can be adjusted for anything from cool water through to very hot, this is an example of **proportional** control. In the event that water in the container is not heating quickly enough, the controller may try to speed up the process by opening up the hot water valve quite wide for a while. This is an example of **derivative** action. If the temperature of the container is settling out too low, despite a good flow of warm water, the controller may open the hot valve more and more as time goes by. This is an example of an **integral** control.

Making a change that is too large when the error is small will lead to overshoot. If the controller were to repeatedly make changes that were too large and repeatedly overshoot the target, the output would *oscillate* around the setpoint in either a constant, growing, or decaying *sinusoid*. If the amplitude of the oscillations increase with time then the system is unstable, whereas if they decrease the system is stable. If the oscillations remain at a constant magnitude the system is *marginally stable*.

In the interest of achieving a gradual convergence to the desired temperature (SP), the controller may *damp* the anticipated future oscillations by tempering its adjustments, or reducing the *loop gain*.

If a controller starts from a stable state with zero error (PV = SP), then further changes by the controller will be in response to changes in other measured or unmeasured inputs to the process that affect the process, and hence the PV. Variables that affect the process other than the MV are known as disturbances. Generally controllers are used to reject disturbances and to implement setpoint changes. Changes in feedwater temperature constitute a disturbance to the faucet temperature control process.

In theory, a controller can be used to control any process which has a measurable output (PV), a known ideal value for that output (SP) and an input to the process (MV) that will affect the relevant PV. Controllers are used in industry to regulate *temperature, pressure, flow rate, chemical* composition, *weight, position, speed* and practically every other variable for which a measurement exists.

Control System

The basic idea behind a PID controller is to read a sensor, then compute the desired actuator output by calculating proportional, integral, and derivative responses and summing those three components to compute the output. Before we start to define the parameters of a PID controller, we shall see what a closed loop system is and some of the terminologies associated with it.

Closed Loop System

In a typical control system, the *process variable* is the system parameter that needs to be controlled, such as temperature (°C), pressure (psi), or flow rate (liters/minute). A sensor is used to measure the process variable and provide feedback to the control system. The *set point* is the desired or command value for the process variable, such as 100 degrees Celsius in the case of a temperature control system. At any given moment, the difference between the process variable and the set point is used by the control system algorithm *(compensator)*, to determine the desired actuator output to drive the system (plant). For instance, if the measured temperature process variable is 100 °C and the desired temperature set point is 120 °C, then the *actuator output* specified by the control algorithm might be to drive a heater. Driving an actuator to turn on a heater causes the system to become warmer, and results in an increase in the temperature process variable. This is called a closed loop control system, because the process of reading sensors to provide constant feedback and calculating the desired actuator output is repeated continuously and at a fixed loop rate as illustrated in figure 1.

In many cases, the actuator output is not the only signal that has an effect on the system. For instance, in a temperature chamber there might be a source of cool air that sometimes blows into the chamber and disturbs the temperature. Such a term is referred to as *disturbance*. We usually try to design the control system to minimize the effect of disturbances on the process variable.

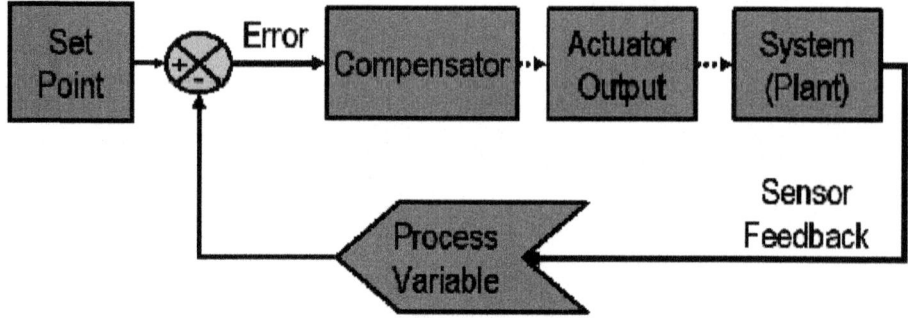

Figure 1 : Block diagram of a typical closed loop system.

Defintion of Terminlogies

The control design process begins by defining the performance requirements. Control system performance is often measured by applying a step function as the set point command variable, and then measuring the response of the process variable. Commonly, the response is quantified by measuring defined waveform characteristics. Rise Time is the amount of time the system takes to go from 10% to 90% of the steady-state, or final, value. Percent Overshoot is the amount that the process variable overshoots the final value, expressed as a percentage of the final value. Settling time is the time required for the process variable to settle to within a certain percentage (commonly 5%) of the final value. Steady-State Error is the final difference between the process variable and set point. Note that the exact definition of these quantities will vary in industry and academia.

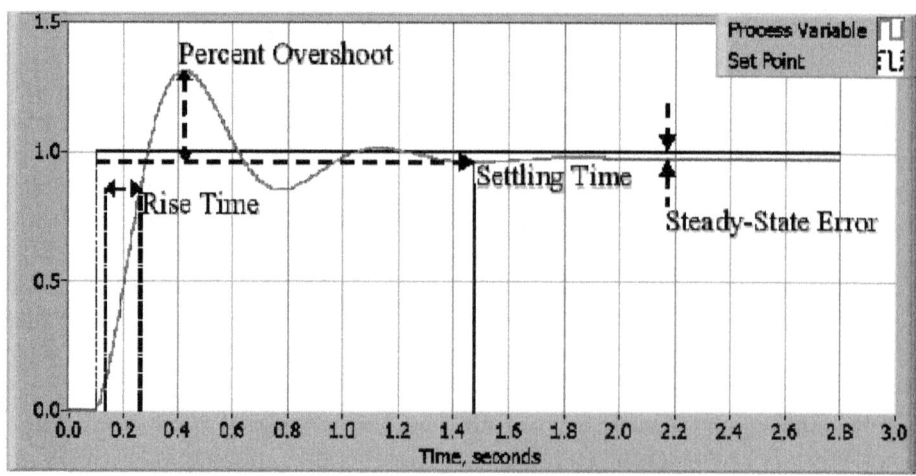

Figure 2 : Response of a typical PID closed loop system.

After using one or all of these quantities to define the performance requirements for a control system, it is useful to define the worst case conditions in which the control system will be expected to meet these design requirements. Often times, there is a disturbance in the system that affects the process variable or the measurement of the process variable. It is important to design a control system that performs satisfactorily during worst case conditions. The measure of how well the control system is able to overcome the effects of disturbances is referred to as the *disturbance rejection* of the control system.

In some cases, the response of the system to a given control output may change over time or in relation to some variable. A *nonlinear system* is a system in which the control parameters that produce a desired response at one operating point might not produce a satisfactory response at another operating point. For instance, a chamber partially filled with fluid will exhibit a much faster response to heater output when nearly empty than it will when nearly full of fluid. The measure of how well the control system will tolerate disturbances and nonlinearities is referred to as the *robustness* of the control system.

Some systems exhibit an undesirable behavior called *deadtime*. Deadtime is a delay between when a process variable changes, and when that change can be observed. For instance, if a temperature sensor is placed far away from a cold water fluid inlet valve, it will not measure a change in temperature immediately if the valve is opened or closed. Deadtime can also be caused by a system or output actuator that is slow to respond to the control command, for instance, a valve that is slow to open or close. A common source of deadtime in chemical plants is the delay caused by the flow of fluid through pipes.

Loop cycle is also an important parameter of a closed loop system. The interval of time between calls to a control algorithm is the loop cycle time. Systems that change quickly or have complex behavior require faster control loop rates.

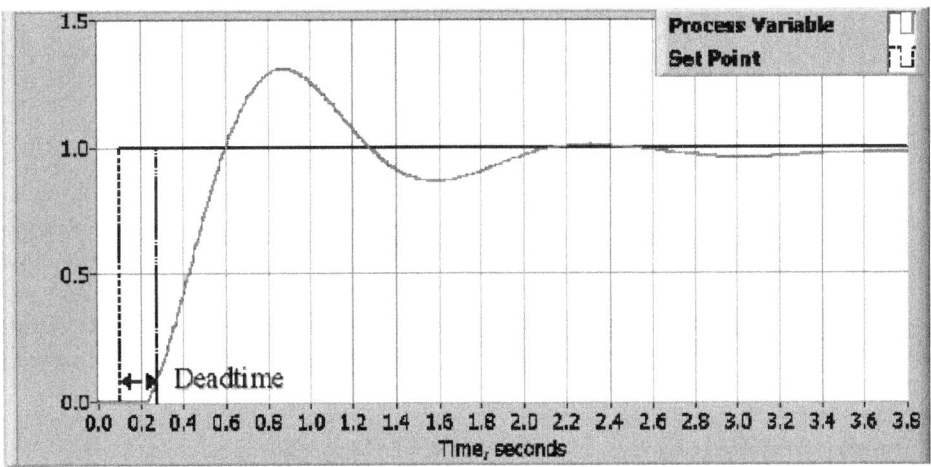

Figure 3 : Response of a closed loop system with deadtime.

Once the performance requirements have been specified, it is time to examine the system and select an appropriate control scheme. In the vast majority of applications, a PID control will provide the required results —

Back to Top

PID Theory

Proportional Response

The proportional component depends only on the difference between the set point and the process variable. This difference is referred to as the Error term. The *proportional gain* (K_c) determines the ratio of output response to the error signal. For instance, if the error term has a magnitude of 10, a proportional gain of 5 would produce a proportional response of 50. In general, increasing the proportional gain will increase the speed of the control system response. However, if the proportional gain is too large, the process variable will begin to oscillate. If K_c is increased further, the oscillations will become larger and the system will become unstable and may even oscillate out of control.

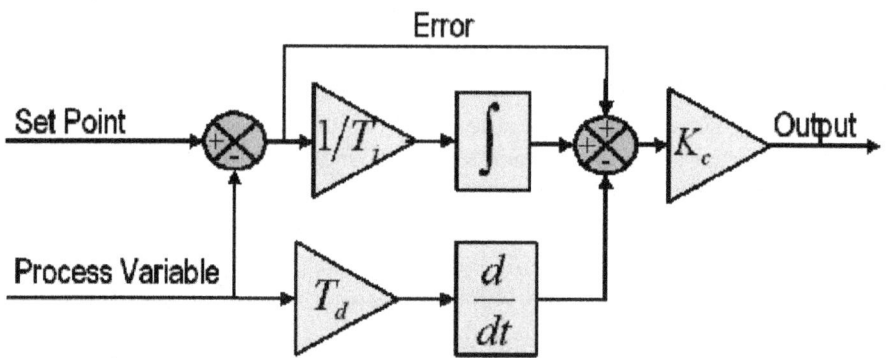

Figure 4 : Block diagram of a basic PID control algorithm.

Integral Response

The integral component sums the error term over time. The result is that even a small error term will cause the integral component to increase slowly. The integral response will continually increase over time unless the error is zero, so the effect is to drive the Steady-State error to zero. Steady-State error is the final difference between the process variable and set point. A phenomenon called integral windup results when integral action saturates a controller without the controller driving the error signal toward zero.

Derivative Response

The derivative component causes the output to decrease if the process variable is increasing rapidly. The derivative response is proportional to the rate of change of the process variable. Increasing the *derivative time* (T_d) parameter will cause the control system to react more strongly to changes in the error term and will increase the speed of the overall control system response. Most practical control systems use very small derivative time (T_d), because the Derivative Response is highly sensitive to noise in the process variable signal. If the sensor feedback signal is noisy or if the control loop rate is too slow, the derivative response can make the control system unstable.

Tuning

The process of setting the optimal gains for P, I and D to get an ideal response from a control system is called *tuning*. There are different methods of tuning of which the "guess and check" method and the Ziegler Nichols method will be discussed.

The gains of a PID controller can be obtained by trial and error method. Once an engineer understands the significance of each gain parameter, this method becomes relatively easy. In this method, the I and D terms are set to zero first and the proportional gain is increased until the output of the loop oscillates. As one increases the proportional gain, the system becomes faster, but care must be taken not make the system unstable. Once P has been set to obtain a desired fast response, the integral term is increased to stop the oscillations. The integral term reduces the steady state error, but increases overshoot. Some amount of overshoot is always necessary for a fast system so that it could respond to changes immediately. The integral term is tweaked to achieve a minimal steady state error. Once the P and I have been set to get the desired fast control system with minimal steady state error, the derivative term is increased until the loop is acceptably quick to its set point. Increasing derivative term decreases overshoot and yields higher gain with stability but would cause the system to be highly sensitive to noise. Often times, engineers need to tradeoff one characteristic of a control system for another to better meet their requirements.

The Ziegler-Nichols method is another popular method of tuning a PID controller. It is very similar to the trial and error method wherein I and D are set to zero and P is increased until the loop starts to oscillate. Once oscillation starts, the critical gain K_c and the period of oscillations P_c are noted. The P, I and D are then adjusted as per the tabular column shown below.

Table 1. Ziegler-Nichols tuning, using the oscillation method.

Control	P	Ti	Td
P	0.5Kc	—	—
PI	0.45Kc	Pc/1.2	—
PID	0.60Kc	0.5Pc	Pc/8

NI LabVIEW and PID

LabVIEW PID toolset features a wide array of VIs that greatly help in the design of a PID based control system. Control output range limiting, integrator anti-windup and bumpless controller output for PID gain changes are some of the salient features of the PID VI. The PID Advanced VI includes all the features of the PID VI along with non-linear integral action, two degree of freedom control and error-squared control.

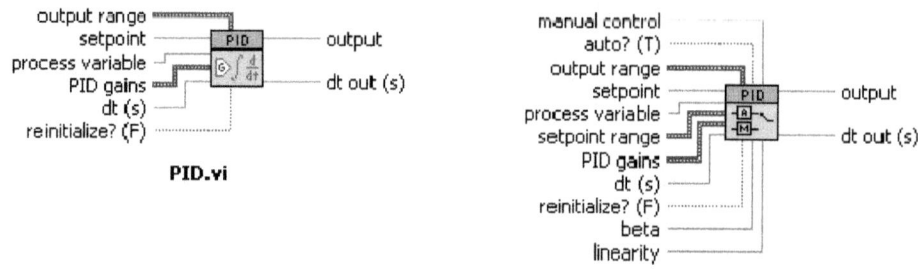

Figure 5 : VIs from the PID controls palette of LabVIEW.

PID palette also features some advanced VIs like the PID Autotuning VI and the PID Gain Schedule VI. The PID Autotuning VI helps in refining the PID parameters of a control system. Once an educated guess about the values of P, I and D have been made, the PID Autotuning VI helps in refining the PID parameters to obtain better response from the control system.

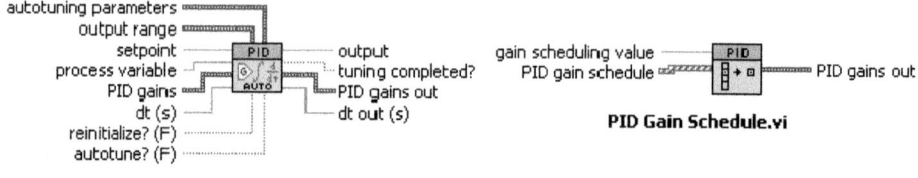

Figure 6 : Advanced VIs from the PID controls palette of LabVIEW.

The reliability of the controls system is greatly improved by using the Lab-VIEW Real Time module running on a real time target. National Instruments provides the new M Series Data Acquisition boards which provide higher accuracy and better performance than an average control system.

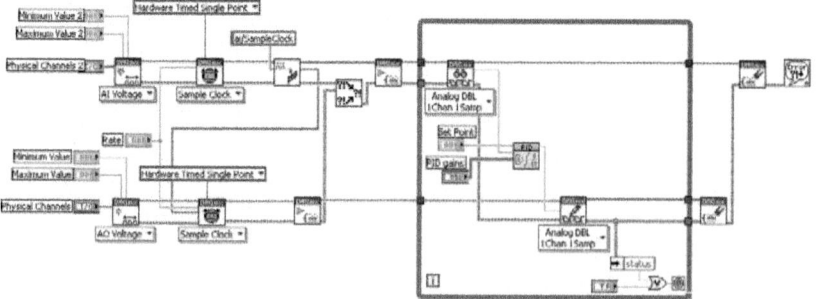

Figure 7 : A typical LabVIEW VI showing PID control with a plug-in NI data acquisition device.

The tight integration of these M Series boards with LabVIEW minimizes the development time involved and greatly increases the productivity of any engineer. Figure 7 shows a typical VI in LabVIEW showing PID control using NI-DAQmx API of M series devices.

Summary

The PID control algorithm is a robust and simple algorithm that is widely used in the industry. The algorithm has sufficient flexibility to yield excellent results in a wide variety of applications and has been one of the main reasons for the continued use over the years. NI LabVIEW and NI plug-in data acquisition devices offer higher accuracy and better performance to make an excellent PID control system.

LOOP TUNING

Tuning a control loop is the adjustment of its control parameters (proportional band/gain, integral gain/reset, derivative gain/rate) to the optimum values for the desired control response. Stability (bounded oscillation) is a basic requirement, but beyond that, different systems have different behavior, different applications have different requirements, and requirements may conflict with one another.

PID tuning is a difficult problem, even though there are only three parameters and in principle is simple to describe, because it must satisfy complex criteria within the *limitations of PID control*. There are accordingly various methods for loop tuning, and more sophisticated techniques are the subject of patents; this section describes some traditional manual methods for loop tuning.

Designing and tuning a PID controller appears to be conceptually intuitive, but can be hard in practice, if multiple (and often conflicting) objectives such as short transient and high stability are to be achieved. Usually, initial designs need to be adjusted repeatedly through computer simulations until the closed-loop system performs or compromises as desired.

Some processes have a degree of *nonlinearity* and so parameters that work well at full-load conditions don't work when the process is starting up from no-load; this can be corrected by *gain scheduling* (using different parameters in different operating regions). PID controllers often provide acceptable control using default tunings, but performance can generally be improved by careful tuning, and performance may be unacceptable with poor tuning.

Stability

If the PID controller parameters (the gains of the proportional, integral and derivative terms) are chosen incorrectly, the controlled process input can be unstable, *i.e.*, its output *diverges*, with or without *oscillation*, and is limited only by saturation or mechanical breakage. Instability is caused by *excess* gain, particularly in the presence of significant *lag*.

Generally, stabilization of response is required and the process must not oscillate for any combination of process conditions and setpoints, though sometimes *marginal stability* (bounded oscillation) is acceptable or desired.

Optimum Behavior

The optimum behavior on a process change or setpoint change varies depending on the application.

Two basic requirements are *regulation* (disturbance rejection–staying at a given setpoint) and *command tracking* (implementing setpoint changes)—these refer to how well the controlled variable tracks the desired value. Specific criteria for command tracking include *rise time* and *settling time*. Some processes must not allow an overshoot of the process variable beyond the setpoint if, for example, this would be unsafe. Other processes must minimize the energy expended in reaching a new setpoint.

Overview of Methods

There are several methods for tuning a PID loop. The most effective methods generally involve the development of some form of process model, then choosing P, I, and D based on the dynamic model parameters. Manual tuning methods can be relatively inefficient, particularly if the loops have response times on the order of minutes or longer.

The choice of method will depend largely on whether or not the loop can be taken "offline" for tuning, and on the response time of the system. If the system can be taken offline, the best tuning method often involves subjecting the system to a step change in input, measuring the output as a function of time, and using this response to determine the control parameters.

Table : Choosing a tuning method

Method	Advantages	Disadvantages
Manual tuning	No math required; online.	Requires experienced personnel.
Ziegler–Nichols	Proven method; online.	Process upset, some trial-and-error, very aggressive tuning.
Software tools	Consistent tuning; online or offline — can employ computer-automated control system design (CAutoD) techniques; may include valve and sensor analysis; allows simulation before downloading; can support non-steady-state (NSS) tuning.	Some cost or training involved.
Cohen–Coon	Good process models.	Some math; offline; only good for first-order processes.

Manual Tuning

If the system must remain online, one tuning method is to first set K_i and K_d values to zero. Increase the K_p until the output of the loop oscillates, then the K_p should be set to approximately half of that value for a "quarter amplitude decay" type response. Then increase K_i until any offset is corrected in sufficient time for the process. However, too much K_i will cause instability. Finally, increase K_d, if required, until the loop is acceptably quick to reach its reference after a load disturbance. However, too much K_d will cause excessive response and overshoot. A fast PID loop tuning usually overshoots slightly to reach the setpoint more quickly; however, some systems cannot accept overshoot, in which case an *over-damped* closed-loop system is required, which will require a K_p setting significantly less than half that of the K_p setting that was causing oscillation.

Table : Effects of *increasing* a parameter independently

Parameter	Rise time	Overshoot	Settling time	Steady-state error	Stability
K_p	Decrease	Increase	Small change	Decrease	Degrade
K_i	Decrease	Increase	Increase	Eliminate	Degrade
K_d	Minor change	Decrease	Decrease	No effect in theory	Improve if K_d small

Ziegler–Nichols Method

Another heuristic tuning method is formally known as the *Ziegler–Nichols method*, introduced by *John G. Ziegler* and *Nathaniel B. Nichols* in the 1940s. As in the method above, the K_i and K_d gains are first set to zero. The proportional gain is increased until it reaches the ultimate gain, K_u, at which the output of the loop starts to oscillate. K_u and the oscillation period P_u are used to set the gains as shown :

Table : Ziegler–Nichols method

Control Type	K_p	K_i	K_d
P	$0.50\,K_u$	—	—
PI	$0.45\,K_u$	$1.2\,K_p/P_u$	—
PID	$0.60\,K_u$	$2\,K_p/P_u$	$K_p P_u/8$

These gains apply to the ideal, parallel form of the PID controller. When applied to the standard PID form, the integral and derivative time parameters T_i and T_d are only dependent on the oscillation period P_u.

PID Tuning Software

Most modern industrial facilities no longer tune loops using the manual calculation methods shown above. Instead, PID tuning and loop optimization software are used to ensure consistent results. These software packages will gather the data, develop process models, and suggest optimal tuning. Some software packages can even develop tuning by gathering data from reference changes.

Mathematical PID loop tuning induces an impulse in the system, and then uses the controlled system's frequency response to design the PID loop values. In loops with response times of several minutes, mathematical loop tuning is recommended, because trial and error can take days just to find a stable set of loop values. Optimal values are harder to find. Some digital loop controllers offer a self-tuning feature in which very small setpoint changes are sent to the process, allowing the controller itself to calculate optimal tuning values.

Other formulas are available to tune the loop according to different performance criteria. Many patented formulas are now embedded within PID tuning software and hardware modules.

Advances in automated PID Loop Tuning software also deliver algorithms for tuning PID Loops in a dynamic or Non-Steady State (NSS) scenario. The software will model the dynamics of a process, through a disturbance, and calculate PID control parameters in response.

Modifications to the PID Algorithm

The basic PID algorithm presents some challenges in control applications that have been addressed by minor modifications to the PID form.

Integral Windup

One common problem resulting from the ideal PID implementations is *integral windup*, where a large change in setpoint occurs (say a positive change) and the integral term accumulates an error larger than the maximal value for the regulation variable (windup), thus the system overshoots and continues to increase as this accumulated error is unwound. This problem can be addressed by :

- Initializing the controller integral to a desired value
- Increasing the setpoint in a suitable ramp
- Disabling the integral function until the PV has entered the controllable region
- Preventing the integral term from accumulating above or below pre-determined bounds
- Back-calculating the integral term to constrain the process output within feasible bounds.

Overshooting from Known Disturbances

For example, a PID loop is used to control the temperature of an electric resistance furnace where the system has stabilized. Now when the door is opened and something cold is put into the furnace the temperature drops below the setpoint. The integral function of the controller tends to compensate this error by introducing another error in the positive direction. This overshoot can be avoided by freezing of the integral function after the opening of the door for the time the control loop typically needs to reheat the furnace.

PI Controller

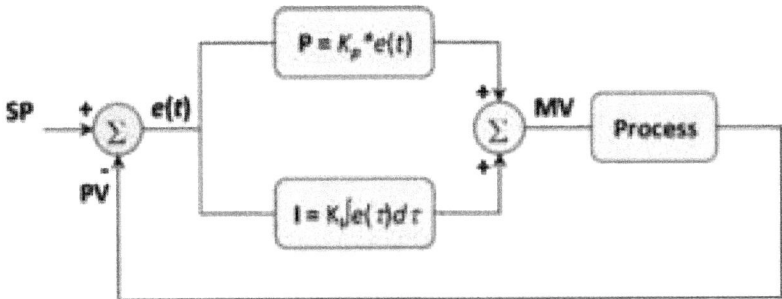

Figure : Basic block of a PI controller.

A **PI Controller** (proportional-integral controller) is a special case of the PID controller in which the derivative (D) of the error is not used.

The controller output is given by

$$K_p \Delta + K_I \int \Delta dt$$

where Δ is the error or deviation of actual measured value (PV) from the setpoint (SP).

$$\Delta = SP - PV.$$

A PI controller can be modelled easily in software such as *Simulink* or *Xcos* using a "flow chart" box involving *Laplace* operators :

$$C = \frac{G(1 + \tau s)}{\tau s}$$

where

$G = K_p =$ proportional gain

$G/\tau = K_I =$ integral gain.

Setting a value for G is often a trade off between decreasing overshoot and increasing settling time.

The lack of derivative action may make the system more steady in the steady state in the case of noisy data. This is because derivative action is more sensitive to higher-frequency terms in the inputs.

Without derivative action, a PI-controlled system is less responsive to real (non-noise) and relatively fast alterations in state and so the system will be slower to reach setpoint and slower to respond to perturbations than a well-tuned PID system may be.

Deadband

Many PID loops control a mechanical device (for example, a valve). Mechanical maintenance can be a major cost and wear leads to control degradation in the form of either *stiction* or a *deadband* in the mechanical response to an input signal. The rate of mechanical wear is mainly a function of how often a device is activated to make a change. Where wear is a significant concern, the PID loop may have an output *deadband* to reduce the frequency of activation of the output (valve). This is accomplished by modifying the controller to hold its output steady if the change would be small (within the defined deadband range). The calculated output must leave the deadband before the actual output will change.

Set Point Step Change

The proportional and derivative terms can produce excessive movement in the output when a system is subjected to an instantaneous step increase in the error, such as a large setpoint change. In the case of the derivative term, this is due to taking the derivative of the error, which is very large in the case of an instantaneous step change. As a result, some PID algorithms incorporate the following modifications :

Derivative of the Process Variable

In this case the PID controller measures the derivative of the measured *process variable* (PV), rather than the derivative of the error. This quantity is always continuous (*i.e.*, never has a step change as a result of changed setpoint). For this technique to be effective, the derivative of the PV must have the opposite sign of the derivative of the error, in the case of negative feedback control.

Set Point Ramping

In this modification, the setpoint is gradually moved from its old value to a newly specified value using a linear or first order differential ramp function. This avoids the *discontinuity* present in a simple step change.

Set Point Weighting

Set point weighting uses different multipliers for the error depending on which element of the controller it is used in. The error in the integral term must be the true control error to avoid steady-state control errors. This affects the controller's set point response. These parameters do not affect the response to load disturbances and measurement noise.

Limitations of PID Control

While PID controllers are applicable to many control problems, and often perform satisfactorily without any improvements or only coarse tuning, they can perform poorly in some applications, and do not in general provide *optimal control*. The fundamental difficulty with PID control is that it is a feed*back* system, with *constant* parameters, and no direct knowledge of the process, and thus overall performance is reactive and a compromise. While PID control is the best controller in an observer without a model of the process, better performance can be obtained by overtly modeling the actor of the process without resorting to an observer.

PID controllers, when used alone, can give poor performance when the PID loop gains must be reduced so that the control system does not overshoot, oscillate or *hunt* about the control setpoint value. They also have difficulties in the presence of non-linearities, may trade-off regulation versus response time, do not react to changing process behavior (say, the process changes after it has warmed up), and have lag in responding to large disturbances.

The most significant improvement is to incorporate feed-forward control with knowledge about the system, and using the PID only to control error. Alternatively, PIDs can be modified in more minor ways, such as by changing the parameters (either *gain scheduling* in different use cases or adaptively modifying them based on performance), improving measurement (higher sampling rate, precision, and accuracy, and low-pass filtering if necessary), or cascading multiple PID controllers.

Linearity

Another problem faced with PID controllers is that they are linear, and in particular symmetric. Thus, performance of PID controllers in non-linear systems (such as *HVAC systems*) is variable. For example, in temperature control, a common use case is active heating (via a heating element) but passive cooling (heating off, but no cooling), so overshoot can only be corrected slowly–it cannot be forced downward. In this case the PID should be tuned to be overdamped, to prevent or reduce overshoot, though this reduces performance (it increases settling time).

Noise in Derivative

A problem with the derivative term is that small amounts of measurement or process *noise* can cause large amounts of change in the output. It is often helpful to filter the measurements with a *low-pass filter* in order to remove higher-frequency noise components. However, low-pass filtering and derivative control can cancel each other out, so reducing noise by instrumentation is a much better choice. Alternatively, a nonlinear *median filter* may be used, which improves the filtering efficiency and practical performance. In some case, the differential band can be turned off in many systems with little loss of control. This is equivalent to using the PID controller as a *PI controller*.

Improvements

Feed-forward

The control system performance can be improved by combining the *feedback* (or closed-loop) control of a PID controller with *feed-forward* (or open-loop) control. Knowledge about the system (such as the desired acceleration and inertia) can be fed forward and combined with the PID output to improve the overall system performance. The feed-forward value alone can often provide the major portion of the controller output. The PID controller can be used primarily to respond to whatever difference or *error* remains between the setpoint (SP) and the actual value of the process variable (PV). Since the feed-forward output is not affected by the process feedback, it can never cause the control system to oscillate, thus improving the system response and stability.

For example, in most motion control systems, in order to accelerate a mechanical load under control, more force or torque is required from the prime mover, motor, or actuator. If a velocity loop PID controller is being used to control the speed of the load and command the force or torque being applied by the prime mover, then it is beneficial to take the instantaneous acceleration desired for the load, scale that value appropriately and add it to the output of the PID velocity loop controller. This means that whenever the load is being accelerated or decelerated, a proportional amount of force is commanded from the prime mover regardless of the feedback value. The PID loop in this situation uses the feedback information to change the combined output to reduce the remaining difference between the process setpoint and the feedback value. Working together, the combined open-loop feed-forward controller and closed-loop PID controller can provide a more responsive, stable and reliable control system.

Bumpless Operation

PID controllers are often implemented with a "bumpless" initialization feature that recalculates an appropriate integral accumulator term to maintain a consistent process output through parameter changes.

Other Improvements

In addition to feed-forward, PID controllers are often enhanced through methods such as PID *gain scheduling* (changing parameters in different operating conditions), *fuzzy logic* or *computational verb logic*. Further practical application issues can arise from instrumentation connected to the controller. A high enough sampling rate, measurement precision, and measurement accuracy are required to achieve adequate control performance. Another new method for improvement of PID controller is to increase the degree of freedom by using fractional order. The order of the integrator and differentiator add increased flexibility to the controller.

Cascade Control

One distinctive advantage of PID controllers is that two PID controllers can be used together to yield better dynamic performance. This is called cascaded PID control. In cascade control there are two PIDs arranged with one PID controlling the setpoint of another. A PID controller acts as outer loop controller, which controls the primary physical parameter, such as fluid level or velocity. The other controller acts as inner loop controller, which reads the output of outer loop controller as setpoint, usually controlling a more rapid changing parameter, flowrate or acceleration. It can be mathematically proven that the working frequency of the controller is increased and the time constant of the object is reduced by using cascaded PID controllers.

For example, a temperature-controlled circulating bath has two PID controllers in cascade, each with its own thermocouple temperature sensor. The outer controller controls the temperature of the water using a thermocouple located far from the heater where it accurately reads the temperature of the bulk of the water. The error term of this PID controller is the difference between the desired bath temperature and measured temperature. Instead of controlling the heater directly, the outer PID controller sets a heater temperature goal for the inner PID controller. The inner PID controller controls the temperature of the heater using a thermocouple attached to the heater. The inner controller's error term is the difference between this heater temperature setpoint and the measured temperature of the heater. Its output controls the actual heater to stay near this setpoint.

The proportional, integral and differential terms of the two controllers will be very different. The outer PID controller has a long time constant — all the water in the tank needs to heat up or cool down. The inner loop responds much more quickly. Each controller can be tuned to match the physics of the system *it* controls — heat transfer and thermal mass of the whole tank or of just the heater — giving better total response.

Alternative Nomenclature and PID Forms

Ideal Versus Standard PID Form

The form of the PID controller most often encountered in industry, and the one most relevant to tuning algorithms is the *standard form*. In this form the K_p gain is applied to the I_{out} and D_{out} terms, yielding :

$$MV(t) = K_p\left(e(t) + \frac{1}{T_i}\int_0^t e(\tau)d\tau + T_d\frac{d}{dt}e(t) \right)$$

where

T_i is the *integral time*

T_d is the *derivative time.*

In this standard form, the parameters have a clear physical meaning. In particular, the inner summation produces a new single error value which is compensated for future and past errors. The addition of the proportional and derivative components effectively predicts the error value at T_d seconds (or samples) in the future, assuming that the loop control remains unchanged. The integral component adjusts the error value to compensate for the sum of all past errors, with the intention of completely eliminating them in T_i seconds (or samples). The resulting compensated single error value is scaled by the single gain K_p.

In the ideal parallel form, shown in the controller theory section

$$MV(t) = K_p e(t) + K_i \int_0^t e(\tau)d\tau + K_d\frac{d}{dt}e(t)$$

the gain parameters are related to the parameters of the standard form through $K_i = \dfrac{K_p}{T_i}$ and $K_d = K_p T_d$. This parallel form, where the parameters are treated as simple gains, is the most general and flexible form. However, it is also the form where the parameters have the least physical interpretation and is generally reserved for theoretical treatment of the PID controller. The standard form, despite being slightly more complex mathematically, is more common in industry.

Reciprocal Gain

In many cases, the manipulated variable output by the PID controller is a dimensionless fraction between 0 and 100% of some maximum possible value, and the translation into real units (such as pumping rate or watts of heater power) is both outside the PID controller proper, and not known with any accuracy. The process variable, however, is in dimensioned units such as temperature. It is common in this case to express the gain K_p not as "output per degree", but rather in the form of a temperature $1/K_p$ which is "degrees per full output". This is the range over which the output changes from 0 to 1 (0% to 100%).

Basing Derivative Action on PV

In most commercial control systems, derivative action is based on PV rather than error. This is because the digitized version of the algorithm produces a large unwanted spike when the SP is changed. If the SP is constant then changes in PV will be the same as changes in error. Therefore this modification makes no difference to the way the controller responds to process disturbances.

$$MV(t) = K_p \left(e(t) + \frac{1}{T_i} \int_0^t e(\tau)d\tau - T_d \frac{d}{dt} PV(t) \right)$$

Basing Proportional Action on PV

Most commercial control systems offer the option of also basing the proportional action on PV. This means that only the integral action responds to changes in SP. While at first this might seem to adversely affect the time that the process will take to respond to the change, the controller may be retuned to give almost the same response - largely by increasing K_p. The modification to the algorithm does not affect the way the controller responds to process disturbances, but the change in tuning has a beneficial effect. Often the magnitude and duration of the disturbance will be more than halved. Since most controllers have to deal frequently with process disturbances and relatively rarely with SP changes, properly tuned the modified algorithm can dramatically improve process performance.

$$MV(t) = K_p \left(-PV(t) + \frac{1}{T_i} \int_0^t e(\tau)d\tau - T_d \frac{d}{dt} PV(t) \right)$$

Tuning methods such as Ziegler-Nichols and Cohen-Coon will not be reliable when used with this algorithm. King describes an effective chart-based method.

Laplace Form of the PID Controller

Sometimes it is useful to write the PID regulator in *Laplace transform* form :

$$G(s) = K_p + \frac{K_i}{s} + K_d s = \frac{K_d s^2 + K_p s + K_i}{s}$$

Having the PID controller written in Laplace form and having the *transfer function* of the controlled system makes it easy to determine the closed-loop transfer function of the system.

PID Pole Zero Cancellation

The PID equation can be written in this form :

$$G(s) = K_d \frac{s^2 + \frac{K_p}{K_d} s + \frac{K_i}{K_d}}{s}$$

When this form is used it is easy to determine the closed loop transfer function.

$$H(s) = \frac{1}{s^2 + 2\zeta\omega_0 s + \omega_0^2}$$

If

$$\frac{K_i}{K_d} = \omega_0^2$$

$$\frac{K_p}{K_d} = 2\zeta\omega_0$$

Then

$$G(s)H(s) = \frac{K_d}{s}$$

While this appears to be very useful to remove unstable poles, it is in reality not the case. The closed loop transfer function from disturbance to output still contains the unstable poles.

Series/Interacting Form

Another representation of the PID controller is the series, or *interacting* form

$$G(s) = K_c \frac{(\tau_i s + 1)}{\tau_i s}(\tau_d s + 1)$$

where the parameters are related to the parameters of the standard form through

$K_p = K_c \cdot \alpha$, $T_i = \tau_i \cdot \alpha$, and

$$T_d = \frac{\tau_d}{\alpha}$$

with

$$\alpha = 1 + \frac{\tau_d}{\tau_i} \,.$$

This form essentially consists of a PD and PI controller in series, and it made early (analog) controllers easier to build. When the controllers later became digital, many kept using the interacting form.

Discrete implementation

The analysis for designing a digital implementation of a PID controller in a *micro-controller* (MCU) or *FPGA* device requires the standard form of the PID controller to be *discretized*. Approximations for first-order derivatives are made by backward *finite differences*. The integral term is discretised, with a sampling time Δt, as follows,

$$\int_0^{t_k} e(\tau)d\tau = \sum_{i=1}^{k} e(t_i)\Delta t$$

The derivative term is approximated as,

$$\frac{de(t_k)}{dt} = \frac{e(t_k) - e(t_{k-1})}{\Delta t}$$

Thus, a *velocity algorithm* for implementation of the discretized PID controller in a MCU is obtained by differentiating $u(t)$, using the numerical definitions of the first and second derivative and solving for $u(t_k)$ and finally obtaining :

$$u(t_k) = u(t_{k-1}) + K_p\left[\left(1 + \frac{\Delta t}{T_i} + \frac{T_d}{\Delta t}\right)e(t_k) + \left(-1 - \frac{2T_d}{\Delta t}\right)e(t_{k-1}) + \frac{T_d}{\Delta t}e(t_{k-2})\right]$$

s.t. $T_i = K_p/K_i$, $T_d = K_d/K_p$

Pseudocode

Here is a simple software loop that implements a PID algorithm :

```
previous_error = 0
integral = 0
start:
  error = setpoint - measured value
  integral = integral + error*dt
  derivative = (error - previous_error)/dt
  output = Kp*error + Ki*integral + Kd*derivative
  previous_error = error
  wait(dt)
  goto start
```

In this example, two variables that will be maintained within the loop are *initialized* to zero, then the loop begins. The current *error* is calculated by subtracting the *measured value* (the process variable or PV) from the current *setpoint* (SP). Then, *integral* and *derivative* values are calculated and these and the *error* are combined with three preset gain terms — the proportional gain, the integral gain and the derivative gain — to derive an *output* value. In the real world, this is *D to A converted* and passed into the process under control as the manipulated variable (or MV). The current error is stored elsewhere for re-use in the next differentiation, the program then waits until dt seconds have passed since start, and the loop begins again, *reading in* new values for the PV and the setpoint and calculating a new value for the error.

Chapter 4

PROPORTIONAL CONTROL-THE SIMPLEST PID CONTROLLER

THE P-ONLY CONTROL ALGORITHM

The simplest algorithm in the PID family is a proportional or P-Only controller. Like all automatic controllers, it repeats a measurement-computation-action procedure at every loop sample time, T, following the logic flow shown in the block diagram below :

General Control Loop Block Diagram

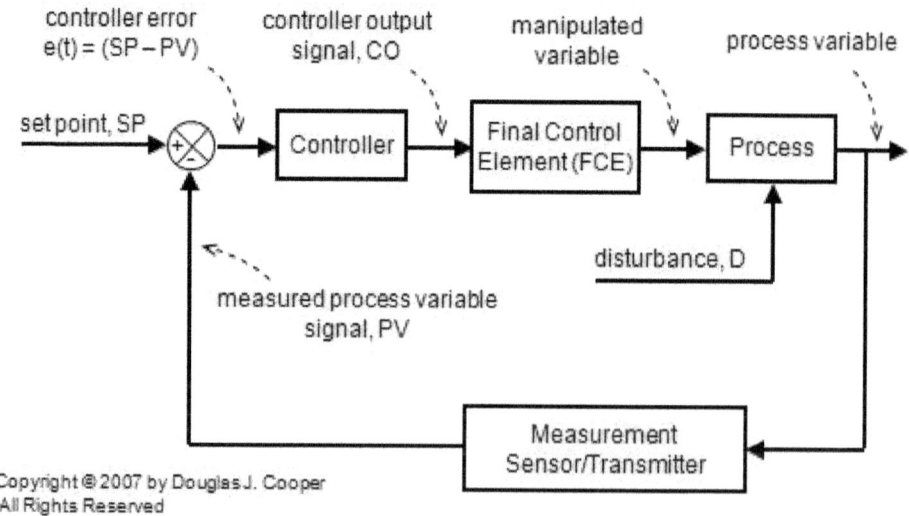

Starting at the far right of the control loop block diagram above :

- A sensor measures and transmits the current value of the process variable, PV, back to the controller (the 'controller wire in')
- Controller error at current time t is computed as set point minus measured process variable, or $e(t)$ = SP–PV
- The controller uses this e(t) in a control algorithm to compute a new controller output signal, CO
- The CO signal is sent to the final control element (*e.g.* valve, pump, heater, fan) causing it to change (the 'controller wire out')
- The change in the final control element (FCE) causes a change in a manipulated variable
- The change in the manipulated variable (*e.g.* flow rate of liquid or gas) causes a change in the PV.

The goal of the controller is to make $e(t)$ = 0 in spite of unplanned and un-measured disturbances. Since $e(t)$ = SP–PV, this is the same as saying a controller seeks to make PV = SP.

The P-only Algorithm

The P-only controller computes a CO action every loop sample time T as :

$$CO = CO_{bias} + Kc \cdot e(t)$$

Where :

CO_{bias} = controller bias or null value

Kc = controller gain, *a tuning parameter*

$e(t)$ = controller error = SP–PV

SP = set point

PV = measured process variable

Design Level of Operation

Real processes display a nonlinear behavior, which means their apparent process gain, time constant and/or dead time changes as operating level changes and as major disturbances change. Since controller design and tuning is based on these Kp, Tp and Θp values, controllers should be designed and tuned for a pre-defined level of operation.

When designing a cruise control system for a car, for example, would it make sense for us to perform bump tests to generate dynamic data when the car is travelling twice the normal speed limit while going down hill on a windy day? Of course not.

Bump test data should be collected as close as practical to the design PV when the disturbances are quiet and near their typical values. Thus, the design level of operation for a cruise control system is when the car is travelling at highway speed on flat ground on a calm day.

Definition : The design level of operation (DLO) is where we expect the SP and PV will be during normal operation while the important disturbances are quiet and at their expected or typical values.

Understanding Controller Bias

Let's suppose the P-Only control algorithm shown above is used for cruise control in an automobile and CO is the throttle signal adjusting the flow of fuel to the engine.

Let's also suppose that the speed SP is 70 and the measured PV is also 70 (units can be mph or kph depending on where you live in the world). Since PV = SP, then $e(t) = 0$ and the algorithm reduces to :

$$CO = CO_{bias} + Kc \cdot (0) = CO_{bias}$$

If CO_{bias} is zero, then when set point equals measurement, the above equation says that the throttle signal, CO, is also zero. This makes no sense. Clearly if the car is travelling 70 kph, then some baseline flow of fuel is going to the engine.

This baseline value of the CO is called the bias or null value. In this example, CO_{bias} is the flow of fuel that, in manual mode, causes the car to travel the design speed of 70 kph when on flat ground on a calm day.

Definition : CO_{bias} is the value of the CO that, in manual mode, causes the PV to steady at the DLO while the major disturbances are quiet and at their normal or expected values.

A P-Only controller bias (sometimes called null value) is assigned a value as part of the controller design and remains fixed once the controller is put in automatic.

Controller Gain, Kc

The P-Only controller has the advantage of having only one adjustable or tuning parameter, Kc, that defines how active or aggressive the CO will move in response to changes in controller error, $e(t)$.

For a given value of $e(t)$ in the P-Only algorithm above, if Kc is small, then the amount added to CO_{bias} is small and the controller response will be slow or sluggish. If Kc is large, then the amount added to CO_{bias} is large and the controller response will be fast or aggressive.

Thus, Kc can be adjusted or tuned for each process to make the controller more or less active in its actions when measurement does not equal set point.

P-only Controller Design

All controllers from the family of PID algorithms (P-Only, PI, PID) should be designed and tuned using our proven recipe :

1. Establish the design level of operation (the normal or expected values for set point and major disturbances).
2. Bump the process and collect controller output (CO) to process variable (PV) dynamic process data around this design level.
3. Approximate the process data behavior with a first order plus dead time (FOPDT) dynamic model.
4. Use the model parameters from step 3 in rules and correlations to complete the controller design and tuning.

The Internal Model Control (IMC) tuning correlations that work so well for PI and PID controllers cannot be derived for the simple P-only controller form. The next best choice is to use the widely-published integral of time-weighted absolute error (ITAE) tuning correlation :

$$\text{Moderate P-only : Kc} = \frac{0.2}{K_p}\left(\frac{T_p}{\theta_p}\right)^{1.22}$$

This correlation is useful in that it reliably yields a moderate Kc value. In fact, some practitioners find that the ITAE Kc value provides a response performance so predictably modest that they automatically start with an aggressive P-only tuning, defined here as two and a half times the ITAE value :

Aggressive P-only : Kc = 2.5 (Moderate Kc)

Reverse Acting, Direct Acting and Control Action

Time constant, Tp, and dead time, Θp, cannot affect the sign of Kc because they mark the passage of time and must always be positive. The above tuning correlation thus implies that Kc must always have the same sign as the process gain, Kp.

When CO increases on a process that has a positive Kp, the PV will increase in response. The process is direct acting. Given this CO to PV relationship, when in automatic mode (closed loop), if the PV starts drifting too high above set point, the controller must decrease CO to correct the error.

This "opposite to the problem" reaction is called *negative feedback* and forms the basis of stable control.

A process with a positive Kp is direct acting. With negative feedback, the controller must be reverse acting for stable control. Conversely, when Kp is negative (a reverse acting process), the controller must be direct acting for stable control.

Since Kp and Kc always have the same sign for a particular process and stable control requires negative feedback, then :

- Direct acting process (Kp and Kc positive) → use a reverse acting controller
- Reverse acting process (Kp and Kc negative) → use a direct acting controller.

In most commercial controllers, a positive value of the Kc is always entered. The sign (or action) of the controller is then assigned by specifying that the controller is either reverse or direct acting to indicate a positive or negative Kc respectively.

If the wrong control action is entered, the controller will quickly drive the final control element (*e.g.*, valve, pump, compressor) to full on/open or full off/closed and remain there until the proper control action entry is made.

Proportional Band

Some manufacturers use different forms for the same tuning parameter. The popular alternative to Kc found in the marketplace is proportional band, PB.

In many industry applications, both the CO and PV are expressed in units of percent. Given that a controller output signal ranges from a minimum (CO_{min}) to maximum (CO_{max}) value, then :

$PB = (CO_{max} - CO_{min})/Kc$

When CO and PV have units of percent and both range from 0% to 100%, the much published conversion between controller gain and proportional band results :

$PB = 100/Kc$

Many case studies on this site assign engineering units to the measured PV because plant software has made the task of unit conversions straightforward. If this is true in your plant, take care when using these conversion formula.

Implementation Issues

Implementation of a P-Only controller is reasonably straightforward, but this simple algorithm exhibits a phenomenon called "offset." In most industrial applications, offset is considered an unacceptable weakness. We explore P-Only control, offset and other issues for the heat exchanger and the gravity drained tanks processes.

P-ONLY CONTROL OF THE HEAT EXCHANGER SHOWS OFFSET

Here we investigate the capabilities of the P-Only controller on our heat exchanger process and highlight some key features and weaknesses of this simple algorithm.

As with all controller implementations, best practice is to follow the four-step design and tuning recipe as we proceed with the study :

Step 1 : Design Level of Operation (DLO)

Real processes display a nonlinear behavior. That is, their process gain (Kp), time constant (Tp) and/or dead time (Θp) changes as operating level changes and as major disturbances change. Since the rules and correlations we use are based on these Kp, Tp and Θp values, controllers should be designed and tuned for a specific level of operation.

The first step in the controller design recipe is to specify our design level of operation (DLO). This includes stating where we expect the set point, SP, and measured process variable, PV, to be during normal operation. Hopefully, these will be the same values as this is the point of a controller.

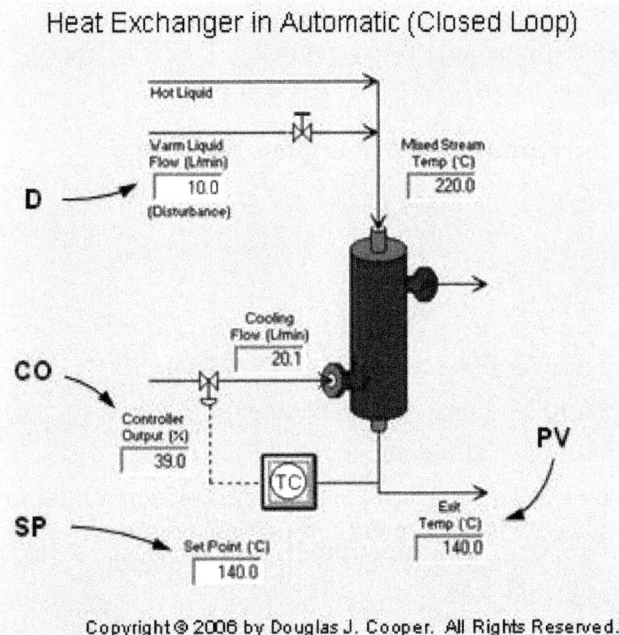

Heat Exchanger in Automatic (Closed Loop)

We also should have some sense of the range of values the SP and PV might assume so we can explore the nature of the process dynamics across that range.

For the heat exchanger, we specify that the SP and PV will normally be at 138°C, and during production, they may range from 138 to 140°C. Thus, we can state :

- Design PV and SP = 138°C with range of 138 to 140°C

We also should know normal or typical values for our major disturbances and be reasonably confident that they are quiet so we may proceed with a bump test. As shown in the graphic above, the heat exchanger process has only one major disturbance variable (D), a side stream labelled Warm Liquid Flow. We specify that the expected or design value for this stream is :

- Expected warm liquid flow disturbance, D = 10L/min.

  We assume that D remains quiet and at this normal design value throughout the study.

Step 2 : Collect Data at the DLO

The next step in the design recipe is to collect dynamic process data as near as practical to our design level of operation. We have previously collected and documented heat exchanger step test data that matches our design conditions.

Step 3 : Fit an FOPDT Model to the Design Data

Here we document a first order plus dead time (FOPDT) model approximation of the heat exchanger step test data from step 2 :

- Process gain (how far), $K_p = -0.53°C/\%$
- Time constant (how fast), $T_p = 1.3$ min
- Dead time (how much delay), $\Theta p = 0.8$ min

Step 4 : Use the Parameters to Complete the Design

The P-Only controller computes a controller output (CO) action every loop sample time T as :

$$CO = CO_{bias} + Kc·e(t)$$

where :

CO_{bias} = controller bias or null value

Kc = controller gain, a tuning parameter

$e(t)$ = controller error defined as SP–PV

- *Computing controller error, e(t)* : Set point (SP) is something we enter into the controller. The PV measurement comes from our sensor (our wire in). With SP and PV values known, controller error can be computed at every loop sample time T as : $e(t) = SP – PV$.

- *Determining Bias Value* : CO_{bias} is the value of the CO that, in manual mode, causes the PV to steady at the DLO while the major disturbances are quiet and at their normal or expected values.

The plot below shows that CO_{bias} can be located with an ordered search. That is, we move CO up and down while in manual mode until the PV settles at the design value of 138°C while the major disturbances (trace not shown) are quiet and at their normal or expected values.

Such a manipulation of our process may be impractical or impossible in production situations. The plot is useful, however, because it helps us visualize how the baseline (bias) value of the CO is linked to the design PV.

When we explore PI control of the heat exchanger, we will discuss how commercial controllers use a bumpless transfer method to automatically provide a value for CO_{bias}.

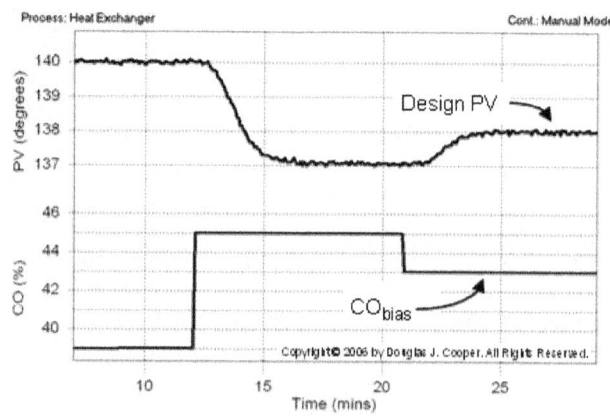

The plot above shows that when CO is held constant at 43% with the disturbances at their normal values, the PV steadies at the design value of 138°C. Thus :

- $CO_{bias} = 43\%$
- *Computing Controller Gain* : For the simple P-Only controller, we compute Kc with the integral of time-weighted absolute error (ITAE) tuning correlation :

Moderate P-Only : $Kc = \dfrac{0.2}{K_p}\left(\dfrac{T_p}{\theta_p}\right)^{1.22}$

This correlation is useful in that it reliably yields a moderate Kc value.

Aside : Dead time, θp, is in the denominator in the correlation, so it cannot equal zero. Otherwise, Kc will approach infinity, a fairly useless result.

Consider that all controllers measure, act, then wait until next sample time before measuring again. This "measure, act, wait" procedure has a delay (or dead time) of one sample time, T, built naturally into its structure.

Thus, by definition, the minimum dead time (θp, min) in a control loop is the loop sample time, T. Dead time can certainly be larger than sample time and it usually is, but it cannot be smaller.

Whether by software or graphical analysis, if we compute a θp that is less than T, we must set $\theta p = T$ everywhere in our tuning correlations. More information about the importance of sample time to controller design and tuning.

- *Best Practice Rule* :

  when using the FOPDT model for controller tuning, θp, min = T

(In the unreal world of pure theory, a true first order process with zero dead time is unconditionally stable under P-Only control. It would not even oscillate, let alone go unstable, at infinite Kc. The tuning correlation is therefore valid even at the theoretical extreme.) Using our FOPDT model values from step 3, we compute :

$$Kc = \dfrac{0.2}{-0.53}\left(\dfrac{1.3}{0.8}\right)^{1.22} = -0.7\% \,/\,°C$$

And our moderate P-only controller becomes :

- $CO = 43\% - 0.7 \cdot e(t)$

Implement and Test

To explore how controller gain, Kc, impacts P-only controller behavior, we test the controller with this ITAE controller gain value. Since the Kc value tends to be moderate, we also study more active or aggressive controller behavior when we double Kc and then double it again :

- $2Kc = -1.4\%/°C$
- $4Kc = -2.8\%/°C$

The performance of the P-only controller in tracking set point changes is pictured below for the ITAE Kc and its multiples. Note that the warm liquid disturbance flow, though not shown, remains constant at 10 L/min throughout the study.

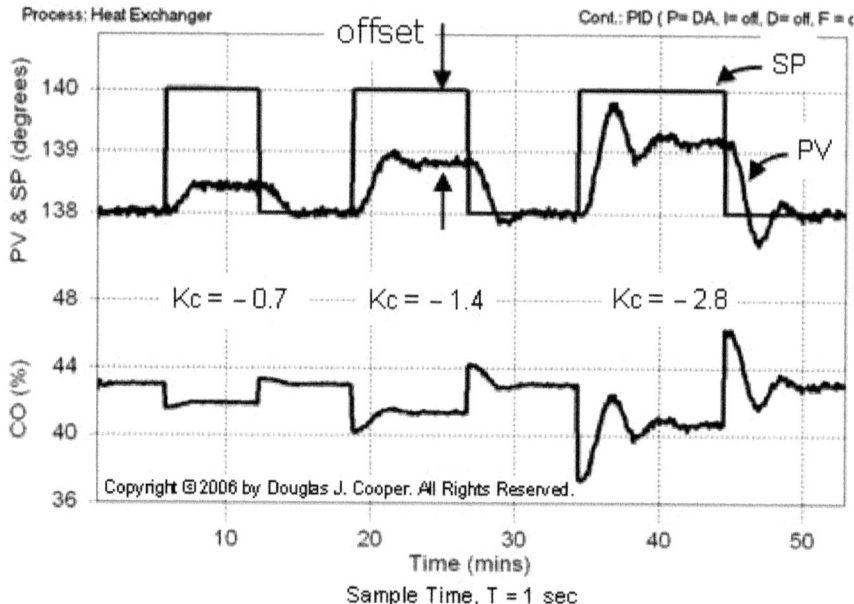

As shown in the figure, whenever the set point is at the design level of 138°C, then PV equals SP.

Each of the three times the SP is stepped away from the DLO, however, the PV settles out at a value short of the set point. The simple P-only controller is not able to eliminate this "offset," or sustained error between the PV and SP. We talk more about offset below.

- *Kc and Controller Activity* : The plot above shows the performance of the P-only controller using three different values of Kc.

One point of this study is to highlight that as Kc increases, the activity of the controller output, CO, increases. The CO trace at the bottom of the plot shows this increasingly active behavior, seen as more dramatic moves in response to the same set point step, as Kc increases across the plot.

Thus, we establish that controller gain, Kc, is responsible for the general, and especially the initial, activity in a controller response. This "response activity related to Kc" behavior carries over to PI and PID controllers.

We also see that as Kc increases across the plot, the offset (difference between SP and final PV) decreases but the oscillatory nature of the response increases.

Offset-the Big Disadvantage of P-only Control

The biggest advantage of P-only control is that there is only one tuning parameter to adjust, so it is relatively easy to achieve a "best" final tuning. The disadvantage is that this simple control algorithm permits offset.

Offset occurs in most processes under P-only control when the set point and/or disturbances are at any value other than that used to determine CO_{bias}.

To understand why offset occurs, let's work our way through the P-only equation :

$$CO = 43\% - 0.7 \cdot e(t)$$

and recognize that :

- When PV equals SP, then error is zero : $e(t) = 0$
- If e(t) is zero, then CO equals the CO_{bias} value of 43%
- If CO is steady at 43%, then the PV settles to 138°C. We know this is true because that's how CO_{bias} was determined in the first place.
- The only way CO can be different from the CO_{bias} value of 43% is if something is added or subtracted from the 43%
- The only way we have something to add or subtract from the 43% is if the error $e(t)$ is not zero
- If $e(t)$ is not zero, then PV cannot equal SP, and we have offset.

Possible Applications?

If P-only controllers permit offset, do they have any place in the process world? Actually, yes.

One example is a surge or swing tank designed to smooth flows between two units. It does not matter what specific liquid level the tank maintains. The level can be at 63% or 36% and we are happy. Just as long as the tank never empties completely or fills so much that it overflows.

A P-only controller can serve this function. Put the set point at a level of 50% and let the offset happen. We can have the controller implemented quickly and keep it tuned with little effort.

P-ONLY DISTURBANCE REJECTION OF THE GRAVITY DRAINED TANKS

Here we investigate the capabilities of the P-only controller for liquid level control of the gravity drained tanks process. Our objective in this study is disturbance rejection (or regulatory control) performance.

Gravity Drained Tanks Process

The measured process variable (PV) is liquid level in the lower tank. The controller output (CO) adjusts the flow into the upper tank to maintain the PV at set point (SP).

The disturbance (D) is a pumped flow out of the lower tank. It's draw rate is adjusted by a different process and is thus beyond our control. Because it runs through a pump, D is not affected by liquid level, though the pumped flow rate drops to zero if the tank empties.

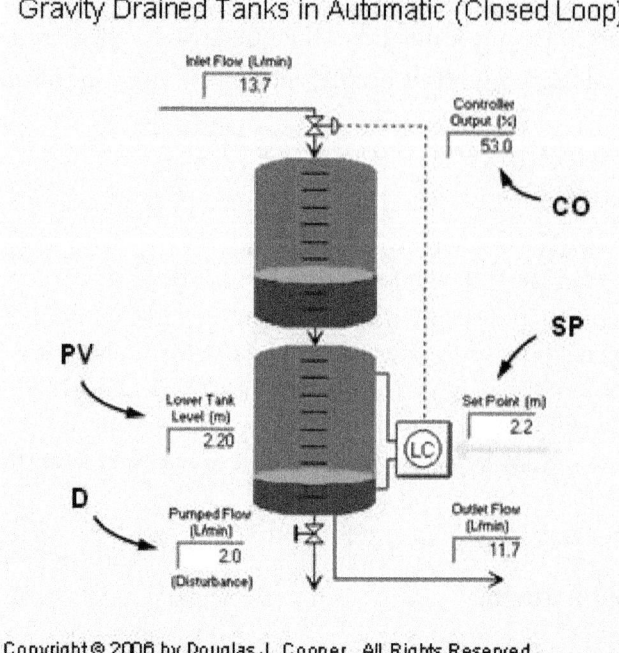

Gravity Drained Tanks in Automatic (Closed Loop)

We begin by summarizing the previously discussed results of steps 1 through 3 of our design and tuning recipe as we proceed with our P-only control investigation :

Step 1 : Determine the Design Level of Operation (DLO)

Our primary objective is to reject disturbances as we control liquid level in the lower tank. As discussed here, our design level of operation (DLO) for this study is :

- Design PV and SP = 2.2 m with range of 2.0 to 2.4 m
- Design D = 2 L/min with occasional spikes up to 5 L/min.

Step 2 : Collect Process Data Around the DLO

When CO, PV and D are steady near the design level of operation, we bump the CO far enough and fast enough to force a clear dynamic response in the PV that dominates the signal and process noise. As detailed here, we performed two different open loop (manual mode) dynamic tests, a step test and a doublet test.

Step 3 : Fit a FOPDT Model to the Dynamic Process Data

The third step of the recipe is to describe the overall dynamic behavior of the process with an approximating first order plus dead time (FOPDT) dynamic model. We define the model parameters and present details of the model fit of step test data here. A model fit of doublet test data using commercial software confirms these values :

- Process gain (how far), Kp = 0.09 m/%
- Time constant (how fast), Tp = 1.4 min
- Dead time (how much delay), Θp = 0.5 min.

Step 4 : Use the FOPDT Parameters to Complete the Design

Following the heat exchanger P-only study, the P-only control algorithm computes a CO action every loop sample time T as :

$CO = CO_{bias} + Kc \cdot e(t)$

Where :

CO_{bias} = controller bias or null value

Kc = controller gain, a tuning parameter

$e(t)$ = controller error, defined as SP–PV.

- *Sample Time, T* : Best practice is to set the loop sample time, T, at one-tenth the time constant or faster (*i.e.*, $T \leq 0.1Tp$). Faster sampling may provide modestly improved performance. Slower sampling can lead to significantly degraded performance.

   In this study, $T \leq (0.1)(1.4 \text{ min})$, so T should be 8 seconds or less. We meet this specification with the common vendor sample time option :

   - Sample time, T = 1 sec.

- *Control Action (Direct/Reverse)* : The gravity drained tanks has a positive Kp. That is, when CO increases, PV increases in response. When in automatic mode (closed loop), if the PV is too high, the controller must decrease the CO to correct the error (read more here). Since the controller must move in the direction opposite of the problem, we specify :

   - Controller is reverse acting.

- *Dead Time Issues* : If dead time is greater than the process time constant (Θp > Tp), control becomes increasingly problematic and a Smith predictor can offer benefit. For this process, the dead time is smaller than the time constant, so :

   - Dead time is small and thus not a concern.

- *Computing Controller Error, e(t)* : Set point, SP, is manually entered into a controller. The measured PV comes from the sensor (our wire in). Since SP and PV are known values, then at every loop sample time, T, controller error can be directly computed as :

   - Error, $e(t) = SP–PV$.

- *Determining Bias Value, CO_{bias}* : CO_{bias} is the value of CO that, in manual mode, causes the PV to remain steady at the DLO when the major disturbances are quiet and at their normal or expected values. Our doublet plots establish that when CO is at 53%, the PV is steady at the design value of 2.2 m, thus :

   - Controller bias, CO_{bias} = 53%.

- *Controller Gain, Kc* : For the simple P-Only controller form, we use the integral of time-weighted absolute error (ITAE) tuning correlation :

$$\textit{Moderate P-only} : Kc = \frac{0.2}{Kp}\left(\frac{T_p}{\theta_p}\right)^{1.22}$$

Aside : Regardless of the values computed in the FOPDT fit, best practice is to set Θp no smaller than sample time, T (or $\Theta p \geq T$) in the control rules and correlations. In this gravity drained tanks study, our FOPDT fit produced a Θp much larger than T, so the "dead time greater than sample time" rule is met.

Using our FOPDT model values from step 3, we compute :

$$Kc = \frac{0.2}{0.09}\left(\frac{1.4}{0.5}\right)^{1.22} = 8\% / m$$

And our moderate P-only controller becomes :

- P-only controller : $CO = 53\% + 8 \cdot e(t)$

Implement and Test

To explore how controller gain impacts P-only performance, we test the controller with the above Kc = 8%/m. Since the correlation tends to produce moderate performance values, we also explore increasingly aggressive or active P-only tuning by doubling Kc (2Kc = 16%/m) and then doubling it again (4Kc = 32%/m).

The ability of the P-only controller to reject step changes in the pumped flow disturbance, D, is pictured below for the ITAE value of Kc and its multiples. Note that the set point remains constant at 2.2 m throughout the study.

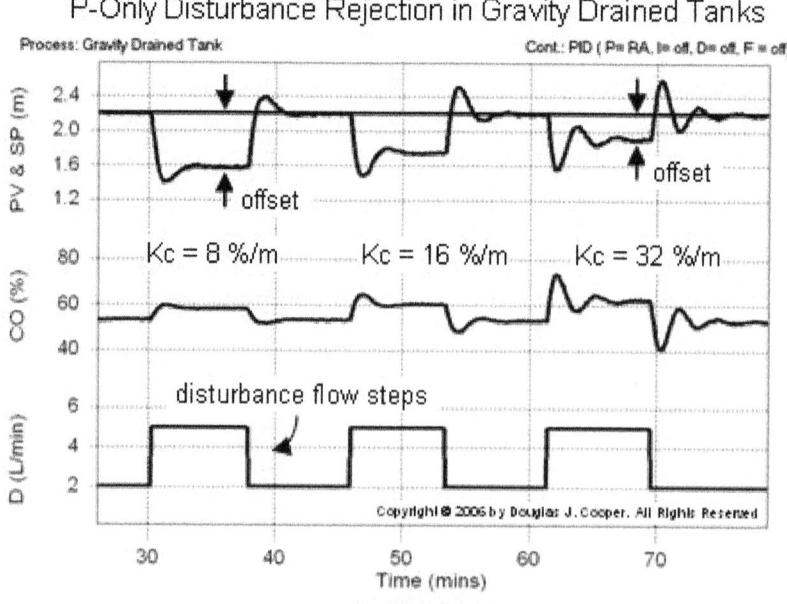

As shown in the figure above, whenever the pumped flow disturbance, D, is at the design level of 2 L/min (*e.g.*, when time is less than 30 min) then PV equals SP.

The three times that D is stepped away from the DLO, however, the PV shifts away from the set point. The simple P-Only controller is not able to eliminate this "offset," or sustained error between the PV and SP. This behavior reinforces that both set point and disturbances contribute to defining the design level of operation for a process.

The figure shows that as Kc increases across the plot :

- The activity of the controller output, CO, increases,
- The offset (difference between SP and final PV) decreases, and
- The oscillatory nature of the response increases.

Offset, or the sustained error between SP and PV when the process moves away from the DLO, is a big disadvantage of P-Only control. Yet there are appropriate uses for this simple controller.

While not our design objective, presented below is the set point tracking ability of the controller when the disturbance flow is held constant :

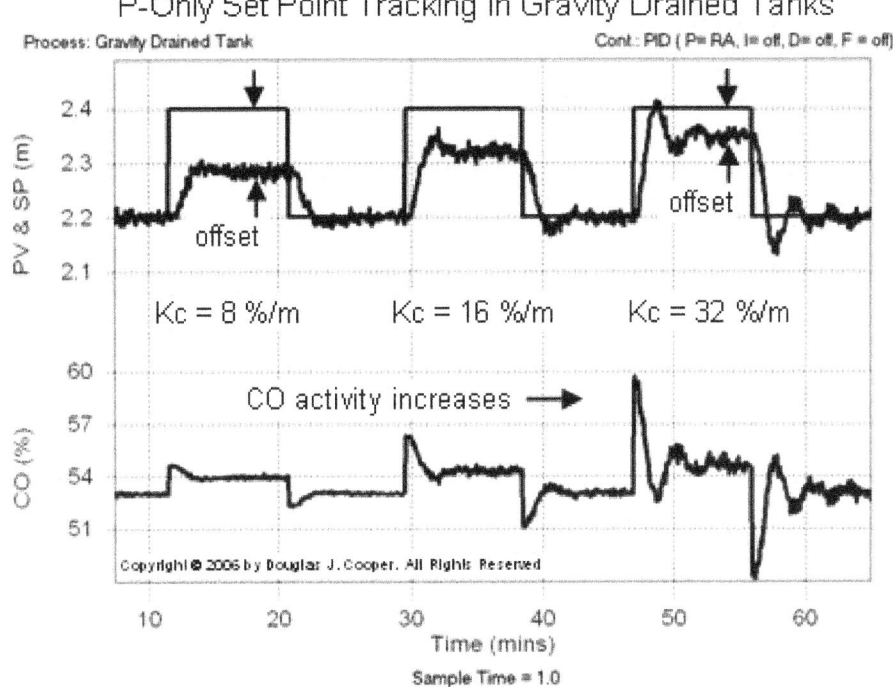

The figure shows that as Kc increases across the plot, the same performance observations made above apply here : the activity of CO increases, the offset decreases, and the oscillatory nature of the response increases.

Aside : It may appear that the random noise in the PV measurement signal is different in the two plots above, but it is indeed the same. Note that the span of the PV axis in the two plots differs by a factor of four. The narrow span of the set point tracking plot greatly magnifies the signal traces, making the noise more visible.

Proportional Band

Different manufacturers use different forms for the same tuning parameter. The popular alternative to controller gain found in the marketplace is proportional band, PB.

If the CO and PV have units of percent and both can range from 0 to 100%, then the conversion between controller gain and proportional band is :

PB = 100/Kc

Thus, as Kc increases, PB decreases. This reverse thinking can challenge our intuition when switching among manufacturers.

Many examples on this site assign engineering units to the measured PV because plant software has made the task of unit conversions straightforward. If this is true in your plant, take care when using this formula.

Integral Action

Integral action has the benefit of eliminating offset but presents greater design challenges.

Chapter 5

INTEGRAL ACTION AND PI CONTROL

Like the P-only controller, the Proportional-Integral (PI) algorithm computes and transmits a controller output (CO) signal every sample time, T, to the final control element (*e.g.*, valve, variable speed pump). The computed CO from the PI algorithm is influenced by the controller tuning parameters and the controller error, *e(t)*.

PI controllers have two tuning parameters to adjust. While this makes them more challenging to tune than a P-only controller, they are not as complex as the three parameter PID controller.

Integral action enables PI controllers to eliminate offset, a major weakness of a P-only controller. Thus, PI controllers provide a balance of complexity and capability that makes them by far the most widely used algorithm in process control applications.

THE PI ALGORITHM

While different vendors cast what is essentially the same algorithm in different forms, here we explore what is variously described as the dependent, ideal, continuous, position form :

$$CO = CO_{bias} + Kc \cdot e(t) + \frac{K_c}{T_i} \int e(t)dt$$

Where : CO = controller output signal (the wire out)

CO_{bias} = controller bias or null value; set by bumpless transfer as explained below

e(t) = current controller error, defined as SP–PV

SP = set point

PV = measured process variable (the wire in)

Kc = controller gain, a tuning parameter

Ti = reset time, a tuning parameter.

The integral mode of the controller is the last term of the equation. Its function is to integrate or continually sum the controller error, $e(t)$, over time.

Some things we should know about the reset time tuning parameter, Ti :

- It provides a separate weight to the integral term so the influence of integral action can be independently adjusted.
- It is in the denominator so smaller values provide a larger weight to (*i.e.* increase the influence of) the integral term.
- It has units of time so it is always positive.

FUNCTION OF THE PROPORTIONAL TERM

As with the P-only controller, the proportional term of the PI controller, $Kc\,e(t)$, adds or subtracts from CO_{bias} based on the size of controller error $e(t)$ at each time t.

As $e(t)$ grows or shrinks, the amount added to CO_{bias} grows or shrinks immediately and proportionately. The past history and current trajectory of the controller error have no influence on the proportional term computation.

The plot below illustrates this idea for a set point response. The error used in the proportional calculation is shown on the plot :

- At time $t = 25$ min, $e(25) = 60–56 = 4$
- At time $t = 40$ min, $e(40) = 60–62 = -2$.

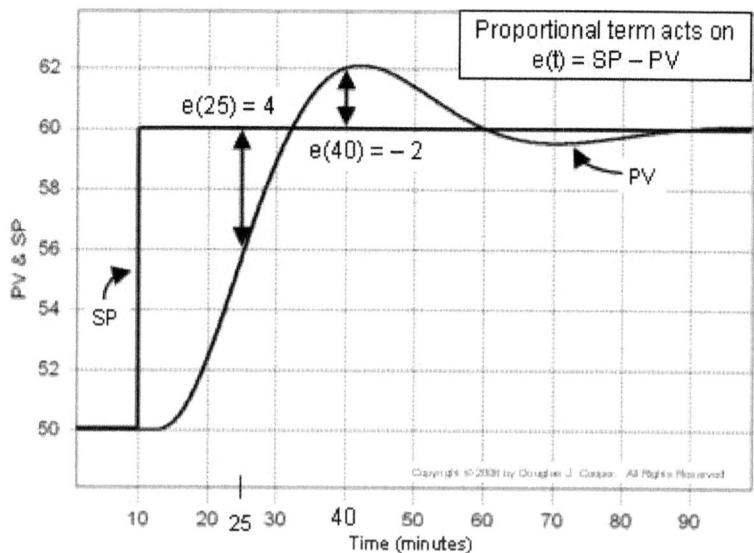

Recalling that controller error $e(t) = SP–PV$, rather than viewing PV and SP as separate traces as we do above, we can compute and plot e(t) at each point in time t.

Below is the identical data to that above only it is recast as a plot of $e(t)$ itself. Notice that in the plot above, PV = SP = 50 for the first 10 min, while in the error plot below, $e(t) = 0$ for the same time period.

This plot is useful as it helps us visualize how controller error continually changes size and sign as time passes.

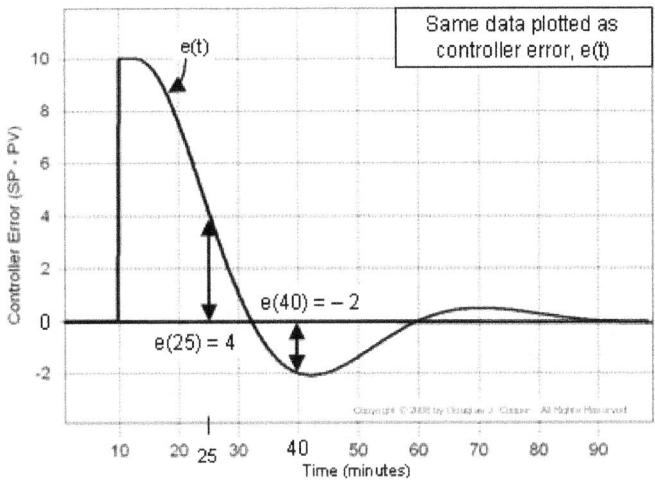

FUNCTION OF THE INTEGRAL TERM

While the proportional term considers the current size of $e(t)$ only at the time of the controller calculation, the integral term considers the history of the error, or how long and how far the measured process variable has been from the set point over time.

Integration is a continual summing. Integration of error over time means that we sum up the complete controller error history up to the present time, starting from when the controller was first switched to automatic.

Controller error is $e(t)$ = SP–PV. In the plot below, the integral sum of error is computed as the shaded areas between the SP and PV traces.

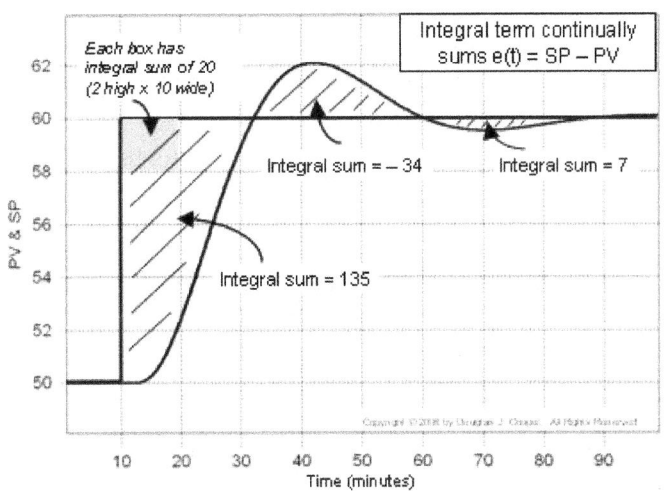

Each box in the plot has an integral sum of 20 (2 high by 10 wide). If we count the number of boxes (including fractions of boxes) contained in the shaded areas, we can compute the integral sum of error.

So when the PV first crosses the set point at around $t = 32$, the integral sum has grown to about 135. We write the integral term of the PI controller as :

$$\frac{K_c}{T_i} \int_0^{32} e(t)dt = \frac{K_c}{T_i}(135)$$

Since it is controller error that drives the calculation, we get a direct view the situation from a controller error plot as shown below :

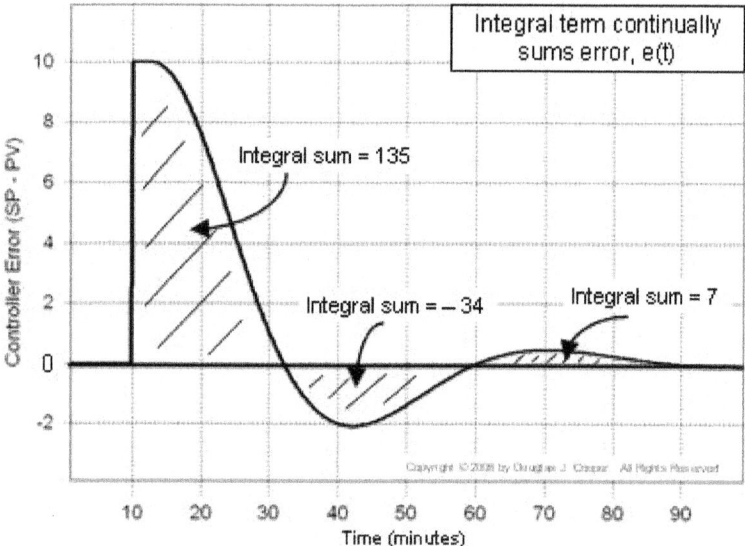

Note that the integral of each shaded portion has the same sign as the error. Since the integral sum starts accumulating when the controller is first put in automatic, the total integral sum grows as long as $e(t)$ is positive and shrinks when it is negative.

At time $t = 60$ min on the plots, the integral sum is 135–34 = 101. The response is largely settled out at $t = 90$ min, and the integral sum is then 135–34 + 7 = 108.

INTEGRAL ACTION ELIMINATES OFFSET

The previous sentence makes a subtle yet very important observation. The response is largely complete at time $t = 90$ min, yet the integral sum of all error is not zero.

In this example, the integral sum has a final or residual value of 108. It is this residual value that enables integral action of the PI controller to eliminate offset.

Most processes under P-only control experience offset during normal operation. Offset is a sustained value for controller error (*i.e.*, PV does not equal SP at steady state).

We recognize from the P-only controller :

$$CO = CO_{bias} + K_c \cdot e(t)$$

that CO will always equal CO_{bias} unless we add or subtract something from it.

The only way we have something to add or subtract from CO_{bias} in the P-only equation above is if $e(t)$ is not zero. It $e(t)$ is not steady at zero, then PV does not equal SP and we have offset.

However, with the PI controller :

$$CO = CO_{bias} + K_c \cdot e(t) + \frac{K_c}{T_i} \int e(t)dt$$

we now know that the integral sum of error can have a final or residual value after a response is complete. This is important because it means that $e(t)$ can be zero, yet we can still have something to add or subtract from CO_{bias} to form the final controller output, CO.

So as long as there is any error (as long as $e(t)$ is not zero), the integral term will grow or shrink in size to impact CO. The changes in CO will only cease when PV equals SP (when $e(t) = 0$) for a sustained period of time.

At that point, the integral term can have a residual value as just discussed. This residual value from integration, when added to CO_{bias}, essentially creates a new overall bias value that corresponds to the new level of operation.

In effect, integral action continually resets the bias value to eliminate offset as operating level changes.

CHALLENGES OF PI CONTROL

There are challenges in employing the PI algorithm :

- The two tuning parameters interact with each other and their influence must be balanced by the designer.
- The integral term tends to increase the oscillatory or rolling behavior of the process response.

Because the two tuning parameters interact with each other, it can be challenging to arrive at "best" tuning values. The value and importance of our design and tuning recipe increases as the controller becomes more complex.

INITIALIZING THE CONTROLLER FOR BUMPLESS TRANSFER

When we switch any controller from manual mode to automatic (from open loop to closed loop), we want the result to be uneventful. That is, we do not want the switchover to cause abrupt control actions that impact or disrupt our process.

We achieve this desired outcome at switchover by initializing the controller integral sum of error to zero. Also, the set point and controller bias value are initialized by setting :

- SP equal to the current PV
- CO_{bias} equal to the current CO.

With the integral sum of error set to zero, there is nothing to add or subtract from CO_{bias} that would cause a sudden change in the current controller output. With the set point equal to the measured process variable, there is no error to drive a change in our CO. And with the controller bias set to our current CO value, we are prepared by default to maintain current operation.

Thus, when we switch from manual mode to automatic, we have "bumpless transfer" with no surprises. This is a result everyone appreciates.

RESET TIME VERSUS RESET RATE

Different vendors cast their control algorithms in slightly different forms. Some use proportional band rather than controller gain. Also, some use reset rate, T_r, instead of reset time. These are simply the inverse of each other :

$$T_r = 1/T_i$$

No matter how the tuning parameters are expressed, the PI algorithms are all equally capable.

But it is critical to know your manufacturer before you start tuning your controller because parameter values must be matched to your particular algorithm form. Commercial software for controller design and tuning will automatically address this problem for you.

PI CONTROL OF THE HEAT EXCHANGER

Here we test the capabilities of the PI controller on the heat exchanger process. Our focus is on design, implementation and basic performance issues. Along the way we will highlight some strengths and weaknesses of this popular algorithm.

As with all controller implementations, best practice is to follow our proven four-step design and tuning recipe as we proceed with this case study.

Step 1 : Design Level of Operation (DLO)

Real processes display a nonlinear behavior. That is, their process gain, time constant and/or dead time changes as operating level changes and as major disturbances change. Since controller design and tuning is based on these process Kp, Tp and Θp values, controllers should be designed and tuned for a specific level of operation.

Thus, the first step in our controller design recipe is to specify our design level of operation (DLO). This includes stating :

- Where we expect the set point, SP, and measured process variable, PV, to be during normal operation.

- The range of values the SP and PV might assume so we can explore the nature of the process dynamics across that range.

We will track along with the same design conditions used in the P-only control study to permit a direct comparison of performance and capability. As in that study, we specify :

- Design PV and SP = 138°C with range of 138 to 140°C.

We also should know normal or typical values for our major disturbances and be reasonably confident that they are quiet so we may proceed with a bump test. The heat exchanger process has only one major disturbance variable, and consistent with the previous study :

- Expected warm liquid flow disturbance = 10 L/min.

Step 2 : Collect Data at the DLO

The next step in the design recipe is to collect dynamic process data as near as practical to our design level of operation. We have previously collected and documented heat exchanger step test data that matches our design conditions.

Step 3 : Fit an FOPDT Model to the Design Data

Here we document a first order plus dead time (FOPDT) model approximation of the step test data from step 2 :

- Process gain (how far), Kp = –0.53°C/%
- Time constant (how fast), Tp = 1.3 min
- Dead time (how much delay), Θp = 0.8 min.

Step 4 : Use the Parameters to Complete the Design

One common form of the PI controller computes a controller output (CO) action every loop sample time T as :

$$CO = CO_{bias} + K_c \cdot e(t) + \frac{K_c}{T_i} \int e(t)dt$$

Where :

CO = controller output signal (the wire out)

CO_{bias} = controller bias or null value; set by bumpless transfer as explained below

$e(t)$ = current controller error, defined as SP–PV

SP = set point

PV = measured process variable (the wire in)

Kc = controller gain, a tuning parameter

Ti = reset time, a tuning parameter.

- *Loop Sample Time, T :* Best practice is to specify loop sample time, T, at 10 times per time constant or faster ($T \le 0.1Tp$). For this study, $T \le 0.13$ min = 8 sec. Faster sampling may provide modestly improved performance, while slower sampling can lead to significantly degraded performance. Most commercial controllers offer an option of T = 1.0 sec, and since this meets our design rule, we use that here.

- *Computing controller error, e(t) :* Set point, SP, is something we enter into the controller. The PV measurement comes from our sensor (our wire in). With SP and PV known, controller error, $e(t) = SP-PV$, can be directly computed at every loop sample time T.

- *Determining Bias Value :* Strictly speaking, CO_{bias} is the value of the CO that, in manual mode, causes the PV to steady at the DLO while the major disturbances are quiet and at their normal or expected values.

- *Bumpless Transfer :* A desirable feature of the PI algorithm is that it is able to eliminate the offset that can occur under P-Only control. The integral term of the PI controller provides this capability by providing updated information that, when combined with the controller bias, keeps the process centered as conditions change.

  Since integral action acts to update (or reset) our bias value over time, CO_{bias} can be initialized in a straightforward fashion to a value that produces no abrupt control actions when we switch to automatic. Most commercial controllers do this with a simple "bumpless transfer" feature. When switching to automatic, they initialize :

  - SP equal to the current PV
  - CO_{bias} equal to the current CO.

  With the set point equal to the measured process variable, there is no error to drive a change in our controller output. And with the controller bias set to our current controller output, we are prepared by default to maintain current operation.

  We will use a controller that employs these bumpless transfer rules when we switch to automatic. Hence, we need not specify any value for CO_{bias} as part of our design.

- *Computing Controller Gain and Reset Time :* Here we use the industry-proven Internal Model Control (IMC) tuning correlations. The first step in using the IMC correlations is to compute Tc, the closed loop time constant. All time constants describe the speed or quickness of a response. The closed loop time constant describes the desired speed or quickness of a controller in responding to a set point change or rejecting a disturbance.

  If we want an active or quickly responding controller and can tolerate some overshoot and oscillation as the PV settles out, we want a small Tc (a short response time) and should choose *aggressive* tuning :

  - *Aggressive :* Tc is the larger of $0.1 \cdot Tp$ or $0.8 \cdot \Theta p$.

Moderate tuning is for a controller that will move the PV reasonably fast while producing little to no overshoot.

- *Moderate* : Tc is the larger of $1 \cdot Tp$ or $8 \cdot \Theta p$.

If we seek a more sluggish controller that will move things in the proper direction, but quite slowly, we choose *conservative* tuning (a big or long Tc).

- *Conservative* : Tc is the larger of $10 \cdot Tp$ or $80 \cdot \Theta p$.

Once we have decided on our desired performance and computed the closed loop time constant, Tc, with the above rules, then the PI correlations for controller gain, Kc, and reset time, Ti, are :

$$Kc = \frac{1}{Kp} \frac{Tp}{(\Theta p + Tc} \text{ and } Ti = Tp$$

Notice that reset time, Ti, is always set equal to the time constant of the process, regardless of desired controller activity.

a. *Moderate Response Tuning* : For a controller that will move the PV reasonably fast while producing little to no overshoot, choose :

Moderate Tc = the larger of $1 \cdot Tp$ or $8 \cdot \Theta p$

$$= \text{larger of } 1(1.3 \text{ min}) \text{ or } 8(0.8 \text{ min})$$

$$= 6.4 \text{ min.}$$

Using this Tc and our model parameters in the tuning correlations above, we arrive at the moderate tuning values :

$$Kc = \frac{1}{-0.53} \frac{1.3}{(0.8 + 6.4)} = -0.34 \frac{\%}{°C} \text{ and } Ti = 1.3 \text{ min} \cdot$$

b. *Aggressive Response Tuning* : For an active or quickly responding controller where we can tolerate some overshoot and oscillation as the PV settles out, specify

Aggressive Tc = the larger of $0.1 \cdot Tp$ or $0.8 \, \Theta p$

$$= \text{larger of } 0.1(1.3 \text{ min}) \text{ or } 0.8(0.8 \text{ min})$$

$$= 0.64 \text{ min}$$

and the aggressive tuning values are :

$$K_c = \frac{1}{-0.53} \frac{1.3}{(0.8 + 6.4)} = -1.7 \frac{\%}{°C} \text{ and } Ti = 1.3 \text{ min}$$

Practitioner's Note : The FOPDT model parameters used in the tuning correlations above have engineering units, so the Kc values we compute also have engineering units. In commercial control systems, controller gain (or proportional band) is normally entered as a dimensionless ($\%/\%$) value.

For commercial implementations, we could :

- Scale the process data before fitting our FOPDT dynamic model so we directly compute a dimensionless K_c.

- Convert the model Kp to dimensionless%/% after fitting the model but before using the FOPDT parameters in the tuning correlations.
- Convert Kc from engineering units into dimensionless%/% after using the tuning correlations.

  CO is already scaled from 0–100% in the above example. Thus, we convert Kc from engineering units into dimensionless%/% using the formula :

  $$Kc(\%/\%) = Kc(\%/units)\frac{PV_{max}\,units - PV_{min}\,units}{100\% - 0\%}$$

  For the heat exchanger, PV_{max} = 250°C and PV_{min} = 0°C. The dimensionless Kc values are thus computed :

  - Moderate Kc = (– 0.34%/°C)·[(250–0°C) ÷ (100–0%)] = –0.85%/%
  - Aggressive Kc = (– 1.7%/°C)·[(250–0°C) ÷ (100–0%)] = –4.2%/%

  We use Kc with engineering units in the remainder and are careful that our PI controller is formulated to accept such values. We would be mindful if we were using a commercial control system, however, to ensure our tuning parameters are cast in the form appropriate for our equipment.

- *Controller Action* : The process gain, Kp, is negative for the heat exchanger, indicating that when CO increases, the PV decreases in response. This behavior is characteristic of a reverse acting process. Given this CO to PV relationship, when in automatic mode (closed loop), if the PV starts drifting above set point, the controller must increase CO to correct the error. Such negative feedback is an essential component of stable controller design.

A process that is naturally reverse acting requires a controller that is direct acting to remain stable. In spite of the opposite labels (reverse acting process and direct acting controller), the details presented above show that both Kp and Kc are negative values.

In most commercial controllers, only positive Kc values can be entered. The sign (or action) of the controller is then assigned by specifying that the controller is either reverse acting or direct acting to indicate a positive or negative Kc, respectively.

If the wrong control action is entered, the controller will quickly drive the final control element (FCE) to full on/open or full off/closed and remain there until a proper control action entry is made.

Implement and Test

The first set point steps to the left show the PI controller performance using the moderate tuning values computed above. The second set point steps to the right show the controller performance using the aggressive tuning values. Note that the warm liquid disturbance flow, though not shown, remains constant at 10 L/min throughout the study.

The asymmetrical behavior of the PV for the set point steps up compared to the steps down is due to the very nonlinear character of the heat exchanger.

If we seek tuning between moderate and aggressive performance, we would average the Kc values from the tuning rules above.

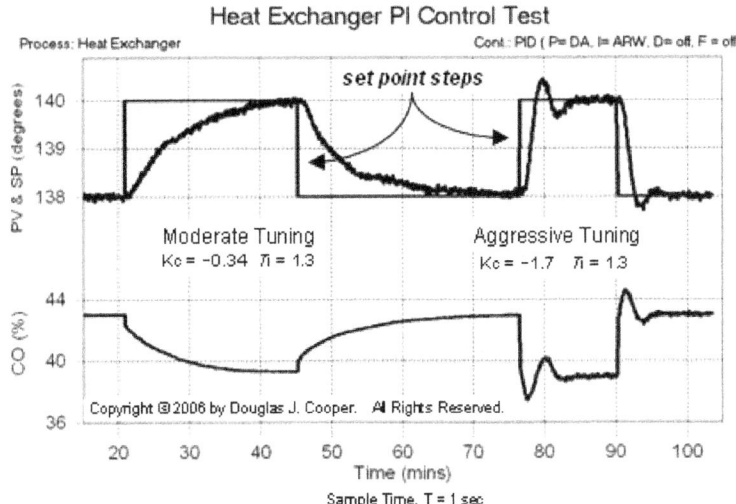

Heat Exchanger PI Control Test

But if we believe we had collected good bump test data (we saw a clear response in the PV when we stepped the CO and the major disturbances were quiet during the test), and the FOPDT model fit appears to be visually descriptive of the data, then we have a good value for *Tp* and that means a good value for *Ti*.

If we are going to fiddle with the tuning, we can tweak Kc and we should leave the reset time alone.

Tuning Recipe Saves Time and Money

The exciting result is that we achieved our desired controller performance based on one bump test and following a controller design recipe. No trial and error was involved. Little off-spec product was produced. No time was wasted.

The method of approximating complex behavior with a FOPDT model and then following a recipe for controller design and tuning has been used successfully on a broad spectrum of processes with streams composed of gases, liquids, powders, slurries and melts. It is a reliable approach that has been proven time and again at diverse plants from a wide range of companies.

THE CHALLENGE OF INTERACTING TUNING PARAMETERS

Many process control practitioners tune by "intuition," fiddling their way to final tuning by a combination of experience and trial-and-error.

Some are quite good at approaching process control as art. Since they are the ones who *define "best" performance* based on the goals of production, the capabilities of the process, the impact on down stream units, and the desires of management, it can be difficult to challenge any claims of success.

To explore the pitfalls of a trial and error approach and reinforce that there is science to controller tuning, we consider the common dependent, ideal form of the *PI controller* :

$$CO = CO_{bias} + Kc \cdot e(t) + \frac{K_c}{T_i} \int e(t)dt$$

Where :

CO = controller output

$e(t)$ = controller error = set point–process variable = SP–PV

Kc = controller gain, a tuning parameter

Ti = reset time, a tuning parameter

For this form, controller activity or aggressiveness increases as Kc increases and as Ti decreases (Ti is in the denominator, so smaller values increase the weighting on the integral action term, thus increasing controller activity).

Since Kc and Ti individually can make a controller more or less aggressive in its response, the two tuning parameters interact with each other. If current controller performance is not what we desire, it is not always clear which value to raise or lower, or by how much.

Example of Interaction Confusion

To illustrate, consider a case where we seek to balance a fairly rapid response to a set point change (a short *rise time*) against a small *overshoot*. While every process application is different, we choose to call the response plot below our desired or base case performance.

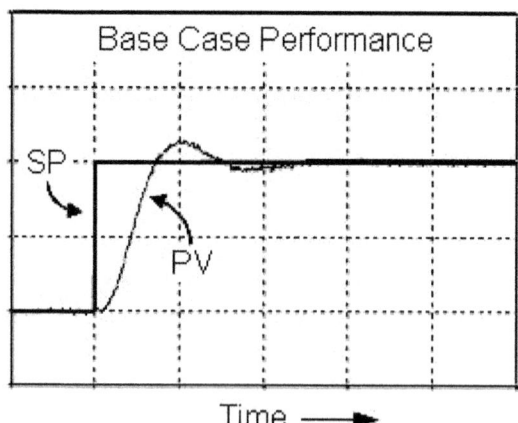

Now consider the two response plots below. These were made using the identical process and controller to that above. The only difference between the base case response above and plot A and plot B below is that different Kc and Ti tuning values were used in each one.

And now the question : what tuning adjustments are required to restore the desired base case performance above starting from each plot below? Or alternatively : how has the tuning been changed from base case performance to produce these different behaviors?

There are no tricks in this question. The "process" is a simple linear second order system with modest dead time. Controller output is not hitting any limits. The scales on the plot are identical. Everything is as it seems, except PI controller tuning is different in each case.

Study the plots for a moment before reading ahead and see if you can figure it out. Each plot has a very different answer.

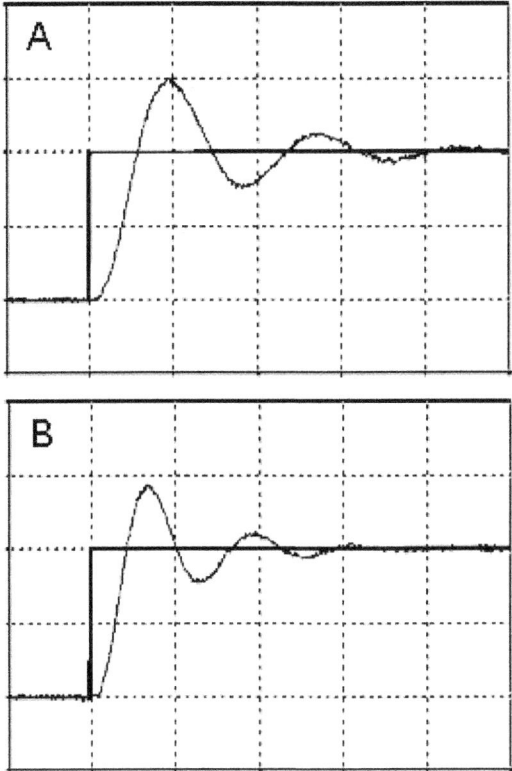

Some Hints

Before we reveal the answer, here is a hint. One plot has been made more active or aggressive in its response by doubling Kc while keeping Ti constant at the original base case value.

The other cuts Ti in half (remember, decreasing Ti makes this PI form more active) while keeping Kc at the base case value :

So we have :

- Base case = Kc and *Ti*
- Plot A or B = 2Kc and *Ti*
- Other Plot B or A = Kc and *Ti*/2

The Answer

Below is a complete tuning map with the base case performance from our challenge problem in the center. The plot shows how performance changes as Kc and *Ti* are doubled and halved from the base case for the dependent, ideal PI controller form.

Impact of Kc and $\overline{Ti}$ on Performance for PI Controller Form: $CO = CO_{bias} + Kc\,e(t) + \frac{Kc}{\overline{Ti}} \int e(t)\,dt$

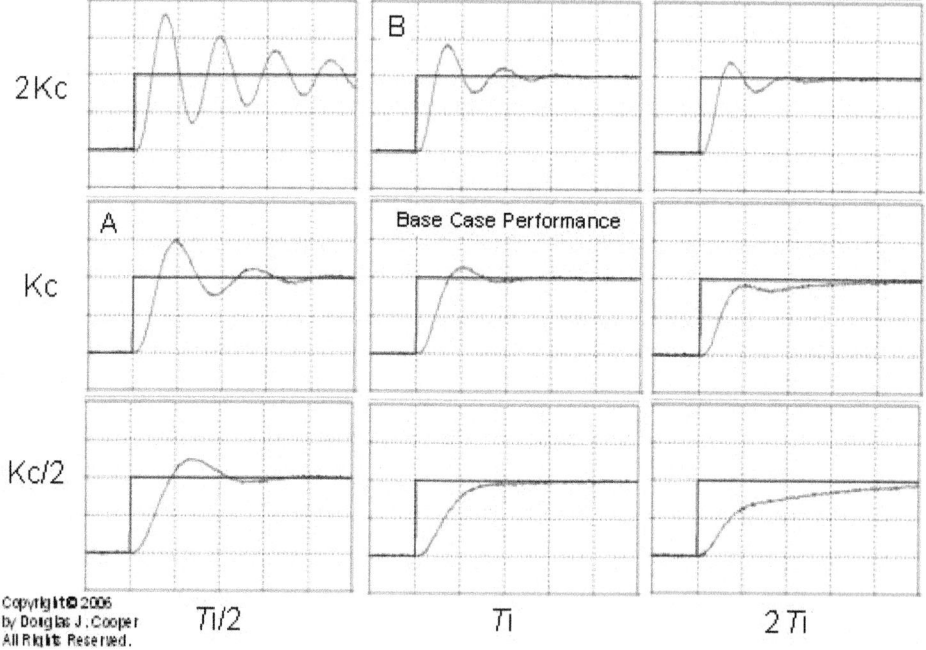

Starting from the center and moving up on the map from the base case performance brings us to plot B. As indicated on the tuning map axis, this direction increases (doubles) controller gain, Kc, thus making the controller more active or aggressive. Moving down on the map from the base case decreases (halves) Kc, making the controller more sluggish in its response.

Moving left on the map from the base case brings us to plot A. As indicated on the tuning map axis, this direction decreases reset time (cuts it in half), again making the controller more active or aggressive. Moving right on the map from the base case increases (doubles) reset time, making the controller more sluggish in its response.

It is clear from the tuning map that the controller is more active or aggressive in its response when Kc increases and Ti decreases, and more sluggish or conservative when Kc decreases and Ti increases.

Building on this observation, it is not surprising that the upper left most plot (2Kc and Ti/2) shows the most active controller response, and the lower right most plot (Kc/2 and 2Ti) is the most conservative or sluggish response.

Back to the question. With what we now know, the answer :

- Base case = Kc and Ti
- Plot B = 2Kc and Ti
- Plot A = Kc and Ti/2.

Interacting Parameters Makes Tuning Problematic

The PI controller has only two tuning parameters, yet it produces very similar looking performance plots located in different places on a tuning map.

If our instincts lead us to believe that we are at plot A when we really are at plot B, then the corrective action we make based on this instinct will compound our problem rather than solve it. This is strong evidence that trial and error is not an efficient or appropriate approach to tuning.

When we consider a PID controller with three tuning parameters, the number of similar looking plots in what would be a three dimensional tuning map increases dramatically. Trial and error tuning becomes almost futile.

We have been exploring a *step by step tuning recipe approach* that produces desired results without the wasted time and off-spec product that results from trial and error tuning.

If we follow this industry-proven methodology, we will improve the safety and profitability of our operation.

Interesting Observation

Before leaving this subject, we make one more very useful observation from the tuning map. This will help build our intuition and may help one day when we are out in the plant.

The right most plot in the center row (Kc, 2Ti) of the tuning map above is reproduced below :

Notice how the PV shows a dip or brief oscillation on its way up to the set point? This is a classic indication that the proportional term is reasonable but the integral term is not getting enough weight in the calculation.

If we cover the right half of the "not enough integral action" plot, the response looks like it is going to settle out with some offset, as would be expected with a P-only controller. When we consider the plot as a whole, we see that as enough time passes, the response completes. This is because the weak integral action finally accumulates enough weight in the calculation to move the PV up to set point.

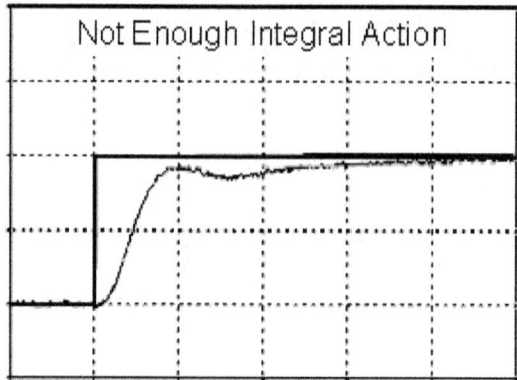

This "oscillates on the way" pattern is a useful marker for diagnosing a lack of sufficient integral action.

PI DISTURBANCE REJECTION IN THE JACKETED STIRRED REACTOR

The control objective for the jacketed reactor is to minimize the impact on reactor operation when the temperature of the liquid entering the cooling jacket changes. As a base case study, we establish here the performance capabilities of a PI controller in achieving this objective.

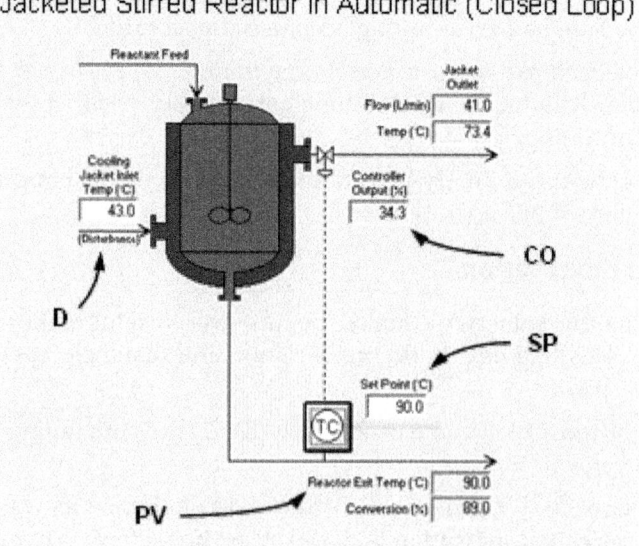

The important variables for this study are labelled in the graphic :

CO = signal to valve that adjusts cooling jacket liquid flow rate (controller output,%)

PV = reactor exit stream temperature (measured process variable,°C)

SP = desired reactor exit stream temperature (set point,°C)

D = temperature of cooling liquid entering the jacket (major disturbance,°C)

We follow our industry proven recipe to design and tune our PI controller :

Step 1 : Design Level of Operation (DLO)

The details of expected process operation and how this leads to our DLO are summarized :

- Design PV and SP = 90°C with approval for brief dynamic (bump) testing of ±2°C.
- Design D = 43°C with occasional spikes up to 50°C.

Step 2 : Collect Process Data Around the DLO

When CO, PV and D are steady near the design level of operation, we bump the process as detailed here to generate CO-to-PV cause and effect response data.

Step 3 : Fit a FOPDT Model to the Dynamic Process Data

We approximate the dynamic behavior of the process by fitting a first order plus dead time (FOPDT) dynamic model to the test data from step 2. The results of the modeling study are presented in detail here and are summarized :

- Process gain (direction and how far), $K_p = -0.5°C/\%$
- Time constant (how fast), $T_p = 2.2$ min
- Dead time (how much delay), $\Theta_p = 0.8$ min.

Step 4 : Use the FOPDT Parameters to Complete the Design

As in the heat exchanger PI control study, we explore what is often called the dependent, ideal form of the PI control algorithm :

$$CO = CO_{bias} + K_c \cdot e(t) + \frac{K_c}{T_i}\int e(t)dt$$

Where :

CO = controller output signal (the wire out)

CO_{bias} = controller bias or null value; set by bumpless transfer

$e(t)$ = current controller error, defined as SP–PV

SP = set point

PV = measured process variable (the wire in)

K_c = controller gain, a tuning parameter

T_i = reset time, a tuning parameter.

Aside : our observations using the dependent ideal PI algorithm directly apply to the other popular PI controller forms. For example, the integral gain for the independent algorithm form, written as :

$$CO = CO_{bias} + Kc \cdot e(t) + Ki \int e(t)dt$$

can be computed as : $Ki = Kc/Ti$. The Kc is the same for both forms, though it is more commonly called the proportional gain for the independent algorithm.

- *Sample Time, T :* Best practice is to set the loop sample time, T, at one-tenth the time constant or faster (*i.e.*, $T \le 0.1Tp$). Faster sampling may provide modestly improved performance, while slower sampling can lead to significantly degraded performance.

  In this study, $T \le 0.1(2.2 \text{ min})$, so T should be 13 seconds or less. We meet this with the sample time option available from most commercial vendors :

  - Sample time, T = 1 sec.

- *Control Action (Direct/Reverse) :* The jacketed stirred reactor process has a negative Kp. That is, when CO increases, PV decreases in response. Since a controller must provide negative feedback, if the process is reverse acting, the controller must be direct acting. That is, when in automatic mode (closed loop), if the PV is too high, the controller must increase the CO to correct the error. Since the controller must move in the same direction as the problem, we specify :

  - Controller is direct acting

- *Dead Time Issues :* If dead time is greater than the process time constant ($\Theta p > Tp$), control becomes increasingly problematic and a Smith predictor can offer benefit. For this process, the dead time is smaller than the time constant, so :

  - Dead time is small and not a concern

- *Computing Controller Error, e(t) :* Set point, SP, is manually entered into a controller. The measured PV comes from the sensor (our wire in). Since SP and PV are known values, then at every loop sample time, T, controller error can be directly computed as :

  - Error, $e(t) = SP-PV$

- *Determining Bias Value, CO_{bias} :* CO_{bias} is the value of CO that, in manual mode, causes the PV to steady at the DLO when the major disturbances are quiet and at their normal or expected values. When integral action is enabled, commercial controllers determine the bias value with a bumpless transfer procedure.

  That is, when switching to automatic, the controller initializes the SP to the current value of PV, and CO_{bias} to the current value of CO. By choosing our current operation as our design state (at least temporarily at switchover), there is no corrective action needed by the controller that will bump the process. Thus,

  - Controller bias, CO_{bias} = current CO for a bumpless transfer

- *Controller Gain, Kc, and Reset Time, Ti :* We use our FOPDT model parameters in the industry-proven Internal Model Control (IMC) tuning correlations to compute PI tuning values.

The first step in using the IMC correlations is to compute Tc, the closed loop time constant. Tc describes how active our controller should be in responding to a set point change or in rejecting a disturbance.

The performance implications of choosing Tc have been explored previously for PI control of the heat exchanger and the gravity drained tanks case studies.

In short, the closed loop time constant, Tc, is computed based on whether we seek :

- Aggressive action and can tolerate some overshoot and oscillation in the PV response,
- Moderate action where the PV will move reasonably fast but show little overshoot,
- Conservative action where the PV will move in the proper direction, but quite slowly.

Once this decision is made, we compute Tc with these rules :

- *Aggressive Response* : Tc is the larger of 0.1 ·Tp or 0.8 Θp
- *Moderate Response* : Tc is the larger of 1 ·Tp or 8 Θp
- *Conservative Response* : Tc is the larger of 10 ·Tp or 80 Θp

With Tc computed, the PI controller gain, Kc, and reset time, Ti, are computed as :

$$Kc = \frac{1}{Kp} \frac{Tp}{(\theta p + Tc)} \text{ and } Ti = Tp$$

Notice that reset time, Ti, is always equal to the process time constant, Tp, regardless of desired controller activity.

a. *Moderate Response Tuning* : For a controller that will move the PV reasonably fast while producing little to no overshoot, choose :

Moderate Tc = the larger of 1 ·Tp or 8 Θp

$\qquad$ = larger of 1(2.2 min) or 8(0.8 min)

$\qquad$ = 6.4 min.

Using this Tc and our model parameters in the tuning correlations above, we arrive at the moderate tuning values :

$$Kc = \frac{1}{-0.5} \frac{2.2}{(0.8 + 6.4)} = -0.6 \frac{\%}{°C} \text{ and } Ti = 2.2 \min .$$

b. *Aggressive Response Tuning* : For an active or quickly responding controller where we can tolerate some overshoot and oscillation as the PV settles out, specify :

Aggressive Tc = the larger of 0.1 ·Tp or 0.8 θp

$\qquad$ = larger of 0.1(2.2 min) or 0.8(0.8 min)

$\qquad$ = 0.64 min.

and the aggressive tuning values are :

$$Kc = \frac{1}{-0.5}\frac{2.2}{(0.8+0.64)} = -3.1\frac{\%}{°C} \text{ and } Ti = 2.2\,\text{min}.$$

Practitioner's Note : The FOPDT model parameters used in the tuning correlations above have engineering units, so the Kc values we compute also have engineering units. In commercial control systems, controller gain (or proportional band) is normally entered as a dimensionless (%/%) value.

For commercial implementations, we could :

- Scale the process data before fitting our FOPDT dynamic model so we directly compute a dimensionless Kc.
- Convert the model Kp to dimensionless %/% after fitting the model but before using the FOPDT parameters in the tuning correlations.
- Convert Kc from engineering units into dimensionless %/% after using the tuning correlations.

CO is already scaled from 0–100% in the above example. Thus, we convert Kc from engineering units into dimensionless %/% using the formula :

$$Kc(\%/\%) = Kc(\%/\text{units})\left(\frac{PV_{max}\text{units} - PV_{min}\text{units}}{100\% - 0\%}\right)$$

For the jacketed stirred reactor, PV_{max} = 250°C and PV_{min} = 0°C. The dimensionless Kc values are thus computed :

- Moderate Kc = (– 0.6%/°C)·[(250–0°C) ÷ (100–0%)] = –1.5%/%
- Aggressive Kc = (– 3.1%/°C)·[(250–0°C) ÷ (100–0%)] = –7.8%/%

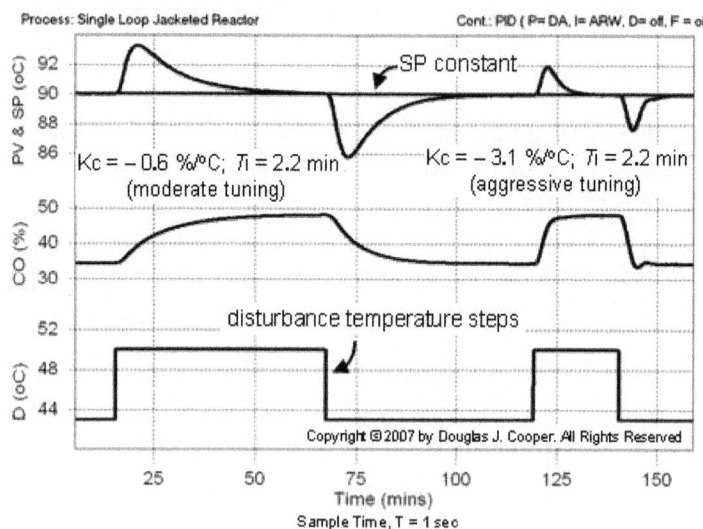

PI Disturbance Rejection in Jacketed Stirred Reactor

Implement and Test

The ability of the PI controller to reject changes in the cooling jacket inlet temperature, D, is pictured below for the moderate and aggressive tuning values computed above. Note that the set point remains constant at 90°C throughout the study.

As expected, the aggressive controller shows a more energetic CO action, and thus, a more active PV response.

While not our design objective, presented below is the set point tracking ability of the PI controller when the disturbance temperature is held constant.

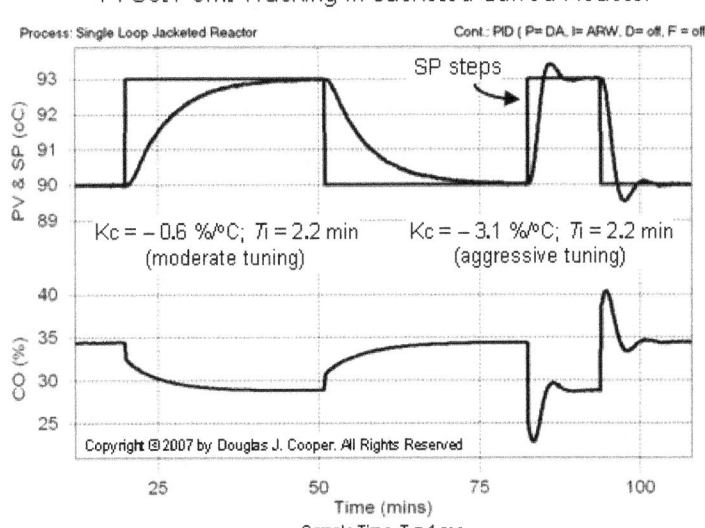

PI Set Point Tracking in Jacketed Stirred Reactor

The plot shows that set point tracking performance matches the descriptions used above for choosing Tc :

- Use aggressive action if we seek a fast response and can tolerate some over-shoot and oscillation in the PV response.
- Use moderate action if we seek a reasonably fast response but seek little to no overshoot in the PV response.

Important => Ti Always Equals Tp

As stated above, the rules provide a constant reset time, Ti, regardless of our desired performance. So if we believe we have collected a good process data set, and the FOPDT model fit looks like a reasonable approximation of this data, then we have a good estimate of the process time constant and $Ti = Tp$ regardless of desired performance.

If we are going to tweak the tuning, Kc should be the only value we adjust. For example, if we seek a performance between moderate and aggressive, we average the Kc values while Ti remains constant.

INTEGRAL (RESET) WINDUP, JACKETING LOGIC AND THE VELOCITY PI FORM

A valve cannot open more than all the way : A pump cannot go slower than stopped. Yet an improperly programmed control algorithm can issue such commands.

Herein lies the problem of integral windup (also referred to as reset windup or integral saturation). It is a problem that has been around for decades and was solved long ago. We discuss why it occurs and how to prevent it to help those who choose to write their own control algorithm.

The PI Algorithm

To increase our comfort level with the idea that different vendors cast the same PI algorithm in different forms, we choose the independent, continuous, position PI form for this discussion :

$$CO = CO_{bias} + Kc \cdot e(t) + Ki \int e(t)dt$$

Where :

CO = controller output signal (the wire out)

CO_{bias} = controller bias or null value

$e(t)$ = current controller error, defined as SP–PV

SP = set point

PV = measured process variable (the wire in)

Kc = proportional gain, a tuning parameter

Ki = integral gain, a tuning parameter.

Note that Kc is the same parameter in both the dependent and independent forms, though it is more typically called controller gain in the dependent form.

Every procedure and observation we have previously discussed about PI controllers applies to both forms. Both even use the same tuning correlations. To tune Ki, we compute Kc and Ti for the dependent form and then divide ($Ki = Kc/Ti$).

Integral (Reset) Windup

As shown below, integration of error means that we continually sum controller error up to the present time.

The integral sum starts accumulating when the controller is first put in automatic and continues to change as long as controller error exists.

If an error is large enough and/or persists long enough, it is mathematically possible for the integral term to grow very large (either positive or negative) :

$$CO = CO_{bias} + Kc \cdot e(t) + Ki \underbrace{\int e(t)dt}_{\substack{\text{can grow} \\ \text{very large}}}$$

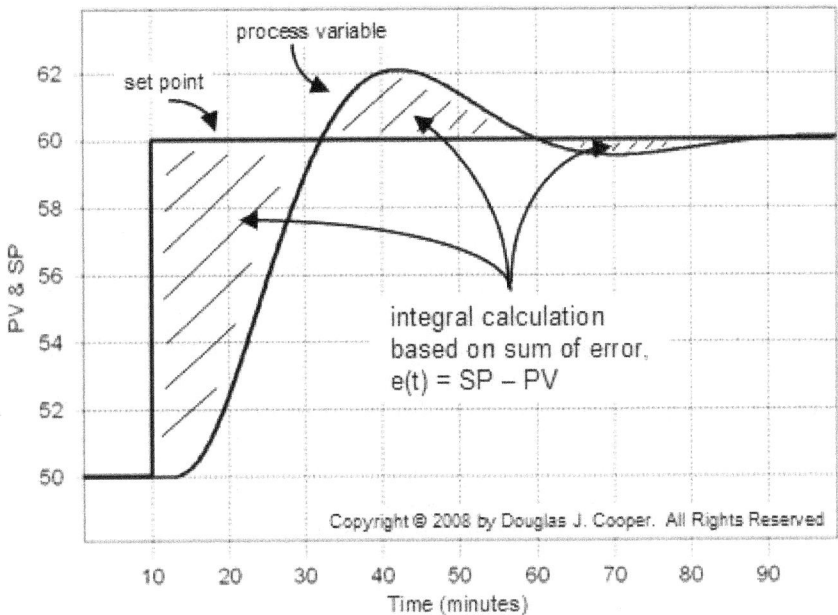

This large integral, when combined with the other terms in the equation, can produce a CO value that causes the final control element (FCE) to saturate. That is, the CO drives the FCE (*e.g.* valve, pump, compressor) to its physical limit of fully open/on/maximum or fully closed/off/minimum.

And if this extreme value is still not sufficient to eliminate the error, the simple mathematics of the controller algorithm, if not jacketed with protective logic, permits the integral term to continue growing.

If the integral term grows unchecked, the equation above can command the valve, pump or compressor to move to 110%, then 120% and more. Clearly, however, when an FCE reaches its full 100% value, these last commands have no physical meaning and consequently, no impact on the process.

Control is Lost

Once we cross over to a "no physical meaning" computation, the controller has lost the ability to regulate the process.

When the computed CO exceeds the physical capabilities of the FCE because the integral term has reached a large positive or negative value, the controller is suffering from *windup*. Because windup is associated with the integral term, it is often referred to as *integral windup* or *reset windup*.

To prevent windup from occurring, modern controllers are protected by either :

- Employing extra "jacketing logic" in the software to halt integration when the CO reaches a maximum or minimum value.

- Recasting the controller into a discrete velocity form that, by its very formulation, naturally avoids windup.

  Both alternatives offer benefits but possess some fairly subtle drawbacks that we discuss below.

Visualizing Windup

To better visualize the problem of windup and the benefit of anti-windup protection, consider the plot from our heat exchanger process below.

To the left is the performance of a PI controller with no windup protection. To the right is the performance of the same controller protected by an anti-windup strategy.

For both controllers, the set point is stepped from 200°C up to 215°C and back again. As shown in the lower trace on the plot, the controller moves the CO to 0%, closing the valve completely, yet this is not sufficient to move the PV up to the new set point.

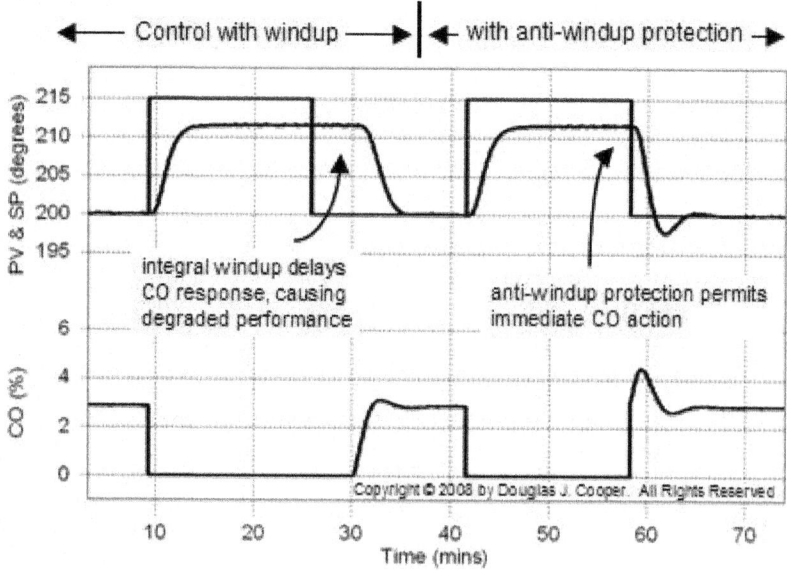

To the left in the plot, the impact of windup is a degraded controller performance. When the set point is stepped back to its original value of 200°C, the windup condition causes a delay in the CO action. This in turn causes a delay in the PV response.

To the right in the plot, anti-windup protection permits the CO, and thus PV, to respond promptly to the command to return to the original SP value of 200°C.

More Details on Windup

The plot below offers more detail. As labelled on the plot :

1. To the left for the Controller with Wind-up case, the SP is stepped up to 215°C. The valve closes completely but is not able to move the PV all the way to the high set point value. Integration is a summing of controller error, and since error persists, the integration term grows very large.

   The sustained error permits the controller to windup (saturate). While it is not obvious from the plot, the PI algorithm is computing values for CO that ask the valve to be open –5%, –8% and more. The control algorithm is just simple math with no ability to recognize that a valve cannot be open to a negative value.

   Note that the chart shows the CO signal bottoming out at 0% while the controller algorithm is computing negative CO values. This misleading information is one reason why windup can be difficult to diagnose as the root cause of a problem from visual inspection of process data trend plots.

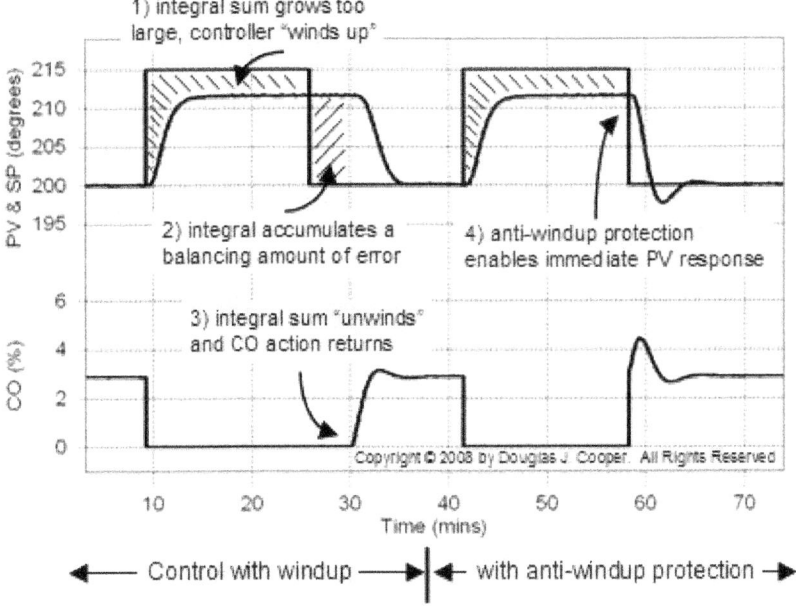

2. When the SP is stepped back to 200°C, it seems as if the CO does not move at first. In reality, the control algorithm started moving the CO when the SP changed, but the values remain in the physically meaningless range of negative numbers.

   So while the valve remains fully closed at 0%, the integral sum is accumulating controller errors of opposite sign. As time passes, the integral term shrinks or "unwinds" as the running sum of errors balance out.

3. When the integral sum of errors shrinks enough, it no longer dominates the CO computation. The CO signal returns from the physically meaningless world of negative values. The valve can finally move in response.

4. To the right in the plot above, the controller is protected from windup. As a result, when the set point is stepped back to 200°C, the CO immediately reacts

with a change that is proportional to the size of the SP change. The PV moves quickly in response to the CO actions as it tracks the SP back to 200°C.

Solution 1 : Jacketing Logic on the Position Algorithm

The PI controller is called the position form because the computed CO is a specific intermediate value between full on/open/maximum and closed/off/minimum. The continuous PI algorithm is specifying the actual position (*e.g.*, 27% open, 64% of maximum) that the final control element (FCE) should assume.

- *Simple logic creates additional problems* : It is not enough to have logic that simply limits or clips the CO if it reaches a maximum (CO_{max}) or minimum (CO_{min}) value because this does nothing to check the growth of the integral sum of errors term.

  In fact, such simple logic was used in the "control with windup" plots just discussed. The CO seems stuck at 0% and we are unaware that the algorithm is actually computing negative valve positions as described in item 1 above.

- *Anti-windup logic outline* : When we switch from manual mode to automatic, we assume that we have initialized the controller using a bumpless transfer procedure. That is, at switchover, the integral sum of error is set to zero, the SP is set equal to the current PV, and the controller bias is set equal to the current CO (implying that $CO_{min} < CO_{bias} < CO_{max}$).

  Thus, there is nothing to cause CO to immediately change and "bump" our process at switch over.

  One approach to creating anti-windup jacketing logic is to artificially manipulate the integral sum of error itself. With our controller properly initialized, the approach is to flip the algorithm around and *back-calculate a value for the integral sum of error* that will provide a desired controller output value ($CO_{desired}$), or :

$$\int e(t)dt = \frac{1}{Ki}\{CO_{desired} - CO_{bias} - Kc \cdot e(t)\}$$

  Note that $CO_{desired}$ can be different in different situations. For example,

- We do not want tuning parameter adjustments to cause sudden CO movements that bump our process. So if tuning values have changed, $CO_{desired}$ is the value of CO from the previous loop calculation cycle.

- If the PI controller computes CO values that are above CO_{max} or below CO_{min}, then we must be concerned about windup and $CO_{desired}$ is set equal to the limiting CO_{max} or CO_{min} value. The anti-windup logic followed at every loop sample time, T, is thus :

1. If tuning parameters have changed since the last loop calculation cycle, then $CO_{desired}$ = current CO. Back calculate the integral sum of error so CO remains unchanged from the previous sample time. This prevents sudden CO bumps due to tuning changes.

2. Update SP and PV for this loop calculation cycle.

3.   Compute : $CO = CO_{bias} + Kc \cdot e(t) + Ki \int e(t)dt$
4.   If $CO > CO_{max}$ or if $CO < CO_{min}$, then the anti-windup (integral desaturation) logic of step 5 is required. Otherwise, proceed to step 6.
5.   If $CO > CO_{max}$, then $CO = CO_{desired} = CO_{max}$. if $CO < CO_{min}$, then $CO = CO_{desired} = CO_{min}$. Back calculate the integral sum of error using our selected $CO_{desired}$ and save it for use in the next control loop calculation cycle.
6.   Implement CO.

Solution 2 : Use the Velocity (Discrete) Controller Form

Rather than computing a CO signal indicating a specific position for our final control element, an alternative is to compute a signal that specifies a change, ΔCO, from current position for the FCE. As explained below, this is called the velocity or discrete controller form.

We employ the dependent algorithm for this presentation, but the derivation that follows can be applied in an analogous fashion to the independent PI form. To derive the discrete velocity form, we must first write the continuous, position form of the PI controller to include the independent variable on the controller output, showing it properly as CO(t) to reflect that it changes with time :

$$CO(t) = CO_{bias} + Kc \cdot e(t) + \frac{Kc}{Ti} \int e(t)dt$$

*   *Deriving the Discrete Velocity Form* : The first step in deriving the discrete form is to take the time derivative of the continuous form. In physics, the time derivative (rate of change) of a position is a velocity. This is why the final form of the PI controller we derive is often called the velocity form.

Taking the derivative of the continuous PI controller above with respect to time yields :

$$\frac{dCO(t)}{dt} = \frac{dCO_{bias}}{dt} + Kc\frac{de(t)}{dt} + \frac{Kc}{Ti}e(t)$$

Since CO_{bias} is a constant and the derivative of a constant is zero, then :

$$\frac{dCO_{bias}}{dt} = 0$$

Removing this term from the equation results in :

$$\frac{dCO(t)}{dt} = Kc\frac{de(t)}{dt} + \frac{Kc}{Ti}e(t)$$

If we assume discrete or finite difference approximations for the continuous derivatives, then the controller becomes :

$$\frac{\Delta CO}{\Delta t} = Kc\frac{\Delta e_i}{\Delta t} + \frac{Kc}{Ti}e_i$$

where e_i is the current controller error, e_{i-1} is the controller error at the previous sample time, T, and $\Delta e_i = e_i - e_{i-1}$.

Recognizing that loop sample time is T = Δt, then the PI controller becomes :

$$\frac{\Delta CO}{T} = Kc\left(\frac{e_i - e_{i-1}}{T}\right) + \frac{Kc}{Ti}e_i$$

Rearranging, we arrive at the *discrete velocity form* of the PI controller :

$$\Delta CO = Kc\left(1 + \frac{T}{Ti}\right)e_i - Kc \cdot e_{i-1} .$$

- *Reason for Anti-Windup Protection :* Discrete velocity algorithms compute a ΔCO that signals the FCE to move a specific distance and direction from its current position. As we can see from the PI controller form above, the computation does not keep track of the current FCE position, nor does it mathematically accumulate any integral sums.

In a sense, the accumulation of integration is stored in the final control element itself. If a long series of ΔCO moves are all positive, for example, the valve, pump or compressor will move toward its maximum value. And once the FCE reaches its maximum limit, any ΔCO commands to move further will have no impact because, a valve cannot open more than all the way and a pump cannot go slower than stopped. It is the physical nature of the FCE itself that provides protection from over-accumulation (*i.e.*, windup).

As long as the CO never reaches CO_{max} or CO_{min}, the continuous position and discrete velocity forms of the PI controller provide identical performance. A properly jacketed continuous position PI controller will also provide windup protection equal to the discrete velocity form. Implicit in these statements is that sample time,T, is reasonably fast and that T and the tuning values (Kc and Ti) are the same when comparing implementations.

- *Concerns with Discrete Velocity PID :* Unfortunately, the usefulness of the discrete velocity form is limited because the method suffers problems when derivative action is included. We find that we must take the derivative of a derivative, yielding a second derivative. A second derivative applied to data that contains even modest noise can produce nonsense results.

Some vendors implement this form anyway and include a signal filter and additional logic sequences to address the problem. Thus, even with the anti-windup benefits of a discrete velocity algorithm, we find the need to jacket the algorithm with protective logic.

Chapter 6

DERIVATIVE ACTION AND PID CONTROL

PID CONTROL AND DERIVATIVE ON MEASUREMENT

Like the PI controller, the Proportional-Integral-Derivative (PID) controller computes a controller output (CO) signal for the final control element every sample time T.

The PID controller is a "three mode" controller. That is, its activity and performance is based on the values chosen for three tuning parameters, one each nominally associated with the proportional, integral and derivative terms.

As we had discussed previously, the PI controller is a reasonably straightforward equation with two adjustable tuning parameters. The number of different ways that commercial vendors can implement the PID form is fairly limited, and they all provide the same performance if properly tuned.

With the addition of a third adjustable tuning parameter, the number of algorithm permutations increases markedly. And there are even different forms of the PID equation itself. This creates added challenges for controller design and tuning.

Here we focus on what a derivative is, how it is computed, and what it means for control. We also explore why derivative on measurement is widely recommended for industrial practice.

The Dependent, Ideal PID Form

A popular way vendors express the dependent, ideal PID controller is :

$$CO = CO_{bias} + Kc \cdot e(t) + \frac{Kc}{Ti} \int e(t)dt + Kc \cdot Td \frac{de(t)}{dt}$$

Where :

CO = controller output signal (the wire out)

CO_{bias} = controller bias; set by bumpless transfer

$e(t)$ = current controller error, defined as SP–PV

SP = set point

PV = measured process variable (the wire in)

Kc = controller gain, a tuning parameter

Ti = reset time, a tuning parameter

Td = derivative time, a tuning parameter.

The first three terms to the right of the equal sign are identical to the PI controller we have already explored in some detail.

The derivative mode of the PID controller is an additional and separate term added to the end of the equation that considers the derivative (or rate of change) of the error as it varies over time.

The Contribution of the Derivative Term

The proportional term considers *how far* PV is from SP at any instant in time. Its contribution to the CO is based on the size of $e(t)$ only at time t. As $e(t)$ grows or shrinks, the influence of the proportional term grows or shrinks immediately and proportionately.

The integral term addresses *how long* and how far PV has been away from SP. The integral term is continually summing $e(t)$. Thus, even a small error, if it persists, will have a sum total that grows over time and the influence of the integral term will similarly grow.

A derivative describes how steep a curve is. More properly, a derivative describes the slope or the rate of change of a signal trace at a particular point in time. Accordingly, the derivative term in the PID equation above considers *how fast*, or the rate at which, error (or PV as we discuss next) is changing at the current moment.

Derivative on PV is Opposite but Equal

While the proportional and integral terms of the PID equation are driven by the controller error, $e(t)$, the derivative computation in many commercial implementations should be based on the value of PV itself.

The derivative of $e(t)$ is mathematically identical to the negative of the derivative of PV everywhere except when set point changes. And when set point changes, derivative on error results in an undesirable control action called *derivative kick*.

Math Note : The mathematical defense that "derivative of $e(t)$ equals the negative derivative of PV when SP is constant" considers that, since $e(t)$ = SP–PV, the equation below follows. That is, derivative of error equals derivative of set point minus process variable.

The derivative of a constant is zero, so when SP is constant, mathematically, the derivative (or slope or rate of change) of the controller error equals the derivative (or slope or rate of change) of the measured process variable, PV, except the sign is opposite.

$$\frac{de(t)}{dt} = \frac{d(\cancel{SP}^0 - PV)}{dt} = -\frac{dPV}{dt}$$

The figures below provide a visual appreciation that the derivative of $e(t)$ is the negative of the derivative of PV.

The top plot shows the measured PV trace after a set point step. The bottom plot shows the $e(t)$ = SP–PV trace for the same event.

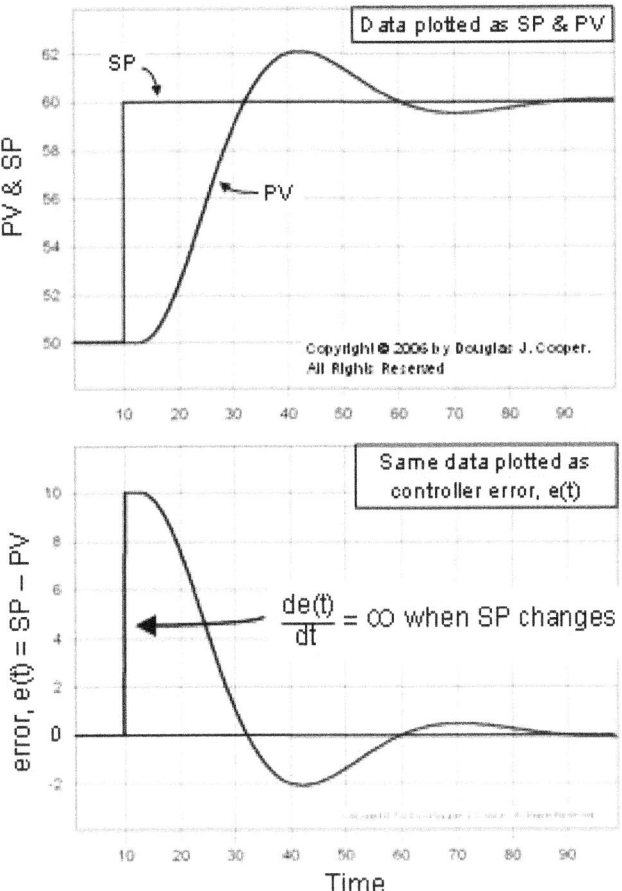

If we compare the two plots after the SP step at time t = 10, we see that the PV trace in the upper plot is an exact reflection of the $e(t)$ trace in the lower plot. The PV trace ascends, peaks and then settles, while in a reflected pattern, the $e(t)$ trace descends, dips and then settles.

Mathematically, this "mirror image" of trace shapes means that the derivatives (or slopes or rates of change) are the same everywhere after the SP step, except they are opposite in sign.

Derivative on PV Used in Practice

While the shape of $e(t)$ and PV are opposite but equal everywhere after the set point step, there is an important difference at the moment the SP changes. The lower plot shows a vertical spike in $e(t)$ at this moment. There is no corresponding spike in the PV plot.

The derivative (or slope) of a vertical spike in the theoretical world approaches infinity. In the real world it is at least a very big number. If Td is large enough to provide any meaningful weight to the derivative term, this huge derivative value will cause a large and sudden manipulation in CO. This large manipulation in CO, referred to as *derivative kick*, is almost always undesirable.

As long as loop sample time, T, is properly specified, the PV trace will follow a gradual and continuous response, avoiding the dramatic vertical spike evident in the $e(t)$ trace.

Because derivative on $e(t)$ is identical to derivative on PV at all times except when the SP changes, and when the set point does change, derivative on error provides information we don't want our controller to use, we substitute the "math note" equation in the yellow box above to obtain the PID with derivative on measurement controller :

$$CO = CO_{bias} + Kc \cdot e(t) + \frac{Kc}{Ti} \int e(t)dt - Kc \cdot Td\frac{dPV}{dt}$$

Derivative on PV Does Not "Kick"

Below we show the heat exchanger case study under PID control using the dependent, ideal algorithm form and moderate tuning values.

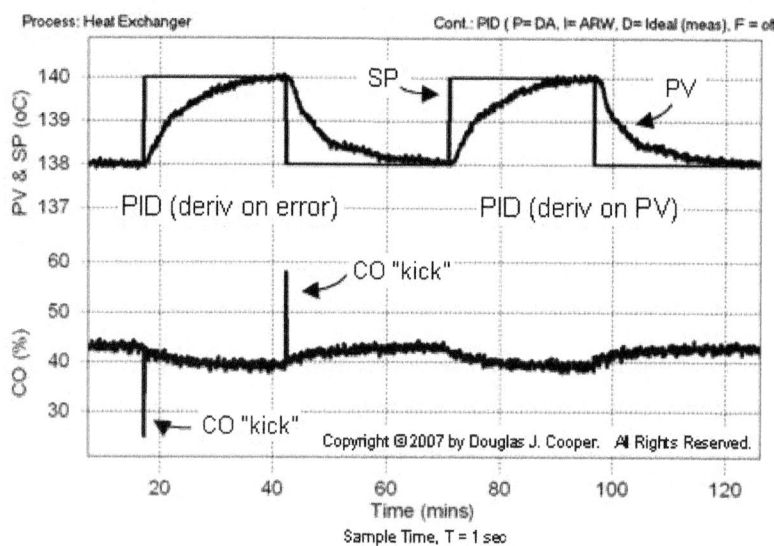

Heat Exchanger Under PID Control
Ideal dependent form: $Kc = -0.47$; $Ti = 1.7$; $Td = 0.31$

The first set point steps to the left in the plot below show loop performance when PID with derivative on error is used. The set points steps to the right present the identical scenario except that PID with derivative on measurement is used.

The "kick" that dominates the CO trace when derivative on error is used is rather dramatic and somewhat unsettling. While it exists for only a brief moment and does not impact performance in this example, we should not assume this will always be true. In any event, such action will eventually take a toll on mechanical final control elements.

Understanding Derivative Action

A rapidly changing PV has a steep slope and this yields a large derivative. This is true regardless of whether a dynamic event has just begun or if it has been underway for some time.

In the plot below, the derivative dPV/dt describes the slope or "steepness" of PV during a process response.

Early in the response, the slope is large and positive when the PV trace is increasing rapidly. When PV is decreasing, the derivative (slope) is negative. And when the PV goes through a peak or a trough, there is a moment in time when the derivative is zero.

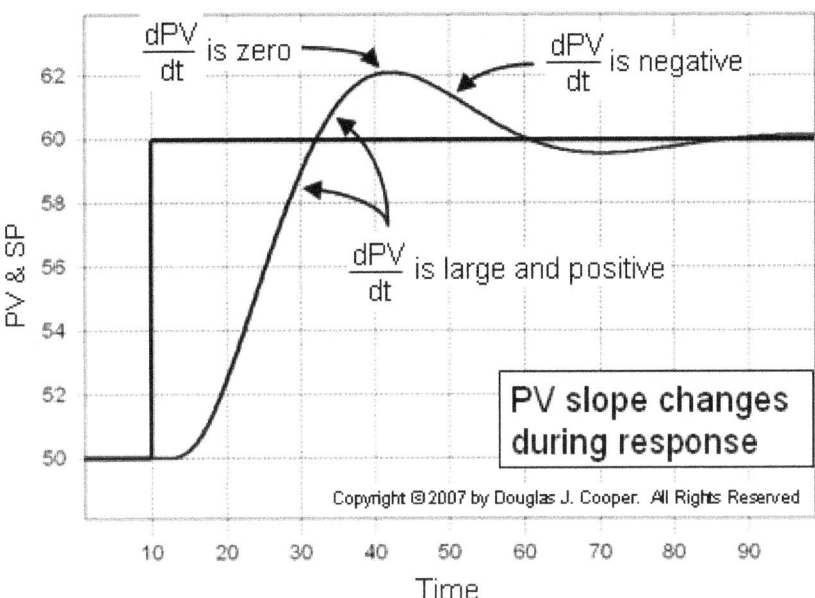

To understand the impact of this changing derivative, let's assume for discussion that :

- Controller gain, Kc, is positive.
- Derivative time, Td (always positive) is large enough to provide meaningful weight to the derivative term. After all, if Td is very small, the derivative

term has little influence, regardless of the slope of the PV. The negative sign in front of the derivative term of the PID with derivative on measurement controller (and given the above assumptions) means that the impact on CO from the derivative term will be opposite to the sign of the slope :

$$CO = CO_{bias} + Kc \cdot e(t) + \frac{Kc}{Ti} \int e(t)dt - Kc \cdot Td\frac{dPV}{dt}$$

Thus, when dPV/dt is large and positive, the derivative term has a large influence and seeks to decrease CO.

Conversely, when dPV/dt is negative, the derivative term seeks to increase CO.

It is interesting to note that the derivative term does not consider whether PV is heading toward or away from the set point (whether $e(t)$ is positive or negative). The only consideration is whether PV is heading up or down and how quickly.

The result is that derivative action seeks to inhibit rapid movements in the PV. This could be an especially useful characteristic when seeking to dampen the oscillations in PV that integral action tends to magnify. Unfortunately, as we will discuss, the potential benefit comes with a price.

THE CHAOS OF COMMERCIAL PID CONTROL

The design and tuning of a three mode PID controller follows the proven recipe we have used with success for P-Only control (*e.g.*, here and here) and PI Control (*e.g.*, here, here and here). The decisions and procedures we established for steps 1–3 of the design and tuning recipe in these previous studies remain unchanged as we move on to the PID algorithm.

Step 4 of the recipe remains the same as well. But it is essential in this step that we match the rules and correlations of step 4 with the particular controller algorithm form we are using.

The challenge arises because the number of PID algorithm forms available from hardware vendors increases markedly when derivative action is included. And unfortunately, these PID algorithms are implemented in many different forms across the commercial market.

The potential for confusion by even a careful practitioner is significant. For example :

- There are three popular PID algorithm forms, and
- Each of these three forms have multiple parameters that are cast in different ways.

As a result, there are literally dozens of possible PID algorithm forms. Matching each controller form with its proper design rules and correlations requires careful attention if performed without the help of software tools.

Common Algorithm Forms

Listed below are the three common PID controller forms. If offered as an option by our vendor (most do offer it), derivative on measured process variable (PV) is the recommended PID form :

- Dependent, ideal PID controller form (derivative on measurement) :

$$CO = CO_{bias} + Kc \cdot e(t) + \frac{Kc}{Ti} \int e(t)dt - Kc \cdot Td \frac{dPV}{dt}$$

- Dependent, interacting form (derivative on measurement) :

$$CO = CO_{bias} + Kc \left(1 + \frac{Td}{Ti} \right) e(t) + \frac{Kc}{Ti} \int e(t)dt - KcTd \frac{dPV}{dt}$$

- Independent PID form (derivative on measurement) :

$$CO = CO_{bias} + Kc \cdot e(t) + Ki \int e(t)dt - kd \frac{dPV}{dt}$$

Where for the above :

CO = controller output signal (the wire out)

CO_{bias} = controller bias; set by bumpless transfer

$e(t)$ = current controller error, defined as SP–PV

SP = set point

PV = measured process variable (the wire in)

Kc = controller gain (also called proportional gain), a tuning parameter

Ki = integral gain, a tuning parameter

Kd = derivative gain, a tuning parameter

Ti = reset time, a tuning parameter

Td = derivative time, a tuning parameter.

Tuning Parameters

Because there has been little standardization on nomenclature, the same tuning parameters can appear under different names in the commercial market. Perhaps more unfortunate, the same parameter can even have a different name within a single company's product line.

A few notes to consider :

1. The dependent forms appear most in products commonly used in the process industries, but the independent form is not uncommon.

2. The majority of DCS and PLC systems now use controller gain, Kc, for their dependent PID algorithms. There are notable exceptions, however, such as Foxboro who uses proportional band (PB = 100/Kc assuming PV and CO both range from 0 to 100%).

3. Reset time, Ti, is slightly more common for the dependent PID algorithms, though it is rarely called that in product documentation. Reset rate, defined as $Tr = 1/Ti$, comes in a close second. Again, the name for this parameter changes with product.

4. Most vendors use derivative time, Td, for their dependent PID algorithms, though few refer to it by that name in their product documentation.

Tune One, Tune Them All

Some good news in all this confusion is that the different forms, if tuned with the proper correlations, will perform exactly the same. No one form is better than another, it is just expressed differently.

In fact, we can show equivalence among the parameters, and thus algorithms, with these relations.

- Proportional

$$Kc, \text{ideal} = Kc, \text{interact} \left(1 + \frac{Td, \text{interact}}{Ti, \text{interact}} \right)$$

  Kc, independent = Kc, ideal

- Integral

$$Ti, \text{ideal} = Ti, \text{interact} \left(1 + \frac{Td, \text{interact}}{Ti, \text{interact}} \right)$$

$$Ki, \text{independent} = \frac{Kc, \text{ideal}}{Ti, \text{ideal}}$$

- Derivative

$$Td, \text{ideal} = \frac{Td, \text{interact}}{\left(1 + \dfrac{Td, \text{interact}}{Ti, \text{interact}} \right)}$$

  Kd, independent = $(Kc, \text{ideal})(Td, \text{ideal})$

Though not presented here, analogous conversion relations can be developed for forms expressed using proportional band and/or reset rate.

Clarity in the Chaos

It is perhaps reasonable to hope that industrial practitioners will have an intuitive understanding of proportional, integral and derivative action. They might know the benefits each term offers and problems each presents. And experienced practitioners will know how design, tune and validate a PID implementation.

Expecting a practitioner to convert that knowledge and intuition over into the confusion of the commercial PID marketplace might not be so reasonable.

Given this, the best solution for those in the real world is to use software that lets us focus on the big picture while the software ensures that details are properly addressed.

Such productivity software should not only provide a "click and go" approach to algorithm and tuning parameter selection, but should also provide this information simply based on our choice of equipment manufacturer and product line.

For example, below is a portion of the controller manufacturer selection available in one commercial software package :

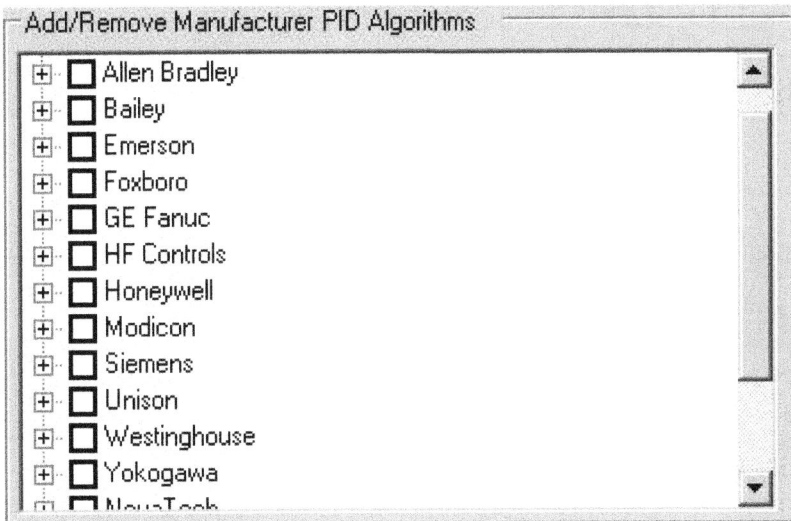

If you select Allen Bradley, Emerson, and Honeywell in the above list, the choice of PID controllers for each company is shown in the next three images :

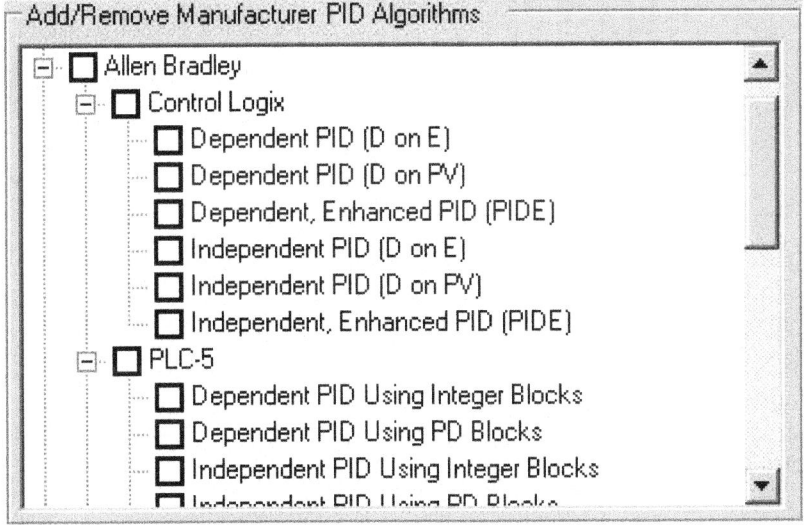

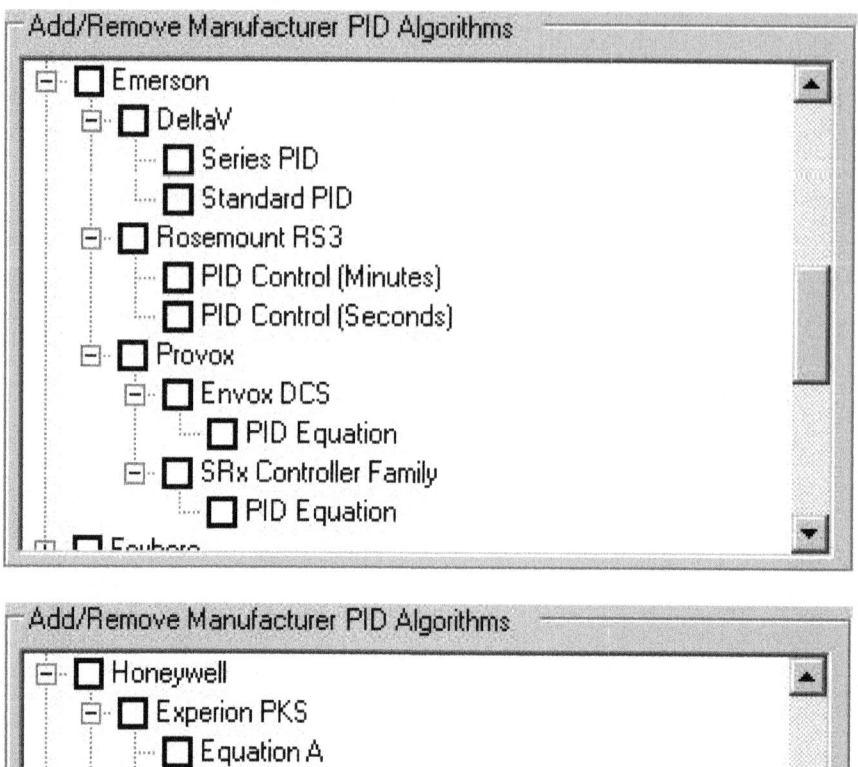

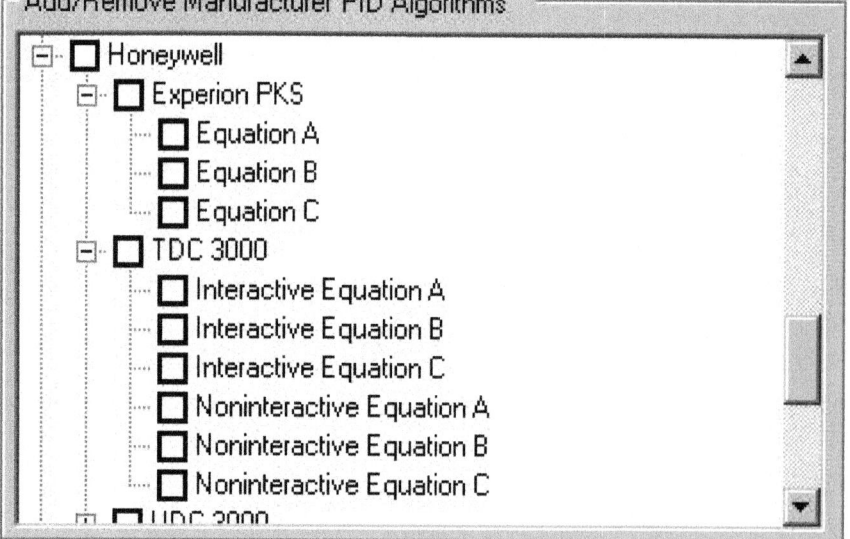

It is clear from these displays that there are different terms and many options for us to select from, all for PID control. And it may not be obvious that the different terms above refer to some version of our "basic three" PID forms.

Too much is at stake in a plant to ask a practitioner to keep track of it all. Software can get us past the details during PID controller design and tuning so we can focus on mission-critical control tasks like improving safety, performance and profitability.

Note : the Laplace domain is a subject that most control practitioners can avoid their entire careers, but it provides is a certain mathematical "elegance."

Below, for example, are the three controller forms assuming derivative on error. Even without familiarity with Laplace, perhaps you will agree the three PID forms indeed look like part of the same family :

- Dependent, ideal (Non-interacting) From

$$\frac{CO(s)}{E(s)} = Kc\left(1+\frac{1}{Ti\,s}+Td\,s\right)$$

- Dependent, Interacting From

$$\frac{CO(s)}{E(s)} = Kc\left(1+\frac{1}{Ti\,s}\right)(Td\,s+1)$$

- Independent From

$$\frac{CO(s)}{E(s)} = Kc\frac{Ki}{s}+Kd\,s$$

MEASUREMENT NOISE DEGRADES DERIVATIVE ACTION

Benefits and Drawbacks

Derivative action has its largest influence when the measured process variable (PV) is changing rapidly (when the *slope of the PV trace is steep*).

The three terms of a properly tuned PID controller thus work together to provide a rapid response to error (proportional term), to eliminate offset (integral term), and to minimize oscillations in the PV (derivative term).

While this sounds great in theory, unfortunately, there are serious drawbacks to including derivative action in our controller.

We have discussed how challenging it can be to balance two *interacting tuning parameters for PI control.*

A PID controller has three tuning parameters (three modes) that all interact and must be balanced to achieve a desired performance. It is often not at all obvious which of the three tuning parameters must be adjusted to correct behavior if performance is considered to be undesirable.

Trial and error tuning is hopeless for any but the most skilled practitioner. Fortunately, our *tuning recipe* provides a quick route to a safe and profitable PID performance.

A second disadvantage relates to the uncertainty in the derivative computation for processes that have noise in the PV signal.

PID Controller Form

For discussion purposes, we use the Dependent, Ideal (Non-interacting) form :

$$CO = CO_{bias} + Kc \cdot e(t) + \frac{Kc}{Ti} \int e(t)dt - Kc \cdot Td\frac{dPV}{dt}$$

The various PID algorithms forms provide identical performance if each algorithm is matched to its proper tuning correlations. Hence, the observations and conclusions presented below are general in nature and are not specific to a particular algorithm form.

Derivative Action Dampens Oscillations

The plot below shows the impact of derivative action on set point response performance. Because noise in the PV signal can impact performance, an idealized noise-free simulation was used to create the plot.

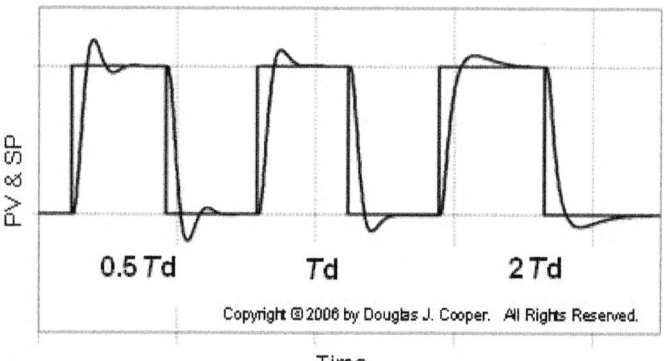

The plot shows the PV response to three pairs of set point (SP) steps. The middle response shows the base case performance of a PID controller tuned using the aggressive correlations referenced above.

For the set point steps to the right and left of the base case, the derivative time, Td, is adjusted while the controller gain, Kc, and reset time, Ti, are kept constant. This let's us isolate the impact of derivative time on performance.

It is apparent that when derivative action is cut in half to the left in the plot, the oscillating nature of the response increases. And when Td is doubled to the right, the increased derivative action inhibits rapid movement in the PV, causing the rise time and settling time to lengthen.

We saw in *this tuning map* on interacting tuning parameters for PI controllers that with only two tuning parameters, performance response plots could look similar.

With the addition of Td, we now have a three dimensional tuning map with a great many similar-looking plots. If we are unhappy with our controller performance, knowing which parameter to adjust and by how much borders on the impossible. With three mode PID control, the orderly approach of our tuning recipe becomes fundamental to success.

Measurement Noise Leads to Controller Output "Chatter"

As discussed in more detail in the *PID control of the heat exchanger* study and summarized here, the side-by-side comparison of PI vs PID control shown below illustrates one unwelcome result of adding derivative action to our controller.

The CO signal trace along the bottom of the plot clearly changes when the derivative term is added. Specifically, derivative action causes the noise (random error) in the PV signal to be amplified and reflected in the controller output.

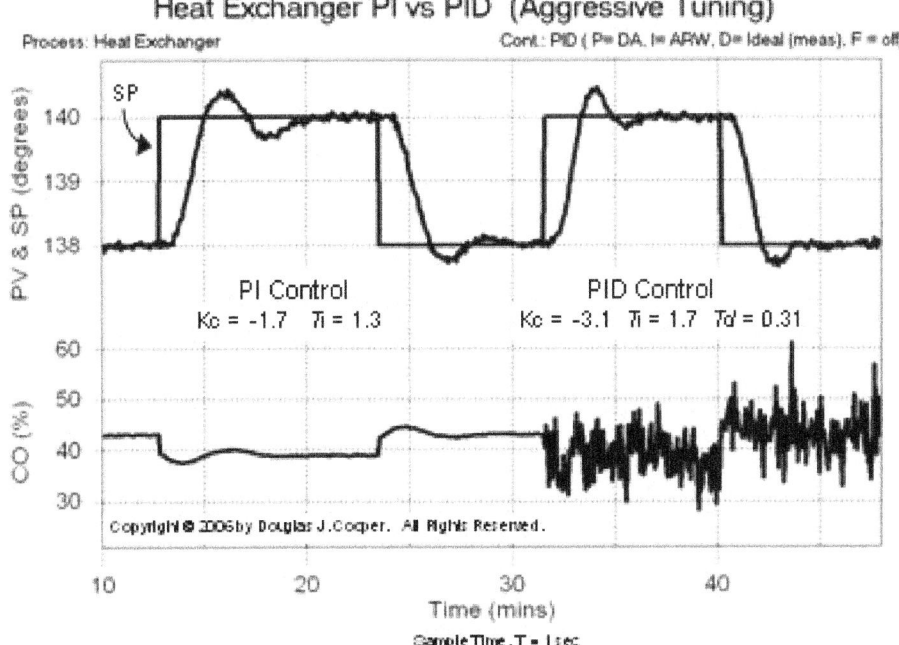

The reason for this extreme CO action or "chatter" is illustrated below. As indicated in the plot, a noisy PV signal produces conflicting derivatives as the slope appears to dramatically alternate direction at every sample.

The consequence of a PV that repeatedly changes from "rapidly increasing slope" to "rapidly decreasing slope" is a derivative term that computes a series of large, alternating CO actions.

The ultimate impact of this alternating derivative computation on the total CO depends on the size of Td (the weight given to the derivative term). As Td grows larger, the "chatter" in the CO signal grows in response.

In any event, extreme control action will increase the wear on a mechanical final control element and lead to an increase in maintenance costs.

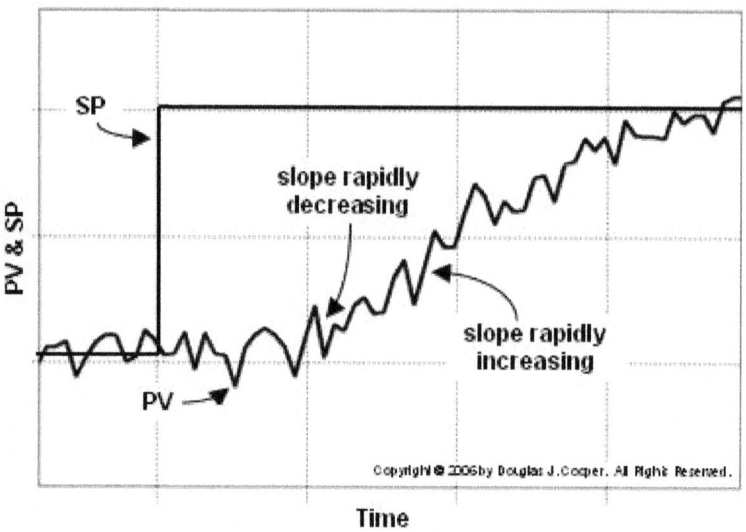

Larger Noise Means Larger Problems

The plot below shows a more subtle problem that measurement noise can cause with derivative action. In particular, this problem arises when the level of operation is near a controller output constraint (either the maximum or minimum CO).

The PID tuning values in the plot are constant throughout the experiment. As indicated on the plot, measurement noise is increased in increments across three set point tracking tests.

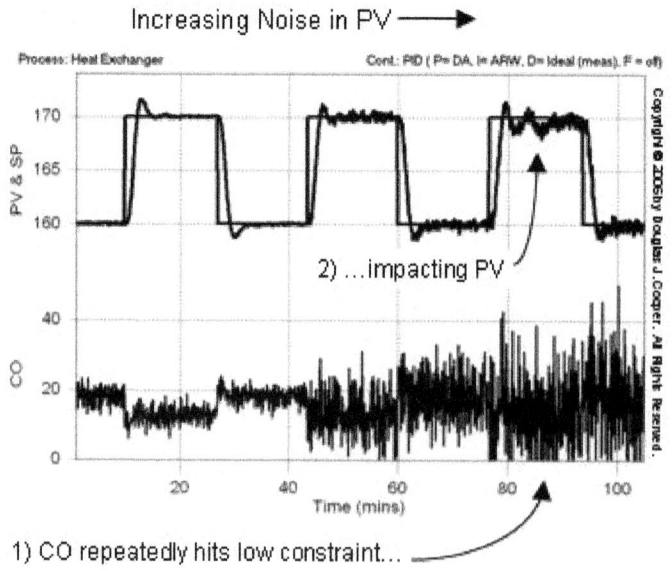

As long as measurement noise causes the derivative to alternate equally between suddenly increasing and suddenly decreasing, and the controller output can reflect this "equality in randomness" unimpeded, then controller performance is reasonably consistent in spite of increasing noise in the PV signal.

If a constraint inhibits the controller output from the "equality in randomness" symmetry, causing it to become skewed or off center, then controller performance degrades.

This is illustrated in the right most set point steps above. For this case, the controller output signal becomes so active that it repeatedly hits the minimum CO value. By constraining CO, the controller output loses its symmetry, causing the PV to wander.

PID DISTURBANCE REJECTION OF THE GRAVITY DRAINED TANKS

Here we investigate the benefits and challenges of derivative action and PID control when disturbance rejection remains our control objective.

As with all controller implementations, we follow our four-step design and tuning recipe. A benefit of this recipe is that steps 1–3 are independent of the controller used, so our previous results from steps 1 and 2 (detailed here) and step 3 (detailed here and here) can be used in this PID study.

We summarize those previous results before proceeding to step 4 and the design and tuning of a PID controller.

Step 1 : Determine the Design Level of Operation (DLO)

The control objective is to reject disturbances as we control liquid level in the lower tank. Our DLO for this study is :
- Design PV and SP = 2.2 m with range of 2.0 to 2.4 m
- Design D = 2 L/min with occasional spikes up to 5 L/min

Step 2 : Collect Process Data Around the DLO

When CO, PV and D are steady near the design level of operation (DLO), we bump the CO as detailed here and force a clear response in the PV that dominates the noise.

Step 3 : Fit a FOPDT Model to the Dynamic Process Data

We approximate the dynamic behavior of the process by fitting test data with a first order plus dead time (FOPDT) dynamic model. A fit of step test data and doublet test data yields these values :
- Process gain (how far), $Kp = 0.09$ m/%
- Time constant (how fast), $Tp = 1.4$ min
- Dead time (how much delay), $\theta p = 0.5$ min

Step 4 : Use the FOPDT Parameters to Complete the Design

The preferred PID algorithm in industrial practice employs derivative on PV, and vendors market this controller in several different forms. Each algorithm form has its own tuning correlations, and if we take care to match algorithm with correlation, they all provide identical capability and performance.

For tuning, we rely on the industry-proven Internal Model Control (IMC) tuning correlations. These require only one specification, the closed loop time constant (Tc), that describes the desired speed or quickness of our controller in responding to a set point (SP) change or rejecting a disturbance (D).

Our PI control study describes what to expect from an aggressive, moderate or conservative controller. Once our desired performance is chosen, the closed loop time constant is computed :

- *Aggressive* : Tc is the larger of $0.1 \cdot Tp$ or $0.8 \cdot \theta p$
- *Moderate* : Tc is the larger of $1 \cdot Tp$ or $8 \theta p$
- *Conservative* : Tc is the larger of $10 \cdot Tp$ or $80 \theta p$.

Because the popular PID forms perform the same if properly tuned, the observations and conclusions we draw from any one algorithms applies to the other forms.

Dependent Ideal PID

Among the most widely used algorithms is the Dependent Ideal (Non-interacting) PID form :

$$CO = CO_{bias} + Kc \cdot e(t) + \frac{Kc}{Ti} \int e(t)dt - Kc \cdot Td \frac{dPV}{dt}$$

Where :

CO = controller output signal (the wire out)

CO_{bias} = controller bias; set by bumpless transfer

$e(t)$ = current controller error, defined as SP–PV

SP = set point

PV = measured process variable (the wire in)

Kc = controller gain, a tuning parameter

Ti = reset time, a tuning parameter

Td = derivative time, a tuning parameter.

- *Design and Tune* : In the P-only study, we had established that for the gravity drained tanks process :
- Sample time, T = 1 sec
- The controller is reverse acting
- Dead time is small compared to Tp and thus not a concern in the design.

After we choose a Tc based on our desired performance, the tuning correlations for the Dependent Ideal PID form are :

$$Kc = \frac{1}{Kp}\left(\frac{Tp+0.5\theta p}{Tc+0.5\theta p}\right), Ti = Tp+0.5\theta p; Td = \frac{Tp\theta p}{2Tp+\theta p}$$

Similar to the PI controller tuning correlations, only controller gain contains Tc, and thus, only Kc changes based on the need for a more or less active controller.

- *Implement and Test* : We first explore an aggressive response tuning for our ideal PID controller :

Aggressive Tc = the larger of 0.1 ·Tp or 0.8 θp

　　　　　　= larger of 0.1 (1.4 min) or 0.8 (0.5 min)

　　　　　　= 0.4 min.

Using this Tc and our Kp, Tp and θp from Step 3 in the tuning correlations above, we compute these aggressive controller gain, reset time and derivative time tuning values :

Aggressive Ideal PID : Kc = 28%/m; Ti = 1.7 min; Td = 0.21 min

The performance of this controller in rejecting changes in the pumped flow disturbance (D) for the gravity drained tanks is shown to the right in plot below. For comparison, the performance of an aggressive PI controller is shown in the plot to the left. Note that the set point (SP) remains constant at 2.2 m throughout the study.

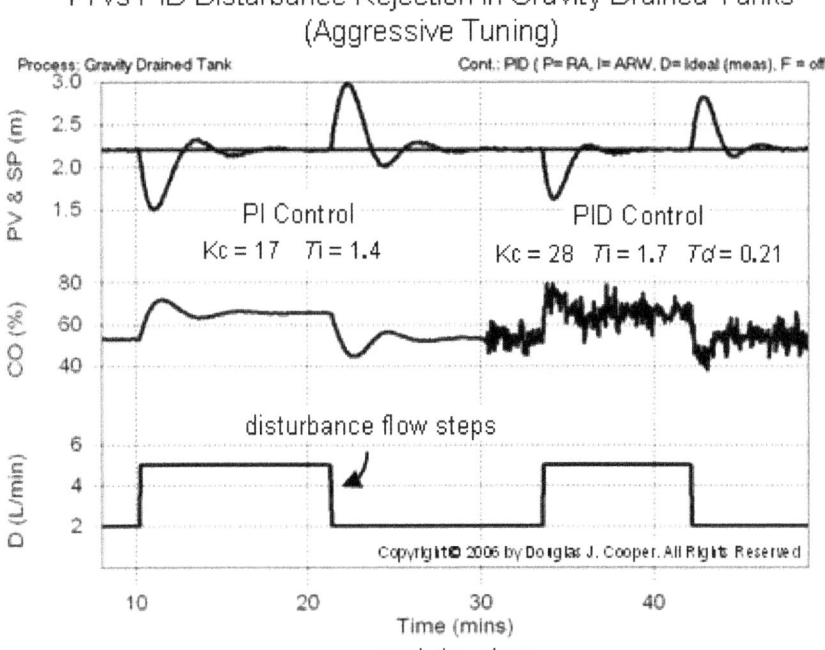

PI vs PID Disturbance Rejection in Gravity Drained Tanks (Aggressive Tuning)

The maximum deviation of the PV from set point during the disturbance rejection event is smaller for the PID controller relative to the PI controller. The PID controller also provides a faster settling time because derivative action tends to reduce the rolling or oscillatory behavior in the PV trace.

Like the heat exchanger PID study, there is an obvious difference in the CO signal trace for the PI vs PID controllers. Derivative action causes the noise (random error) in the PV signal to be amplified and reflected in the control output (CO) signal.

Such extreme control action will cause excessive wear in a valve or other mechanical final control element, requiring increased maintenance. This consequence of noise in the measured PV can be a serious disadvantage with PID control.

Ideal vs Interacting PID

We compare the Dependent Ideal PID form above to the performance of the Dependent Interacting PID form and establish that they are identical in performance if properly tuned. The Dependent Interacting form is written :

$$CO = CO_{bias} + Kc\left(1 + \frac{Td}{Ti}\right)e(t) + \frac{Kc}{Ti}\int e(t)dt - KcTd\frac{dPV}{dt}$$

- *Design and Tune* : We use the same rules above to choose a Tc that reflects our desired performance. The IMC tuning correlations for the Dependent, Interacting form are then :

$$Kc = \frac{1}{Kp}\left(\frac{Tp}{Tc + 0.5\theta p}\right); Ti = Tp; Td = 0.5\theta p$$

As before, only controller gain contains Tc, and thus, only Kc changes based on a desire for a more or less active controller. Sample time remains for this implementation at $T = 1$ sec and the controller remains as reverse acting.

- *Implement and Test* : We choose a moderate response tuning in this example :

Moderate Tc = the larger of $1 \cdot Tp$ or $8\ \theta p$

$$= \text{larger of } 1.0\ (1.4\ \text{min}) \text{ or } 8\ (0.5\ \text{min})$$

$$= 0.4\ \text{min}.$$

Using this Tc and our model parameters in the proper tuning correlations (ideal or interacting), we arrive at these moderate tuning values :

Moderate Ideal PID : Kc = 4.3%/m; Ti = 1.7 min; Td = 0.21 min

Moderate Interacting PID : Kc = 3.7%/m; Ti = 1.4 min; Td = 0.25 min.

As shown in the plot below, moderate tuning provides a reasonably fast disturbance rejection response while producing little or no oscillations as the PV settles.

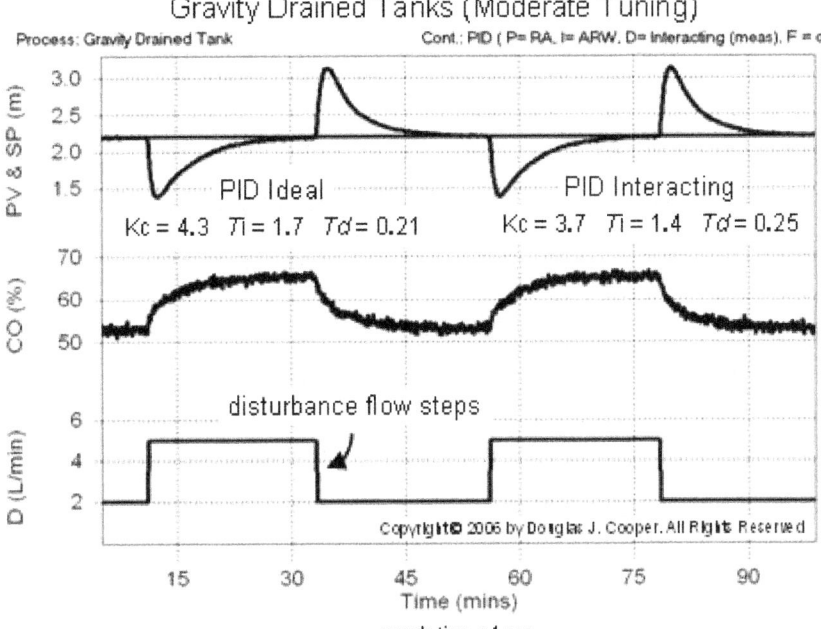

The indistinguishable behavior confirms that the two controllers indeed are identical in capability and performance if tuned with their own correlations.

Aside : Our observations using the dependent ideal and dependent interacting PID algorithms directly apply to the other popular PID controller forms.

For example, the independent PID algorithm form is written :

$$CO = CO_{bias} + Kc \cdot e(t) + Ki \int e(t)dt - Kd\frac{dPV}{dt}$$

The integral and derivative gains in the above independent form can be computed, for example, using the ideal PID correlations as : $Ki = Kc/Ti$ and $Kd = Kc \times Td$.

Because of these mathematical identities, performance and capability observations drawn about one algorithm will apply directly to the other.

Ideal Moderate vs Ideal Aggressive

As shown in the plot below, we compare moderate tuning side-by-side with aggressive tuning for the dependent ideal PID controller. For a different perspective, we make this comparison using a set point tracking objective.

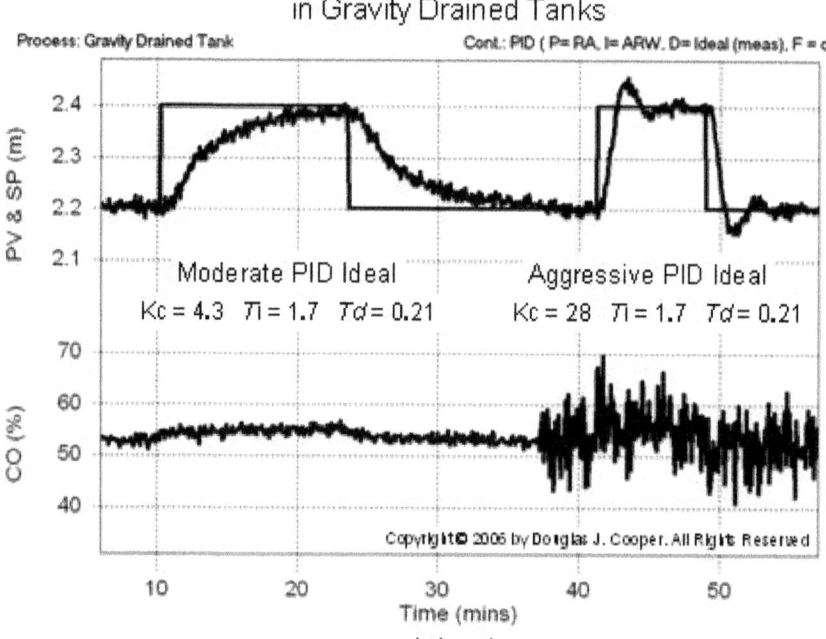

The performance of the two controllers matches the design descriptions provided here. That is, a controller tuned with :

- A moderate Tc will move the PV reasonably fast while producing little to no overshoot.

- An aggressive Tc will move the PV quickly enough to produce some overshoot and then oscillation as the PV settles out.

Need for CO Filtering

The excessive activity in the CO signal can be a problem, and a controller output (CO) signal filter is one solution. An interesting observation from the above plot is that the degree of "chatter" in the CO signal grows as controller gain, Kc, increases.

Chapter 7

SIGNAL FILTERS AND THE PID WITH CONTROLLER OUTPUT FILTER ALGORITHM

USING SIGNAL FILTERS IN OUR PID LOOP

Sources of Noise

Random behavior in the PV measurement arises because of signal noise and process noise.

Signal noise tends to have higher frequency relative to the characteristic dynamics of process control applications (*i.e.*, processes with streams comprised of liquids, gases, powders, slurries and melts). Sources of signal noise include :

- Electrical interference
- Jitter (clock related irregularities such as variations in sample spacing)
- Quantizing of signal samples into overly-broad discrete "buckets" from low resolution or improperly specified instrumentation (*e.g.* too-large measurement span relative to operating range).

Process noise tends to be lower in frequency. This category borders on the philosophical as to what constitutes a disturbance to be controlled versus noise to be filtered.

Bubbles and splashing that randomly corrupts liquid pressure drop measurements is an example of process noise that might benefit from filtering.

A less clear candidate for filtering is a temperature measurement in a poorly-mixed vessel. The mixing patterns can cause lower-frequency random variations in the temperature signal that are unrelated to changes in the bulk vessel temperature.

It is important to emphasize that before we try to filter away a problem, we should first work to understand the source of the random error. Rather than "fix" the noise by hiding it with additional or modified algorithms, we should attempt to reduce or eliminate the problem through normal engineering and maintenance practices.

The Filtered Signal

The plot below shows the random behavior of a raw (unfiltered) PV signal and the smoother trace of a filtered PV signal.

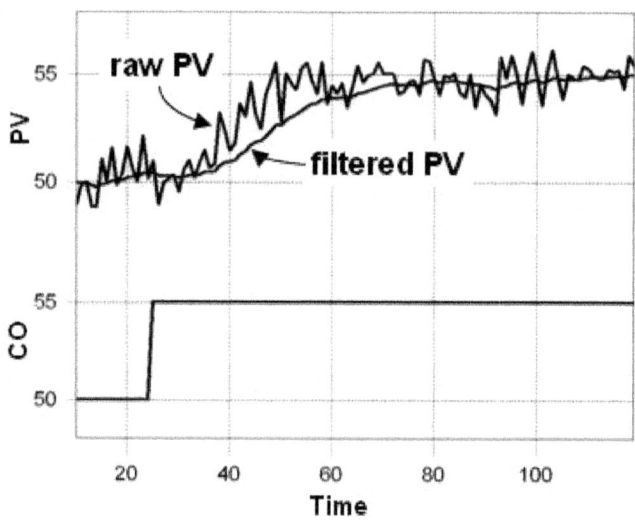

As the above plot illustrates, a filter is able to receive a noisy signal and yield a signal with reduced random variation. A "better" filter design is one that decreases the random variation while retaining more of the true dynamic information of the original signal.

Filters can be analog (hardware) or digital (software); high, low or band pass; linear or nonlinear; designed in the time, Z-transform or frequency domain; and much more. Filters can collect and process data at a rate faster than the control loop sample time, so many data points can go into a single PV sample forwarded to the controller.

Filters Add Delay

The filtered signal in the plot above, though perhaps visually appealing, clearly lags behind the actual dynamic response of the unfiltered signal. More specifically, the filtered signal has an increased dead time and time constant relative to the behavior of the actual process.

Signal filters offer benefit in process control applications because they can temper the large CO moves caused by derivative action of noisy PV measurements.

Yet they add delay in sensing the true state of a process, and this has negative consequences in that as delay increases, the best achievable control performance decreases.

The design challenge is to find the careful balance between signal smoothing and information delay to achieve the controller performance we desire for our application.

External Filters in Control

As shown below, there are three popular places to put external filters in the feedback loop. By "external," we mean that the filters are designed, installed and maintained separately from the controller.

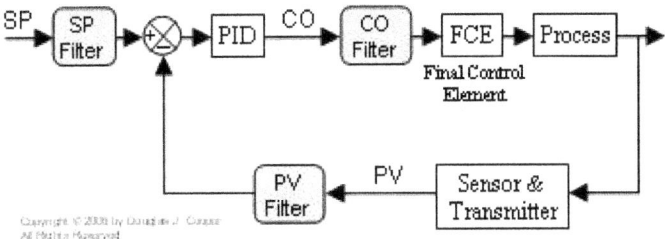

- *Set Point Filters :* Set point filters are not associated with the noisy PV problem, but are included here to make the discussion general.

A set point filter takes a step change in SP, and as shown below, forwards a smooth transition signal to the controller.

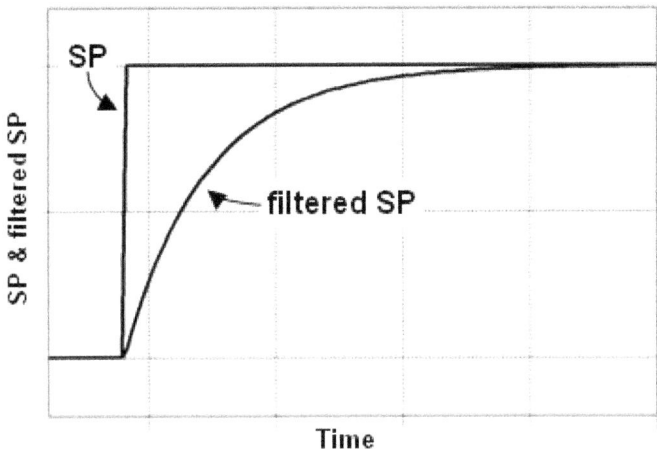

SP filters do not influence the disturbance rejection (regulatory) performance of a controller. Hence, these filters permit a controller to be tuned aggressively to reject disturbances, yet the smoothed SP transition results in a moderate set point tracking (servo) performance from this same aggressive controller.

SP filters are also used to limit overshoot at the top of a set point ramp or step change. If this is our design objective, an alternative is to eliminate the filter and employ a controller that uses proportional on PV rather than proportional on error. For example (compare to PI with proportional on error *here*) :

$$CO = CO_{bias} - Kc\,PV + \frac{Kc}{Ti}\int e(t)dt$$

Finally, and unfortunately, SP filters are occasionally used as a bandage to mask the fact that a controller is simply poorly designed and/or tuned. We all recognize that this is a practice to be avoided.

- *PV Filters* : Signal filters are frequently placed between the sensor transmitter and the controller (or more likely, the multiplexer feeding the controller). In process control applications, these filters should be analog (hardware) devices designed specifically to minimize high frequency electrical interference.

While measurement noise does degrade derivative action in that it leads to chatter in the CO signal, this "noise leads to CO chatter" effect is very modest for proportional action. And interestingly, integral action is unaffected by noise because the constant summing of error literally averages the random variations in the signal.

Since filtering adds delay and this hurts best possible control performance, and because noise is not an issue for proportional and integral action, it is generally poor practice to filter the PV signal external to the controller for anything beyond electrical interference.

The preferred approach is to selectively filter only that signal destined for the derivative computation. This moves the filter inside the controller architecture as discussed below.

- *CO Filters* : While PV filters smooth the signal feeding the controller, CO filters smooth the noise or "chatter" in the CO signal sent to the final control element. Even if PV signal noise does not appears to cause performance problems, a CO filter can offer potential benefits as it reduces fluctuations in the controller output and this reduces wear on the FCE.

If a noisy PV is an issue in our controller, our first attempts should be to locate and correct the problem. If after that exercise, our decision is to design and implement a filter in our feedback loop, CO filters are attractive alternatives.

Internal Filters in Control

For feedback control, filtering need only be applied to the signal feeding the derivative term. As stated before, noise does not present a problem for proportional and integral action. These elements will perform best without the delay introduced from a signal filter.

When we selectively filter just that signal feeding the derivative calculation, the filter becomes part of the controller architecture. Hence, we can still use our design recipe, though the correlations for tuning this four mode form are different from the four mode "PID with CO Filter" form mentioned above.

There are two common architectures that are identical in capability, though different in presentation.

As shown below, we filter the PV signal before feeding it to the derivative mode of the PID algorithm for computation :

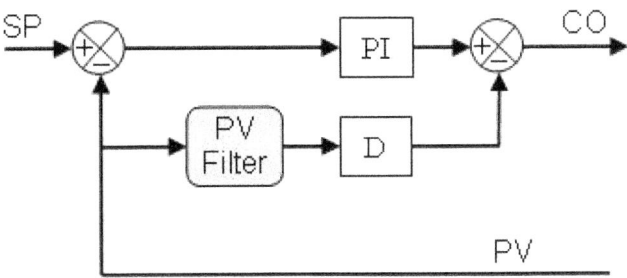

We can also compute the derivative action with the noisy signal and then filter the computed result :

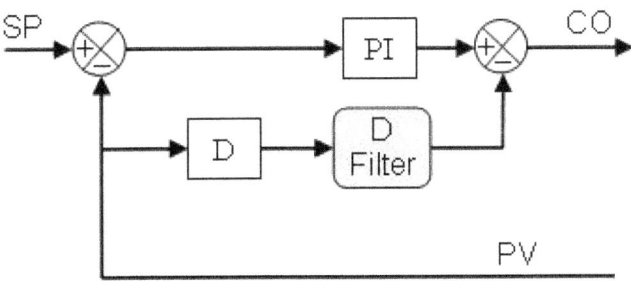

If the same filter form is used (*e.g.*, first-order), we can show mathematically that both options above are identical.

As it turns out, many commercial controllers use this internal derivative filtering form where they implement a first-order filter and fix the filter time at one-tenth of the derivative time value.

PID WITH CONTROLLER OUTPUT (CO) FILTER

The derivative action of a PID controller can cause noise in the measured process variable (PV) to be amplified and reflected as "chatter" in the controller output (CO) signal. Signal filters, implemented as either analog hardware or digital software, offer a popular solution to this problem.

If noise is impacting controller performance, our first attempts should be to locate and correct the underlying fault. Filters are poor cures for a bad design or failing equipment.

If we decide to employ a filter in our loop, an algorithm designed to smooth the controller output signal holds some allure.

The CO Filter Architecture

Below is a loop architecture with a PID controller followed by a controller output filter.

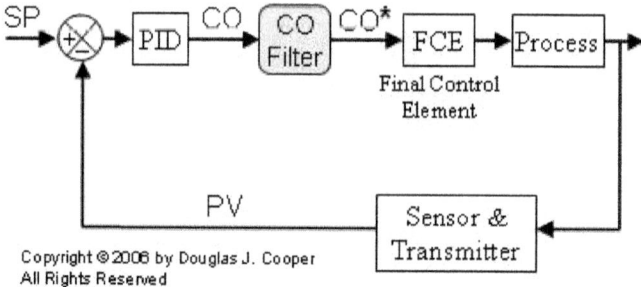

The benefits of this architecture include :

- A CO filter works to limit large controller output moves regardless of the underlying cause. As a result, CO filters can reduce persistent controller output fluctuations that cause wear in a mechanical final control element (FCE).

- A CO filter is a single solution that addresses both a noisy PV measurement problem and computational oddities that may exist in our vendor's particular PID algorithm,

- Perhaps most important, the tuning recipe we have employed so successfully with PI and PID algorithms can be directly applied to a PID with CO filter architecture.

PID Plus External Filter

As shown above, the filter computation is performed after the PID controller has computed the CO. Below, the PID controller output is computed using the non-interacting, dependent, ideal form, but any of the popular algorithms can be used with this "PID plus filter" architecture :

$$CO = CO_{bias} + Kc\ e(t) + \frac{Kc}{Ti}\int e(t)dt + Kc\ Td\frac{de(t)}{dt}$$

Where :

CO = controller output signal (the wire out)

CO_{bias} = controller bias; set by bumpless transfer

$e(t)$ = current controller error, defined as SP–PV

SP = set point

PV = measured process variable (the wire in)

Kc = controller gain, a tuning parameter

Ti = reset time, a tuning parameter

Td = derivative time, a tuning parameter.

The Filtering Algorithm

A first order filter yields a smoothed CO* value as :

$$T_f \frac{dCO*}{dt} + CO* = CO$$

Where :

CO = raw PID controller output signal

CO* = filtered CO signal sent to FCE (*e.g.,* valve)

T_f = filter time constant, a tuning parameter

A comparison of the first order filter above to a general first order plus dead time (FOPDT) model form reveals that :

- The gain (or scaling factor) of the filter is one. That is, the filtered CO* has the same zero, span and units as the CO value from the PID algorithm.
- There is no dead time built into the filter. The CO* forwarded to the FCE is computed immediately after the PID algorithm yields the raw (unfiltered) CO signal.
- The degree of filtering, or how quickly CO* moves toward the unfiltered CO value, is set by T_f, the filter time constant.

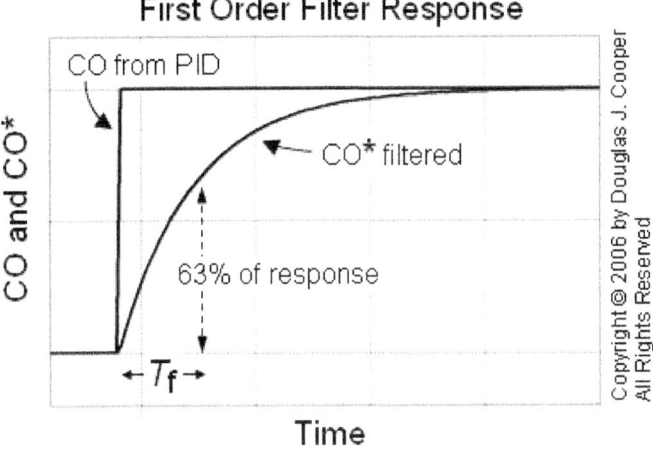

First Order Filter Response

Filtering Adds Delay

The degree of smoothing depends on the size of the filter time constant, T_f :

- A smaller T_f means CO* moves quickly and follows closer to changes in CO, so there is little filtering or smoothing of the signal.
- A larger T_f means CO* responds more slowly to changes in CO, so the filtering or smoothing is greater.

Shown below is a series of CO signals from a PID controller. Also shown is the filtered CO* trace using the same T_f as in the plot above.

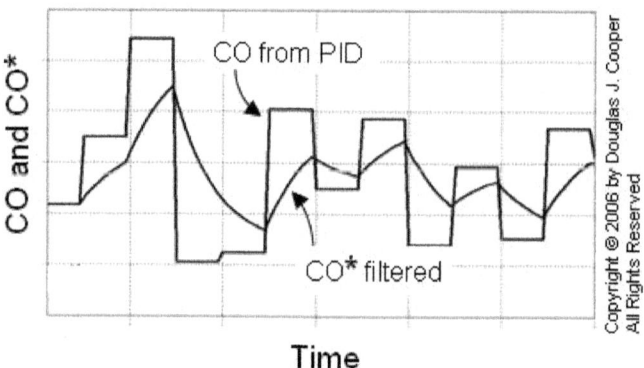

As smoothing (or filter time, T_f) increases, the filtered signal may become more visually appealing, but more filtering means additional information delay in the control loop computation.

As delay increases in a control loop, the best achievable control performance decreases. The design challenge is to find the careful balance between signal smoothing and information delay to achieve the controller performance we desire for our process.

Combining Controller Plus Filter

To use our tuning recipe, we must first combine the controller and filter into a single unified equation. Since both equations above have CO isolated on one side of the equal sign, we can set the two equations equal to yield :

$$T_f \frac{dCO^*}{dt} + CO^* = CO_{bias} + Kc\,e(t) + \frac{Kc}{Ti}\int e(t)dt + KcTd\frac{de(t)}{dt}$$

Moving the left-most term to the right-hand side produces the unified PID with Filter equation :

$$CO^* = CO_{bias} + Kc\,e(t) + \frac{Kc}{Ti}\int e(t)dt + Kc\,Td\frac{de(t)}{dt} - T_f\frac{dCO^*}{dt}$$

Unified PID with Filter Form

For design and tuning purposes going forward, we will use the unified PID with Filter equation form, as if it were represented by the schematic :

As implied by this diagram, we will drop the CO versus CO* nomenclature and simply write the unified PID with Filter equation as :

$$CO = CO_{bias} + Kc\ e(t) + \frac{Kc}{Ti}\int e(t)dt + Kc\ Td\frac{de(t)}{dt} - T_f\frac{dCO}{dt}$$

Filter Time Constant

Many commercial controllers that include some form of a PID with Filter algorithm cast the filter time constant as a fraction of the controller derivative time, or :

$$T_f = \alpha\ Td$$

The unified PID with Filter algorithm then becomes :

$$CO = CO_{bias} + Kc\ e(t) + \frac{Kc}{Ti}\int e(t)dt + Kc\ Td\frac{de(t)}{dt} - \alpha\ Td\frac{dCO}{dt}$$

We will use this αTd form in the example that follows. If your controller output filter uses T_f, the conversion is computed : $T_f = \alpha Td$.

Discrete-Time Implementation

While a CO filter offers potential for benefit in loops with noise and/or delicate mechanical FCEs, if our vendor does not offer the option, we must program the filter ourselves. Fortunately, the code is straightforward.

As above, the first order filtering algorithm is expressed :

$$T_f\frac{dCO*}{dt} + CO* = CO$$

In discrete-time form, we write :

$$T_f\frac{(CO*new - CO*old)}{T} + CO*old = CO$$

or

CO*new = CO*old + (T/T_f)(CO – CO*old)

Where T is the loop sample time and CO is the unfiltered PID signal.

In a computer program, the "new" CO* is computed directly from the "old" CO* value at each loop, so we do not need to keep track of these labels. The filter computation can be programmed in one line as :

COstar = COstar + (T/T_f)(CO–COstar)

PID WITH CO FILTER CONTROL OF THE HEAT EXCHANGER

The same tuning recipe we successfully demonstrated for PI control and PID control design and tuning can be used when a controller output (CO) filter is added to the heat exchanger process control loop.

Here we explore PID with CO Filter control using the unified (controller with internal filter) form. The unified form is identical to a PID with external first-order

CO filter implementation. Hence, the methods we use and observations we make apply equally to both internal and external filter architectures.

We follow the same four-step design and tuning recipe used for all control implementations.

Steps 1-3 of the PID with CO Filter design are identical to our previous PI and PID control case studies. The details for steps 1-3, stated below as summary conclusions, are presented with discussion in the PI control study.

Step 1 : Design Level of Operation (DLO)

- Design PV and SP = 138°C with operation ranging from 138 to 140°C
- Expected warm liquid flow disturbance = 10 L/min.

Step 2 : Fit an FOPDT Model to the Dynamic Data

The first order plus dead time (FOPDT) model approximation of the heat exchanger data from step 2 is :

- Process gain (how far), K_p = –0.53°C/%
- Time constant (how fast), T_p = 1.3 min
- Dead time (how much delay), θ_p = 0.8 min.

Step 3 : Use the Parameters to Complete the Design

Vendors market the PID algorithm in a number of different forms, creating a confusing array of choices for the practitioner. The addition of a CO filter makes a bad situation worse. A filter adds another adjustable (tuning) parameter and significantly increases the number of possible algorithm forms.

Trial and error tuning of a four mode (four tuning parameter) controller with filter while our process is making product is a sure path to waste and expense. With so many interacting variables, we will likely settle for an operation that "isn't horrible" rather than a performance that is near optimal.

The dilemma is real and our tuning recipe is the answer. Yet for success in this final step, it is critical that we match our algorithm with its proper tuning correlations.

The way to do this reliably is with loop tuning software. In fact, a good package will help with all of the steps, from data collection and model fitting through vendor algorithm selection and final performance analysis. When our task list includes maintaining and tuning loops during production, commercial software will pay for itself in days.

- *Sample Time and Bumpless Transfer* : As explained in the PI control study, best practice is to set loop sample time $T \leq 0.1T_p$ (10 times per time constant or faster). For this example, T = 1.0 sec.

Also, like most commercial controllers, we employ bumpless transfer. Thus, when switching to automatic, SP is set equal to the current PV and CO_{bias} is set equal to the current CO.

- *Controller Action* : Controller gain is negative for the heat exchanger, yet most commercial controllers require that a positive value of Kc be entered. The way we indicate a negative sign is to choose the direct acting option during implementation. If the wrong control action is entered, the controller will quickly drive the final control element to full on/open or full off/closed and remain there until a proper control action entry is made.

- *Specify Desired Performance* : We use the industry-proven Internal Model Control (IMC) tuning correlations in this study. IMC correlations employ a closed loop time constant, Tc, that describes the desired speed or quickness of our controller in responding to a set point change or rejecting a disturbance. We must decide whether we seek :

- An *aggressive* controller with a rapid response and some overshoot :

  Tc is the larger of 0.1 ·Tp or 0.8 θp

- A *moderate* controller that will move the PV reasonably fast yet produce little to no overshoot in a set point response :

  Tc is the larger of 1 ·Tp or 8 θp

- A *conservative* controller that will move the PV in the proper direction, but quite slowly :

  Tc is the larger of 10 ·Tp or 80 θp.

Ideal PID with Filter Example

$$CO = CO_{bias} + Kc\, e(t) + \frac{Kc}{Ti}\int e(t)dt + KcTd\frac{de(t)}{dt} - \alpha Td\frac{dCO}{dt}$$

If our vendor offers the option, the preferred algorithm in industrial practice is PID with derivative on measurement (derivative on PV).

While a CO filter can largely address derivative kick, the filter term must be made larger than otherwise necessary to do so. Thus, there remains a performance benefit to derivative on measurement even when using a CO filter.

$$CO = CO_{bias} + Kc\, e(t) + \frac{Kc}{Ti}\int e(t)dt - Kc\, Td\frac{dPV}{dt} - \alpha\, Td\frac{dCO}{dt}$$

The IMC tuning correlations for either of the above PID with CO Filter forms are :

Kc	Ti	Td	α
$\frac{1}{Kp}\left(\frac{Tp+0.5\theta p}{Tc+\theta p}\right)$	$Tp+0.5\theta p$	$\frac{Tp\theta p}{2Tp+\theta p}$	$\frac{Tc(Tp+0.5\theta p)}{Tp(Tc+\theta p)}$

We start our study by choosing an aggressive response tuning :

Aggressive Tc = the larger of 0.1 $\cdot Tp$ or 0.8 θp

$\qquad$ = larger of 0.1(1.3 min) or 0.8(0.8 min)

$\qquad$ = 0.64 min.

Using this Tc and our Kp, Tp and θp from Step 3 in the tuning correlations above yields the aggressive PID w/ Filter tuning values below. Also listed are the PID Ideal and PI controller tuning values from earlier studies :

PID w/ Filter :

Kc = –2.2%/°C Ti = 1.7 min Td = 0.31 min α = 0.6

PID :

Kc = –3.1%/°C Ti = 1.7 min Td = 0.31 min

PI :

Kc = –1.7%/°C Ti = 1.3 min.

Below we compare the performance of these three aggressively tuned controllers side-by-side for the heat exchanger process simulation. Shown are three set point step pairs from 138°C up to 140°C and back again. Though not shown, the disturbance flow rate remains constant at 10 L/min throughout the study.

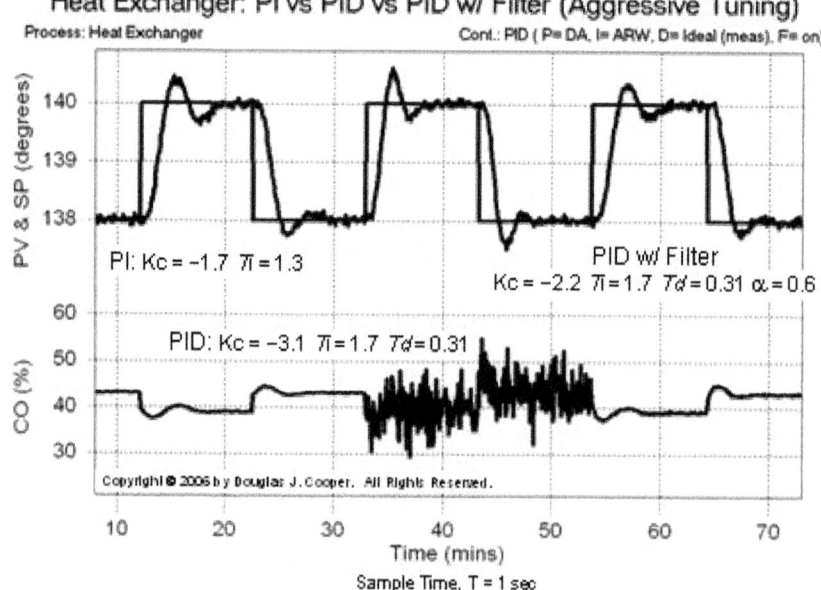

To the left is set point tracking performance for the PI controller. The middle set point steps show the performance of the PID controller. Derivative action enables a slightly faster rise time and settling time, but the derivative action causes the noise in the PV signal to be amplified and reflected as "chatter" in the control output signal.

To the right is the set point tracking performance of the PID w/ CO Filter controller. Indeed, the filter does an impressive job of cleaning up the chatter in the controller output signal without degrading performance.

In truth, however, the four tuning parameter PID w/ Filter performs similar to the two tuning parameter PI controller.

Tune One, Tune Them All

To complete this study, we compare the dependent, ideal, non-interacting form above to the performance of the dependent, interacting form :

$$CO = CO_{bias} + Kc\left(1+\frac{Td}{Ti}\right)e(t)+\frac{Kc}{Ti}\int e(t)dt - KcTd\frac{dPV}{dt} - \alpha\,Td\frac{dCO}{dt}$$

The tuning correlations for the dependent, interacting form are :

Kc	Ti	Td	α
$\dfrac{1}{Kp}\left(\dfrac{Tp}{Tc+\theta p}\right)$	Tp	$0.5\theta p$	$\dfrac{Tc}{Tc+\theta p}$

For variety, we choose moderate response tuning for this comparison :

Moderate Tc = the larger of $1\cdot Tp$ or $8\cdot\theta p$

$\qquad\qquad$ = larger of 1(1.3 min) or 8(0.8 min)

$\qquad\qquad$ = 6.4 min.

Using this Tc and our model parameters in the proper tuning correlations, we arrive at these moderate tuning values :

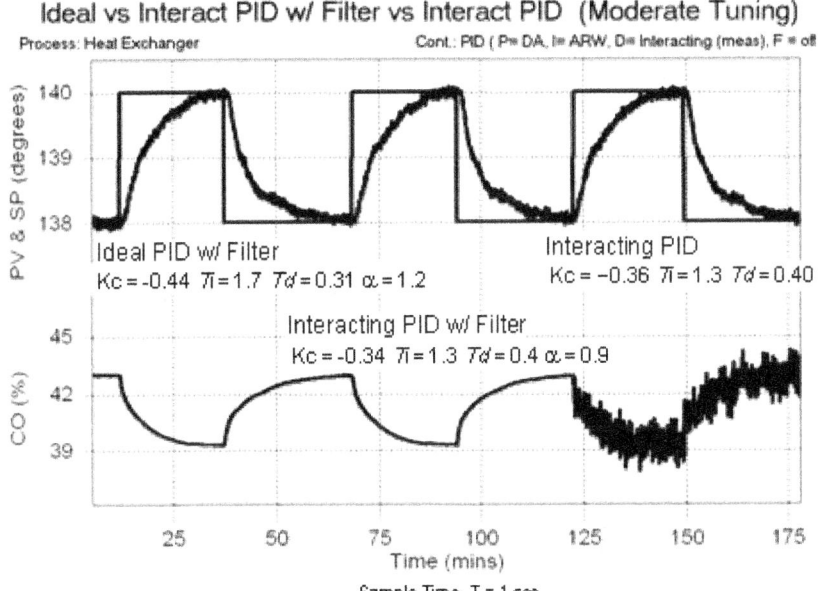

Ideal vs Interact PID w/ Filter vs Interact PID (Moderate Tuning)

Dependent, Interacting :

Kc = –0.34%/°C Ti = 1.3 min Td = 0.40 min α = 0.9

Dependent, Ideal :

Kc = –0.44%/°C Ti = 1.7 min Td = 0.31 min α = 1.2.

As shown in the plot below, we see that moderate tuning provides a reasonably fast PV response while producing no overshoot. But more important, we establish that the interacting form and the ideal form provide identical performance when tuned with their own correlations.

The third set point step shows the performance of a straight PID with no filter. This reinforces the benefits of a CO filter if derivative action is being contemplated.

Observations

Our study of the heat exchanger process has shown that PID controllers provide minor performance improvements over PI controllers. Yet derivative action causes noise in the PV to be reflected as chatter in the CO signal, and this counterbalances the small benefits of the derivative term.

But now we have elevated a difficult two tuning parameter PI controller design into an extremely challenging four parameter PID w/ Filter controller design. And at best, this extra effort still provides only modest performance benefits.

Unless the economic impact of a loop is substantial, many practitioners conclude that the PI controller is the best choice. It is faster to implement, easier to maintain, and provides performance approaching that of the PID w/ Filter controller.

PID WITH CO FILTER DISTURBANCE REJECTION IN THE JACKETED STIRRED REACTOR

Here we explore the performance of a PID with controller output (CO) filter algorithm in meeting this same disturbance rejection objective. We use the unified PID with CO filter controller in this study. The unified form is identical to a PID with external first-order CO filter implementation. Thus, the methods and observations from this investigation apply equally to both controller architectures.

The important variables for the jacketed reactor are (*view a process graphic*) :

CO = signal to valve that adjusts cooling jacket liquid flow rate (controller output,%)

PV = reactor exit stream temperature (measured process variable,°C)

SP = desired reactor exit stream temperature (set point,°C)

D = temperature of cooling liquid entering the jacket (major disturbance,°C).

We follow our *industry proven recipe* to design and tune our PID with CO filter controller. Recall that steps 1-3 of a design remain the same regardless of the controller used. For this process and objective, the results of steps 1-3 are summarized from previous investigations :

Step 1 : Design Level of Operation (DLO)

DLO details are summarized :

- Design PV and SP = 90°C with approval for brief dynamic (bump) testing of ±2°C
- Design D = 43°C with occasional spikes up to 50°C.

Step 2 : Collect Process Data Around the DLO

When CO, PV and D are steady near the design level of operation, we *bump the jacketed stirred reactor* to generate CO-to-PV cause and effect process response data.

Step 3 : Fit a FOPDT model to the dynamic process data

We approximate the dynamic behavior of the process by fitting a first order plus dead time (FOPDT) dynamic model to the test data from step 2. The results *of the modeling study* are summarized :

- Process gain (direction and how far), K_p =–0.5°C/%
- Time constant (how fast), T_p = 2.2 min
- Dead time (how much delay), θ_p = 0.8 min.

Step 4 : Use the FOPDT Parameters to Complete the Design

- *Algorithm Form :* A PID with CO filter controller, regardless of whether the *filter is internal or external*, presents us with a "four adjustable tuning parameter" challenge. As more parameters are included in our controller, the array of *vendor algorithm forms* increases.

  The various controller forms are all capable of delivering a similar, predictable performance, as long as we match our algorithm with its proper tuning correlations. Certainly, a "guess and test" approach to tuning a four-mode controller while our process is making product is a sure path to wasting feedstock and utilities, creating safety and environmental concerns, and putting plant profitability at risk.

  Modern *loop tuning software* will not only guide data analysis and model fitting, but will ensure our tuning matches our vendor's algorithm. Such software will even display expected final performance prior to implementation. If our task list includes maintaining and tuning control loops during production, such software will pay for itself in days.

- *Sample Time :* As *discussed here*, best practice is to set loop sample time to T ≤ 0.1 T_p (10 times per time constant or faster). We meet this design criterion with the widely-available vendor option of T = 1.0 sec.

- *Controller Action :* Kp, and thus Kc, are negative for our jacketed stirred reactor process. Most commercial controllers have us specify a negative Kc by entering a positive value into the controller and then choosing the "direct acting" option. If the wrong control action is entered, the controller will drive the final control element to full on/open or full off/closed and remain there until a proper control action entry is made.

- *Specify Desired Performance :* We use the industry-proven Internal Model Control (IMC) tuning correlations in this study. IMC correlations employ a closed loop time constant, Tc, that describes the desired speed or quickness of our controller in responding to a set point change or rejecting a disturbance.

Our *PI control study* describes what to expect from an aggressive, moderate or conservative controller. Once our desired performance is chosen, the closed loop time constant is computed :

  - *Aggressive performance :* Tc is the larger of 0.1 ·Tp or 0.8 θp
  - *Moderate performance :* Tc is the larger of 1 ·Tp or 8 θp
  - *Conservative performance :* Tc is the larger of 10 ·Tp or 80 θp

- *The Tuning Correlations*

$$CO = CO_{bias} + Kc\, e(t) + \frac{Kc}{Ti}\int e(t)dt + KcTd\frac{de(t)}{dt} - \alpha\, Td\frac{dCO}{dt}$$

If our vendor offers the option, *the preferred algorithm* in industrial practice is PID with derivative on measurement, PV :

$$CO = CO_{bias} + Kc\, e(t) + \frac{Kc}{Ti}\int e(t)dt + KcTd\frac{dPV}{dt} - \alpha\, Td\frac{dCO}{dt}$$

The IMC tuning correlations for either of the above PID with CO filter forms are the same and listed in the chart below :

	Controller Gain Kc	Reset Time Ti	Derive Time Td	Filter Const α
PI	$\dfrac{1}{Kp}\dfrac{Tp}{(8p + Tc)}$	Tp		
Ideal PID	$\dfrac{1}{Kp}\left(\dfrac{Tp + 0.5\theta p}{Tc + 0.5\theta p}\right)$	Tp + 0.5θp	$\dfrac{Tp\theta p}{2Tp + \theta p}$	
PID w/CO Filter	$\dfrac{1}{Kp}\left(\dfrac{Tp + 0.5\theta p}{Tc + \theta p}\right)$	Tp + 0.5θp	$\dfrac{Tp\theta p}{2Tp + \theta p}$	$\dfrac{Tc(Tp + 0.5\theta p)}{Tp(Tc + \theta p)}$

PID with CO Filter Disturbance Rejection Study

In the plots below, we compare the performance of the PID with CO filter controller side-by-side with that of the PI controller and the ideal (unfiltered) PID controller.

Our objective is rejecting the impact on reactor operation when the temperature of cooling liquid entering the reactor jacket changes. We test both moderate and aggressive response tuning for the three controllers.

a. *Moderate Response Tuning :* For a controller that will move the PV reasonably fast while producing little to no overshoot, choose :

Moderate Tc = the larger of $1 \cdot Tp$ or $8 \, \theta p$

$\qquad\qquad$ = larger of $1(2.2 \text{ min})$ or $8(0.8 \text{ min})$

$\qquad\qquad$ = 6.4 min.

Using this Tc and the Kp, the moderate IMC tuning values are :

PI : Kc = –0.61%/°C Ti = 2.2 min

PID : Kc = –0.77%/°C Ti = 2.6 min Td = 0.34 min

PID w/ Filter : Kc = –0.72%/°C Ti = 2.6 min Td = 0.34 min α = 1.1

b. *Aggressive Response Tuning :* For an active or quickly responding controller where we can tolerate some overshoot and oscillation as the PV settles out, specify :

Aggressive Tc = the larger of $0.1 \cdot Tp$ or $0.8 \, \theta p$

$\qquad\qquad$ = larger of $0.1(2.2 \text{ min})$ or $0.8(0.8 \text{ min})$

$\qquad\qquad$ = 0.64 min.

and the aggressive IMC tuning values are :

PI : Kc = –3.1%/°C Ti = 2.2 min

PID : Kc = –5.0%/°C Ti = 2.6 min Td = 0.34 min

PID w/ Filter : Kc = –3.6%/°C Ti = 2.6 min Td = 0.34 min α = 0.5

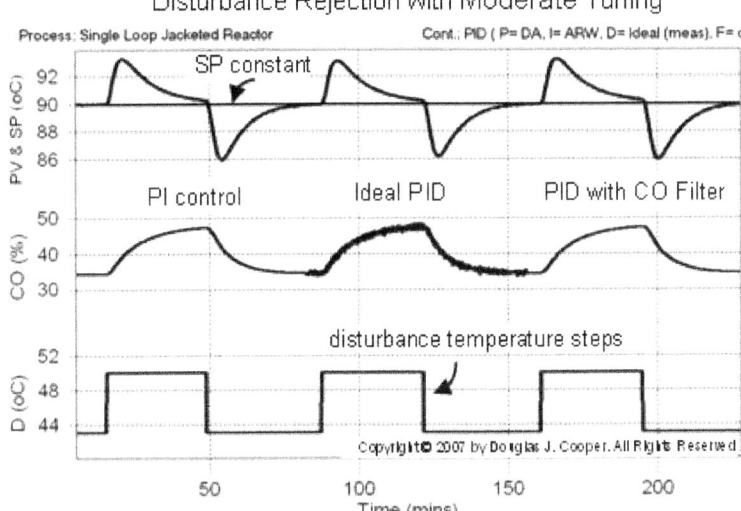

Disturbance Rejection with Moderate Tuning

- *Implement and Test* : A comparison of the three controllers in rejecting a disturbance change in the cooling jacket inlet temperature, D, is shown below. This
  plot shows controller performance when using the moderate tuning values
  computed above. Note that the set point remains constant at 90°C throughout
  the study.

The PI controller performance is shown to the left in the plot above. The ideal
PID performance is in the middle. The plot reveals that the benefit of derivative
action is marginal at best. There is a clear penalty, however, in that derivative
action causes the modest *noise in the PV signal* to be amplified and reflected as
"chatter" in the CO signal.

To the right in the plot above is the performance of the PID with CO filter
controller. The filter is effective in reducing the controller output chatter caused
by the derivative action without degrading performance. In truth, however, the
four tuning parameter PID with filter performs similar to the two tuning parameter PI controller.

The disturbance rejection performance of the controllers when tuned for aggressive action is shown below. Note that the axis scales for the plots both above
and below are the same to permit a visual comparison.

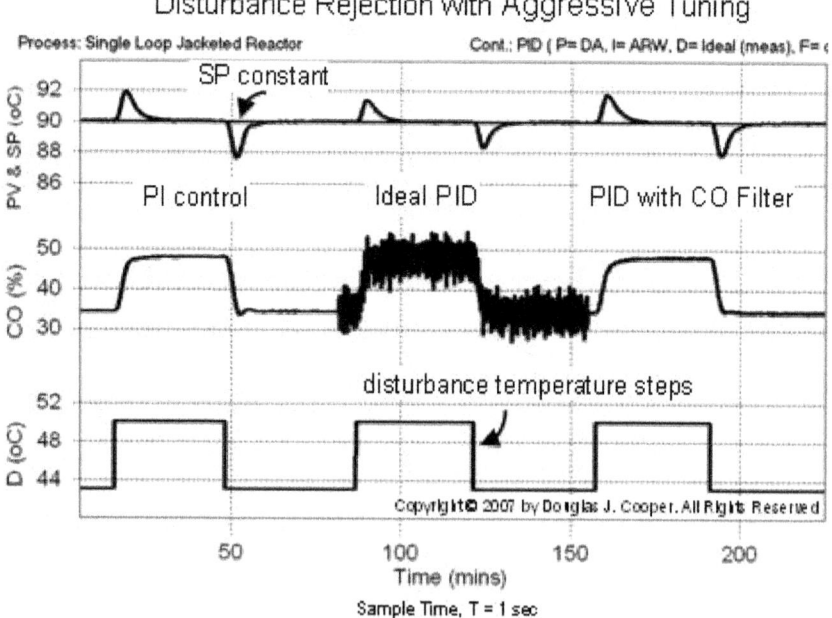

Disturbance Rejection with Aggressive Tuning

The aggressive tuning provides a smaller maximum deviation from set point
and a faster settling time relative to the moderate tuning performance. The only
obvious difference is that as a PID controller (middle of plot) becomes more aggressive in its actions, the CO chatter grows as a problem and filtering solutions
become increasingly beneficial.

But ultimately, just as with the moderate tuning case, the two mode (or two tuning parameter) PI controller compares favorably with the four mode PID with CO filter controller.

While not our design objective, presented below is the set point tracking ability of the aggressively tuned controllers when the disturbance temperature is held constant :

Set Point Tracking with Aggressive Tuning

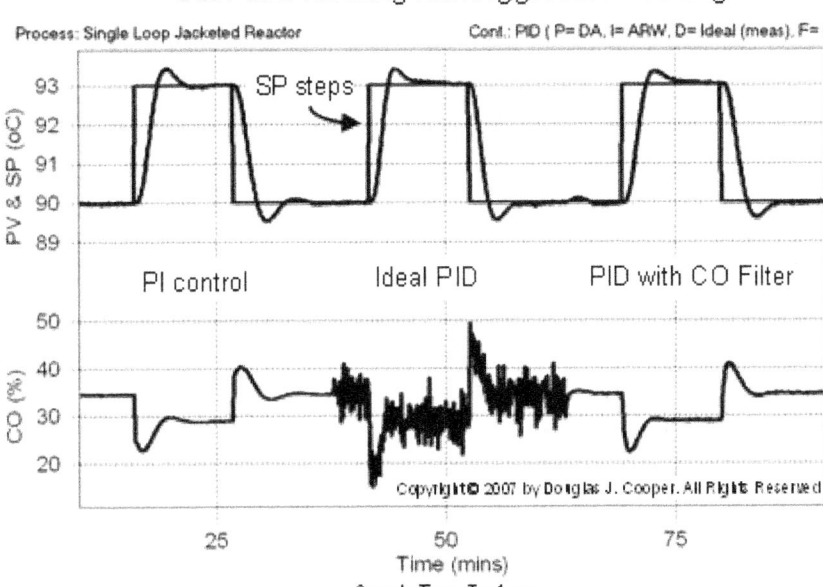

The set point tracking response of the ideal PID controller is marginally better in that it shows a slightly shorter rise time, smaller overshoot and faster settling time. The CO chatter that comes as a price for these minor benefits will likely increase maintenance costs as our final control element (*e.g.*, valve, pump or compressor) wears from this excessive activity.

The four mode PID with CO filter addresses the chatter, but it is not clear that the added complexity is worth the marginal performance benefits.

Thus, many practitioners conclude that the PI controller provides the best balance of complexity and performance. It is faster to implement, easier to maintain, and provides performance approaching that of the PID with CO filter controller.

Chapter 8

FUNDAMENTAL PRINCIPLES OF PROCESS CONTROL

MOTIVATION AND TERMINOLOGY OF AUTOMATIC PROCESS CONTROL

Automatic control systems enable us to operate our processes in a safe and profitable manner. Consider, as on this site, processes with streams comprised of gases, liquids, powders, slurries and melts. Control systems achieve this "safe and profitable" objective by continually measuring *process variables* such as temperature, pressure, level, flow and concentration – and taking actions such as opening valves, slowing down pumps and turning up heaters – all so that the measured process variables are maintained at operator specified set point values.

Safety First

The overriding motivation for automatic control is safety, which encompasses the safety of people, the environment and equipment.

The safety of plant personnel and people in the community are the highest priority in any plant operation. The design of a process and associated control system must always make human safety the primary objective.

The trade off between safety of the environment and safety of equipment is considered on a case by case basis. At the extremes, the control system of a multi-billion dollar nuclear power facility will permit the entire plant to become ruined rather than allow significant radiation to be leaked to the environment.

On the other hand, the control system of a coal-fired power plant may permit a large cloud of smoke to be released to the environment rather than allowing damage to occur to, say, a single pump or compressor worth a few thousand dollars.

The Profit Motive

When people, the environment and plant equipment are properly protected, our control objectives can focus on the profit motive. Automatic control systems offer strong benefits in this regard.

Plant-level control objectives motivated by profit include :

- Meeting final product specifications
- Minimizing waste production
- Minimizing environmental impact
- Minimizing energy use
- Maximizing overall production rate.

It can be most profitable to operate as close as possible to these minimum or maximum objectives. For example, our customers often set our product specifications, and it is essential that we meet them if failing to do so means losing a sale.

Suppose we are making a film or sheet product. It takes more raw material to make a product thicker than the minimum our customers will accept on delivery. Consequently, the closer we can operate to the minimum permitted thickness constraint without going under, the less material we use and the greater our profit.

Or perhaps we sell a product that tends to be contaminated with an impurity and our customers have set a maximum acceptable value for this contaminant. It takes more processing effort (more money) to remove impurities, so the closer we can operate to the maximum permitted impurity constraint without going over, the greater the profit.

Whether it is a product specification, energy usage, production rate, or other objective, approaching these targets ultimately translates into operating the individual process units within the plant as close as possible to predetermined set point values for temperature, pressure, level, flow, concentration and the other measured process variables.

Controllers Reduce Variability

As shown in the plot below, a poorly controlled process can exhibit large variability in a measured process variable (*e.g.*, temperature, pressure, level, flow, concentration) over time.

Suppose, as in this example, the measured process variable (PV) must not exceed a maximum value. And as is often the case, the closer we can run to this operating constraint, the greater our profit (note the vertical axis label on the plot).

To ensure our operating constraint limit is not exceeded, the operator-specified set point (SP), that is, the point where we want the control system to maintain our PV, must be set far from the constraint to ensure it is never violated. Note in the plot that SP is set at 50% when our PV is poorly controlled.

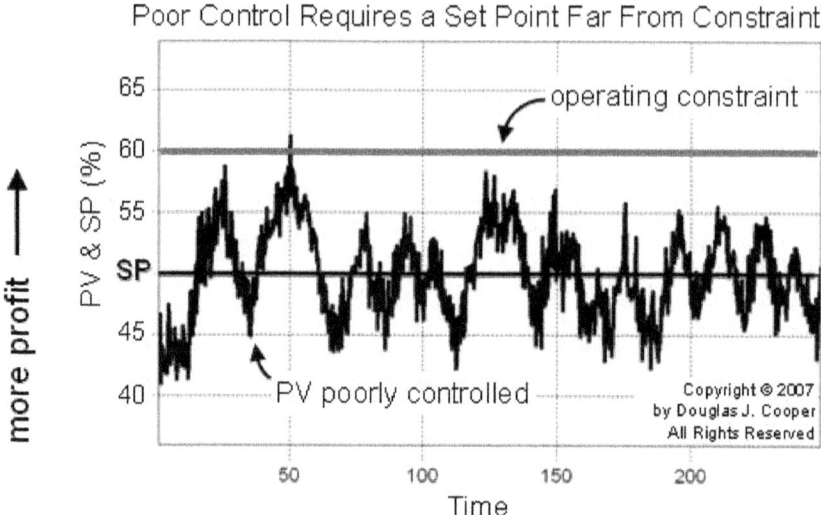

Below we see the same process with improved control. There is significantly less variability in the measured PV, and as a result, the SP can be moved closer to the operating constraint.

With the SP in the plot below moved to 55%, the average PV is maintained closer to the specification limit while still remaining below the maximum allowed value. The result is increased profitability of our operation.

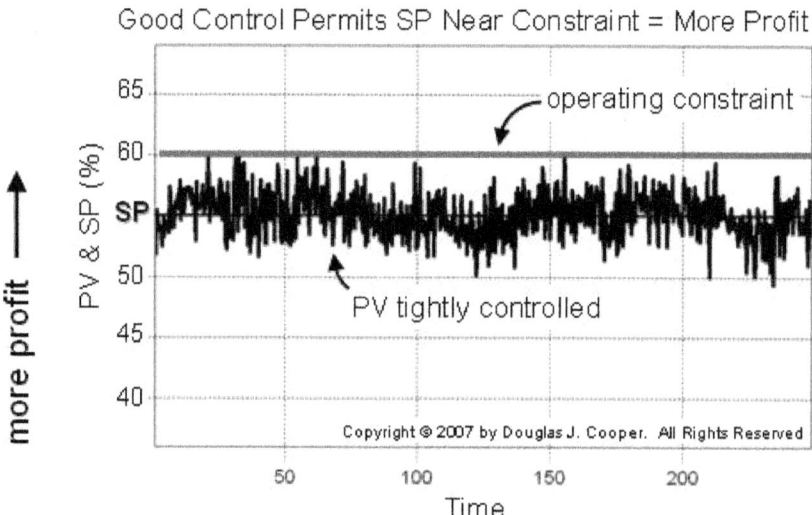

Terminology of Control

We establish the jargon for this site by discussing a home heating control system as illustrated below :

This is a simplistic example because a home furnace is either on or off. Most control challenges have a *final control element* (FCE), such as a valve, pump or compressor, that can receive and respond to a complete range of *controller output* (CO) signals between full on and full off. This would include, for example, a valve that can be open 37% or a pump that can be running at 73%.

For our home heating process, the *control objective* is to keep the *measured process variable* (PV) at the *set point value* (SP) in spite of *unmeasured disturbances* (D).

For our home heating system :

PV = Process variable is house temperature

CO = Controller output signal from thermostat to furnace valve

SP = Set point is the desired temperature set on the thermostat by the home owner

D = Heat loss disturbances from doors, walls and windows; changing outdoor temperature; sunrise and sunset; rain...

To achieve this control objective, the measured process variable is compared to the thermostat set point. The difference between the two is the *controller error*, which is used in a *control algorithm* such as a PID (proportional-integral-derivative) controller to compute a CO signal to the final control element (FCE).

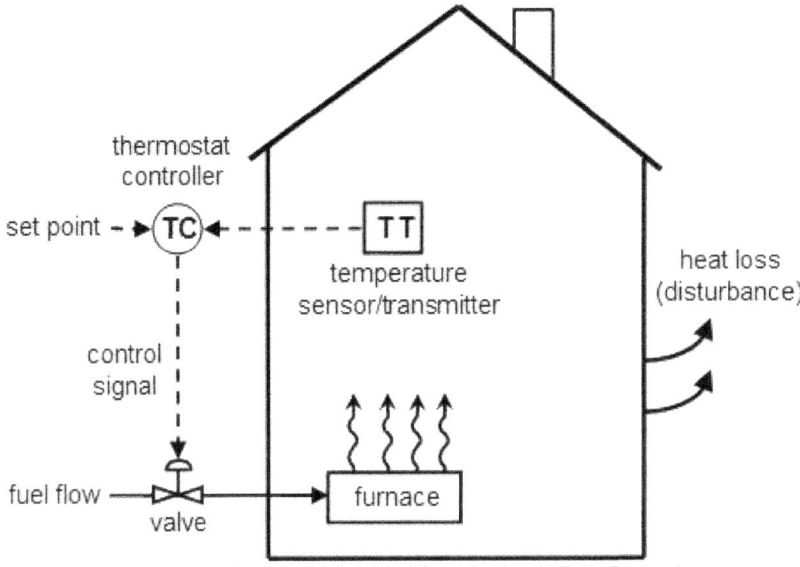

The change in the controller output (CO) signal causes a response in the final control element (fuel flow valve), which subsequently causes a change in the *manipulated process variable* (flow of fuel to the furnace). If the manipulated process variable is moved in the right direction and by the right amount, the measured process variable will be maintained at set point, thus satisfying the control objective.

This example, like all in process control, involves a measurement, computation and action :

- Is the measured temp colder than set point (SP–PV > 0)? Then open the valve.
- Is the measured temp hotter than set point (SP–PV < 0)? Then close the valve.

Note that computing the necessary controller action is based on *controller error*, or the difference between the set point and the measured process variable, *i.e.*

$e(t)$ = SP–PV (error = set point–measured process variable)

In a home heating process, control is an on/off or open/close decision. And as outlined above, it is a straightforward decision to make. The price of such simplicity, however, is that the capability to tightly regulate our measured PV is rather limited.

One situation not addressed above is the action to take when PV = SP (*i.e.*, $e(t) = 0$). And in industrial practice, we are concerned with *variable position* final control elements, so the challenge elevates to computing :

- The direction to move the valve, pump, compressor, heating element...
- How far to move it at this moment
- How long to wait before moving it again
- Whether there should be a delay between measurement and action

THE COMPONENTS OF A CONTROL LOOP

Components of a Control Loop

A controller seeks to maintain the measured process variable (PV) at set point (SP) in spite of unmeasured disturbances (D). The major components of a control system include a sensor, a controller and a final control element. To design and implement a controller, we must :

1. Have identified a process variable we seek to regulate, be able to measure it (or something directly related to it) with a sensor, and be able to transmit that measurement as an electrical signal back to our controller, and
2. Have a final control element (FCE) that can receive the controller output (CO) signal, react in some fashion to impact the process (*e.g.*, a valve moves), and as a result cause the process variable to respond in a consistent and predictable fashion.

Home Temperature Control

As shown below, the home heating control system can be organized as a traditional control loop block diagram. Block diagrams help us visualize the components of a loop and see how the pieces are connected.

A home heating system is simple on/off control with many of the components contained in a small box mounted on our wall. Nevertheless, we introduce the idea of control loop diagrams by presenting a home heating system in the same way we would a more sophisticated commercial control application.

Home Heating Control Loop Block Diagram

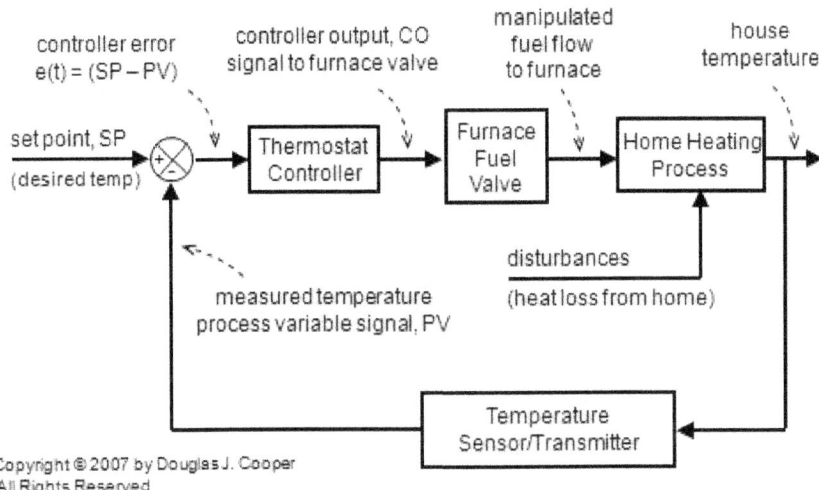

Starting from the far right in the diagram above, our process variable of interest is house temperature. A sensor, such as a thermistor in a modern digital thermostat, measures temperature and transmits a signal to the controller.

The measured temperature PV signal is subtracted from set point to compute controller error, $e(t)$ = SP–PV. The action of the controller is based on this error, $e(t)$.

In our home heating system, the controller output (CO) signal is limited to open/close for the fuel flow solenoid valve (our FCE). So in this example, if e(t) = SP–PV > 0, the controller signals to open the valve. If $e(t)$ = SP–PV < 0, it signals to close the valve. As an aside, note that there also must be a safety interlock to ensure that the furnace burner switches on and off as the fuel flow valve opens and closes.

As the energy output of the furnace rises or falls, the temperature of our house increases or decreases and a feedback loop is complete. The important elements of a home heating control system can be organized like any commercial application :

- *Control Objective* : Maintain house temperature at SP in spite of disturbances
- *Process Variable* : House temperature
- *Measurement Sensor* : Thermistor; or bimetallic strip coil on analog models
- *Measured Process Variable (PV) Signal* : Signal transmitted from the thermistor
- *Set Point (SP)* : Desired house temperature
- *Controller Output (CO)* : Signal to fuel valve actuator and furnace burner
- *Final Control Element (FCE)* : Solenoid valve for fuel flow to furnace
- *Manipulated Variable* : Fuel flow rate to furnace
- *Disturbances (D)* : Heat loss from doors, walls and windows; changing outdoor temperature; sunrise and sunset; rain...

A General Control Loop and Intermediate Value Control

The home heating control loop above can be generalized into a block diagram pertinent to all feedback control loops as shown below.

General Control Loop Block Diagram

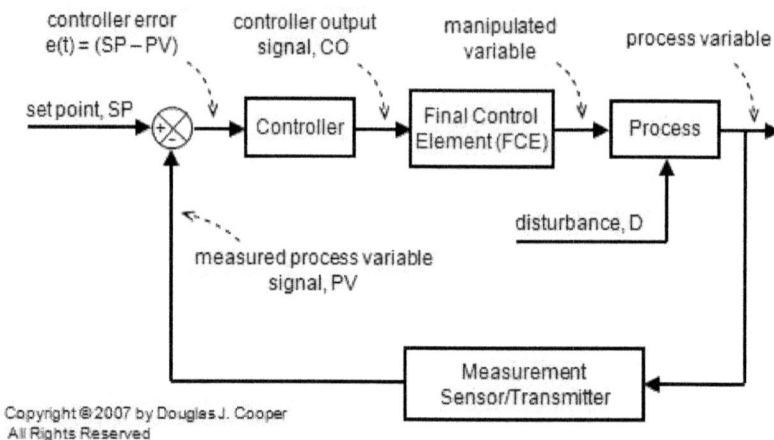

Both diagrams above show a closed loop system based on negative feedback. That is, the controller takes actions that counteract or oppose any drift in the measured PV signal from set point.

While the home heating system is on/off, our focus going forward shifts to intermediate value control loops. An intermediate value controller can generate a full range of CO signals anywhere between full on/off or open/closed. The PI algorithm and PID algorithm are examples of popular intermediate value controllers.

To implement intermediate value control, we require a sensor that can measure a full range of our process variable, and a final control element that can receive and assume a full range of intermediate positions between full on/off or open/closed. This might include, for example, a process valve, variable speed pump or compressor, or heating or cooling element.

Note from the loop diagram that the process variable becomes our official PV only after it has been measured by a sensor and transmitted as an electrical signal to the controller. In industrial applications. These are most often implemented as 4-20 milliamps signals, though commercial instruments are available that have been calibrated in a host of amperage and voltage units.

With the loop closed as shown in the diagrams, we are said to be in automatic mode and the controller is making all adjustments to the FCE. If we were to open the loop and switch to manual mode, then we would be able to issue CO commands through buttons or a keyboard directly to the FCE. Hence :

- Open loop = manual mode
- Closed loop = automatic mode.

Cruise Control and Measuring Our PV

Cruise control in a car is a reasonably common intermediate value control system. For those who are unfamiliar with cruise control, here is how it works.

We first enable the control system with a button on the car instrument panel. Once on the open road and at our desired cruising speed, we press a second button that switches the controller from manual mode (where car speed is adjusted by our foot) to automatic mode (where car speed is adjusted by the controller).

The speed of the car at the moment we close the loop and switch from manual to automatic becomes the set point. The controller then continually computes and transmits corrective actions to the gas pedal (throttle) to maintain measured speed at set point.

It is often cheaper and easier to measure and control a variable directly related to the process variable of interest. This idea is central to control system design and maintenance. And this is why the loop diagrams above distinguish between our "process variable" and our "measured PV signal."

Cruise control serves to illustrate this idea. Actual car speed is challenging to measure. But transmission rotational speed can be measured reliably and inexpensively. The transmission connects the engine to the wheels, so as it spins faster or slower, the car speed directly increases or decreases.

Thus, we attach a small magnet to the rotating output shaft of the car transmission and a magnetic field detector (loops of wire and a simple circuit) to the body of the car above the magnet. With each rotation, the magnet passes by the detector and the event is registered by the circuitry as a "click." As the drive shaft spins faster or slower, the click rate and car speed increase or decrease proportionally.

So a cruise control system really adjusts fuel flow rate to maintain click rate at the set point value. With this knowledge, we can organize cruise control into the essential design elements :

- *Control Objective :* Maintain car speed at SP in spite of disturbances
- *Process Variable :* Car speed
- *Measurement Sensor :* Magnet and coil to clock drive shaft rotation
- *Measured Process Variable (PV) Signal :* "Click rate" signal from the magnet and coil
- *Set Point (SP) :* Desired car speed, recast in the controller as a desired click rate
- *Controller Output (CO) :* Signal to actuator that adjusts gas pedal (throttle)
- *Final Control Element (FCE) :* Gas pedal position
- *Manipulated Variable :* Fuel flow rate
- *Disturbances (D) :* Hills, wind, curves, passing trucks.

The traditional block diagram for cruise control is thus :

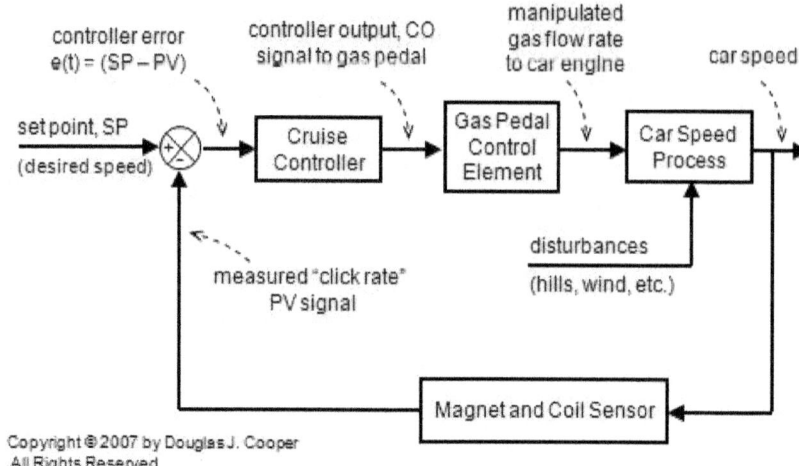

Car Cruise Control Loop Block Diagram

Instruments Should Be Fast, Cheap and Easy

The above magnet and coil "click rate = car speed" example introduces the idea that when purchasing an instrument for process control, there are wider considerations that can make a loop faster, easier and cheaper to implement and maintain. Here is a "best practice" checklist to use when considering an instrument purchase :

- Low cost
- Easy to install and wire
- Compatible with existing instrument interface
- Low maintenance
- Rugged and robust
- Reliable and long lasting
- Sufficiently accurate and precise
- Fast to respond (small time constant and dead time)
- Consistent with similar instrumentation already in the plant.

PROCESS DATA, DYNAMIC MODELING AND A RECIPE
FOR PROFITABLE CONTROL

It is best practice to follow a formal procedure or "recipe" when designing and tuning a PID (proportional-integral-derivative) controller. A recipe-based approach is the fastest method for moving a controller into operation. And perhaps most important, the performance of the controller will be superior to a controller tuned using a guess-and-test or trial-and-error method.

Additionally, a recipe-based approach overcomes many of the concerns that make control projects challenging in a commercial operating environment. Specifically, the recipe-based method causes less disruption to the production schedule, wastes less raw material and utilities, requires less personnel time, and generates less off-spec product.

The Recipe for Success is Short

1. Establish the design level of operation (DLO), defined as the expected values for set point and major disturbances during normal operation
2. Bump the process and collect controller output (CO) to process variable (PV) dynamic process data around this design level
3. Approximate the process data behavior with a first order plus dead time (FOPDT) dynamic model
4. Use the model parameters from step 3 in rules and correlations to complete the controller design and tuning. For now, we introduce some initial thoughts about steps 2 and 4.

Step 2 : Bumping Our Process and Collecting CO to PV Data

From a controller's view, a complete control loop goes from wire out to wire in as shown below. Whenever we mention controller output (CO) or process variable (PV) data anywhere on this site, we are specifically referring to the data signals exiting and entering our controller at the wire termination interface.

Controller's View: Wire Out (CO) to Wire In (PV) Loop

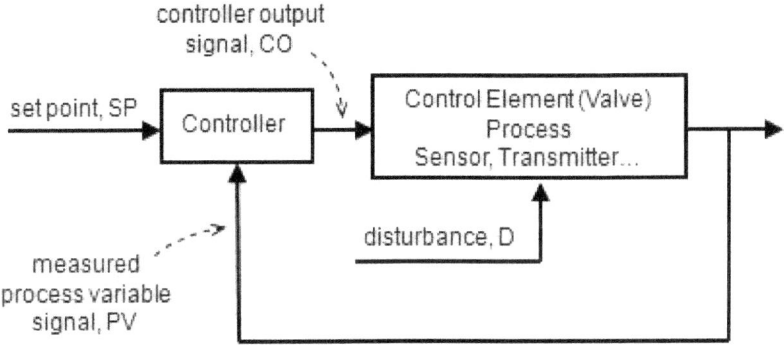

To generate CO to PV data, we bump our process. That is, we step or pulse the CO (or the set point if in automatic mode as discussed here) and record PV data as the process responds. Here are three basic rules we follow in all of our examples :

• *Start with the process at steady state and record everything* : The point of bumping the CO is to learn about the cause and effect relationship between it and the PV. With the plant initially at steady state, we are starting with a clean slate. The dynamic behavior of the process is then clearly isolated as the PV responds. It is important that we start capturing data before we make the initial CO bump and then sample and record quickly as the PV responds.

- *Make sure the PV response dominates the process noise :* When performing a bump test, it is important that the CO moves far enough and fast enough to force a response that clearly dominates any noise or random error in the measured PV signal. If the CO to PV cause and effect response is clear enough to see by eye on a data plot, we can be confident that modern software can model it.

- *The disturbances should be quiet during the bump test :* We desire that the dynamic test data contain PV response data that has been clearly, and in the ideal world exclusively, forced by changes in the CO.

Data that has been corrupted by unmeasured disturbances is of little value for controller design and tuning. The model will then incorrectly describe the CO to PV cause and effect relationship. And as a result, the controller will not perform correctly. If we are concerned that a disturbance event has corrupted test data, it is conservative to rerun the test.

Step 4 : Using Model Parameters for Design and Tuning

The final step of the recipe states that once we have obtained model parameters that approximate the dynamic behavior of our process, we can complete the design and tuning of our PID controller.

We look ahead at this last step because this is where the payoff of the recipe-based approach is clear. To establish the merit, we assume for now that we have determined the design level of operation for our process (step 1), we have collected a proper data set rich in dynamic process information around this design level (step 2), and we have approximated the behavior revealed in the process data with a first order plus dead time (FOPDT) dynamic model (step 3).

Thankfully, we do not need to know what a FOPDT model is or even what it looks like. But we do need to know about the three model parameters that result when we fit this approximating model to process data.

The FOPDT (first order plus dead time) model parameters, listed below, tell us important information about the measured process variable (PV) behavior whenever there is a change in the controller output (CO) signal :

- Process gain, Kp (tells the direction and how far PV will travel)
- Process time constant, Tp (tells how fast PV moves after it begins its response)
- Process dead time, θp (tells how much delay before PV first begins to respond)

We note that the first order plus dead time (FOPDT) dynamic model has the form :

$$Tp \frac{dPV(t)}{dt} + PV(t) = Kp \cdot CO(t - \theta p)$$

Where :

PV(t) = measured process variable as a function of time

CO(t – θp) = controller output signal as a function of time and shifted by θp

θp = process dead time

t = time

The other variables are as listed above this box. It is a first order differential equation because it has one derivative with one time constant, Tp. It is called a first order plus dead time equation because it also directly accounts for a delay or dead time, θp, in the CO(t) to PV(t) behavior.

We study what these three model parameters are and how to compute them, but here is why process gain, Kp, process time constant, Tp, and process dead time, θp, are all important :

- *Tuning :* These three model parameters can be plugged into proven correlations to directly compute P-Only, PI, PID, and PID with CO Filter tuning values. No more trial and error. No more tweaking our way to acceptable control. Great performance can be readily achieved with the step by step recipe listed above.

- *Controller Action :* Before implementing our controller, we must input the proper direction our controller should move to correct for growing errors. Some vendors use the term "reverse acting" and "direct acting." Others use terms like "up-up" and "up-down" (as CO goes up, then PV goes up or down). This specification is determined solely by the sign of the process gain, Kp.

- *Loop Sample Time, T :* Process time constant, Tp, is the clock of a process. The size of Tp indicates the maximum desirable loop sample time. Best practice is to set loop sample time, T, at 10 times per time constant or faster ($T \leq 0.1Tp$). Sampling faster will not necessarily provide better performance, but it is a safer direction to move if we have any doubts. Sampling too slowly will have a negative impact on controller performance. Sampling slower than five times per time constant will lead to degraded performance.

- *Dead Time Problems :* As dead time grows larger than the process time constant (θp > Tp), the control loop can benefit greatly from a model based dead time compensator such as a Smith predictor. The only way we know if θp > Tp is if we have followed the recipe and computed the parameters of a FOPDT model.

- *Model Based Control :* If we choose to employ a Smith predictor, a dynamic feed forward element, a multivariable decoupler, or any other model based controller, we need a dynamic model of the process to enter into the control computer. The FOPDT model from step 2 of the recipe is often appropriate for this task.

Fundamental to Success

With tuning values, loop specifications, performance diagnostics and advanced control all dependent on knowledge of a dynamic model, we begin to see that process gain, Kp; process time constant, Tp; and process dead time, θp; are parameters of fundamental importance to success in process control.

SAMPLE TIME IMPACTS CONTROLLER PERFORMANCE

There are two sample times, T, used in process controller design and tuning.

One is the control loop sample time that specifies how often the controller samples the measured process variable (PV) and then computes and transmits a new controller output (CO) signal.

The other is the rate at which CO and PV data are sampled and recorded during a bump test of our process. Bump test data is used to design and tune our controller prior to implementation.

In both cases, sampling too slow will have a negative impact on controller performance. Sampling faster will not necessarily provide better performance, but it is a safer direction to move if we have any doubts.

Fast and slow are relative terms defined by the process time constant, Tp. Best practice for both control loop sample time and bump test data collection are the same :

Best Practice : Sample time should be 10 times per process time constant or faster ($T \leq 0.1Tp$). We explore this "best practice" rule in a detailed study here. This study employs some fairly advanced concepts, so it is placed further down in the Table of Contents.

Yet perhaps we can gain an appreciation for how sample time impacts controller design and tuning with this thought experiment :

Suppose you see me standing on your left. You close your eyes for a time, open them, and now I am standing on your right. Do you know how long I have been at my new spot? Did I just arrive or have I been there for a while? What path did I take to get there? Did I move around in front or in back of you? Maybe I even jumped over you?

Now suppose your challenge is to keep your hands at your side until I pass by, and just as I do, you are to reach out and touch me. What are your chances with your eyes closed (and loud music is playing so you cannot hear me)?

Now lets say you are permitted to blink open your eyes briefly once per minute. Do you think you will have a better chance of touching me? How about blinking once every ten seconds? Clearly, as you start blinking say, two or three times a second, the task of touching me becomes easy. That's because you are sampling fast enough to see my "process" behavior fully and completely.

Based on this thought experiment, sampling too slow is problematic and sampling faster is generally better.

Keep in mind the "$T \leq 0.1Tp$" rule as we study PID control. This applies both to sampling during data collection, and the "measure and act" loop sample time when we implement our controller.

Chapter 9

CONTROL OF INTEGRATING PROCESSES

RECOGNIZING INTEGRATING (NON-SELF-REGULATING) PROCESS BEHAVIOR

Cruise control of a car is a self-regulating process. If we keep the fuel flow to the engine constant while travelling on flat ground on a windless day, the car will settle out at some constant speed. If we increase the fuel flow rate a fixed amount, the car will accelerate and then steady out at a different constant speed.

If the exchanger cooling rate and disturbance flow rate are held constant at fixed values, the exit temperature will steady at a constant value. If we increase the cooling rate, wait a bit, and then return it to its original value, the exchanger exit temperature will respond during the experiment and then return to its original steady state.

But some processes where the streams are comprised of gases, liquids, powders, slurries and melts do not naturally settle out at a steady state operating level. Process control practitioners refer to these as non-self-regulating, or more commonly, as integrating processes.

Integrating (non-self-regulating) processes can be remarkably challenging to control. After exploring the distinctive behaviors illustrated below, you may come to realize that some of the level, temperature, pressure, pH and other loops you work with have such a character.

Integrating (Non-Self-Regulating) Behavior in Manual Mode

The upper plot below shows the open loop (manual mode) behavior of the more common self-regulating process. In this idealized response, the controller output (CO) and process variable (PV) are initially at steady state. The CO is stepped up and back from this steady state. As shown, the PV responds to the step, but ultimately returns to its original operating level.

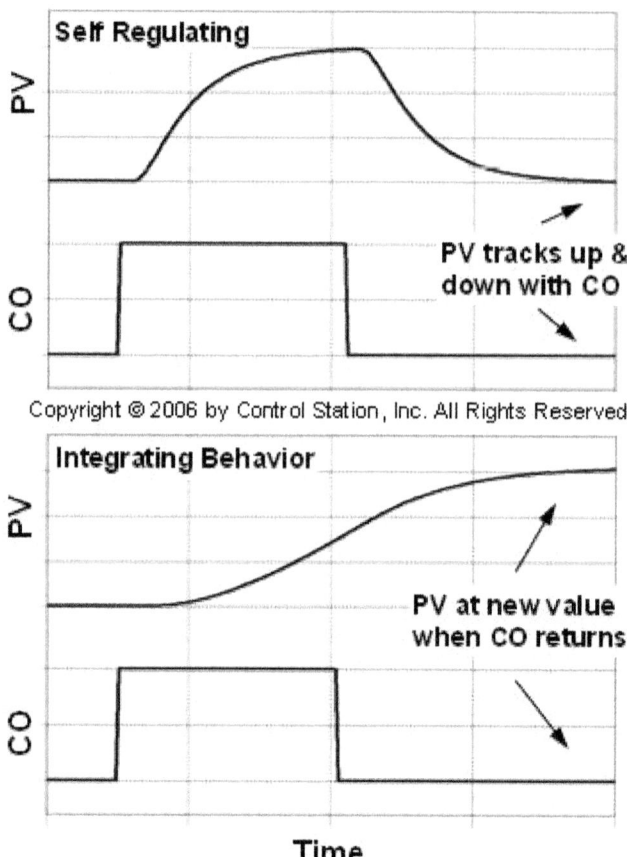

Time

The lower plot above shows the open loop response of an ideal integrating process. The distinctive behavior is that the PV settles at a new operating level when the CO returns to its original value.

In truth, the integrating behavior plot above is misleading in that it implies that for such processes, a steady CO will produce a steady PV. While possible with idealized simulations like that used to generate the plot, such a "balance point" behavior is rarely found in open loop (manual mode) for integrating processes in an industrial operation.

More realistically, if left uncontrolled, the lack of a balance point means the PV of an integrating process will naturally tend to drift up or down, possibly to extreme and even dangerous levels. Consequently, integrating processes are rarely operated in manual mode for very long.

Aside : The behavior shown in the integrating plot can also appear across small regions of operation of a self-regulating process if there is a significant dead-band in the final control element (FCE). This might result, for example, from loose mechanical linkages in a valve. For this investigation, we assume the FCE operates properly.

P-Only Control Behavior is Different

To appreciate the difference in behavior for integrating processes, we first recall P-Only control of a self-regulating process.

As shown below, when the set point (SP) is initially at the design level of operation (DLO) in the first moments of operation, then PV equals SP. Recall that *the DLO* is where we expect the SP and PV to be during normal operation when the major disturbances are at their normal or typical values.

The set point is then stepped up from the DLO on the left half of the plot. The simple P-Only controller is unable to track the changing SP and a steady error, called *offset*, results. The offset grows as each step moves the SP farther away from the DLO.

Midway through the plot, a disturbance occurs. Its size was pre-calculated in this ideal simulation to eliminate the offset. The SP is then stepped back down to the right in the plot and we see that offset shifts, but again grows in a similar and predictable pattern.

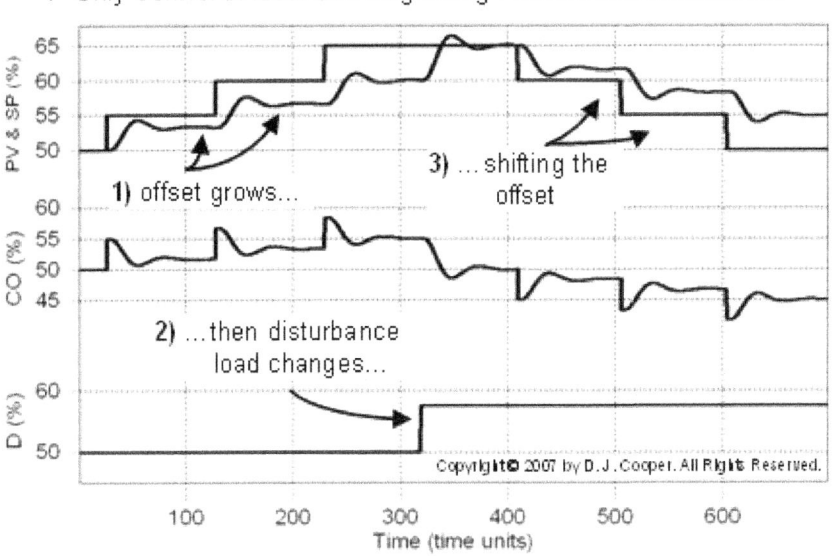

P-Only Control of Ideal Self-Regulating Process and Disturbance

With this as background, we next consider an ideal integrating process simulation under P-Only control as shown below :

Even under simple P-Only control as shown in the left half of the plot, the PV is able to track the SP steps with no offset. This behavior can be quite confusing as it does not fit the expected behavior we have just seen above for the more common self-regulating process.

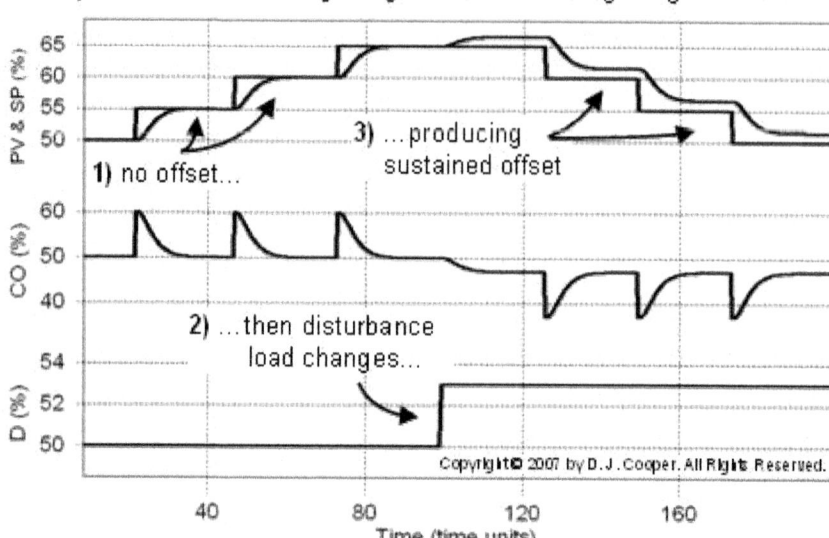

P-Only Control of Ideal Integrating Process with Integrating Disturbance

The reason this happens is that integrating processes have a natural accumulating character. In fact, this is why "integrating process" is used as a descriptor for non-self-regulating processes. Since the process integrates, then it appears that the controller does not need to.

Yet the set point steps to the right in the above plot show that this is not completely correct. Once a disturbance shifts the baseline operation of the process, shown roughly at the midpoint in the above plot, an offset develops and remains constant, even as SP returns to its original design value.

Controller Output Behavior is Telling

If we study the CO trace in the two plots above, we see one feature that distinguishes self-regulating from integrating process behavior.

In the self-regulating process plot above, the average CO value tracks up and then down as the SP steps up and then down.

In the integrating process plot above, the CO spikes with each SP step, but then in a most unintuitive fashion, returns to the same steady value. It is only the change in the disturbance flow that causes the average CO to shift midway through the plot, though it then remains centered around the new value for the remainder of the SP steps.

Pi Control Behavior is Different

The plot below shows the ideal self-regulating process controlled using the popular *dependent ideal PI* algorithm. Reset time, T_i, is held constant throughout the experiment while controller gain, Kc, is doubled and then doubled again.

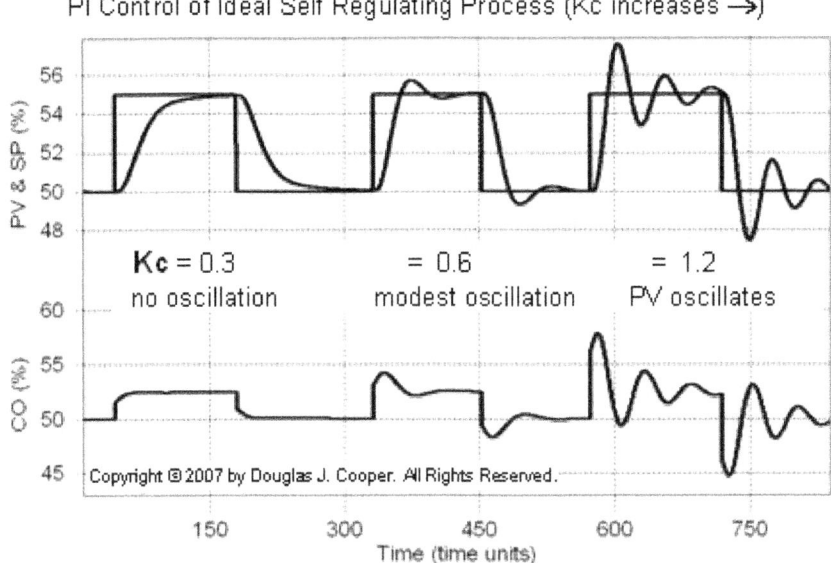

PI Control of Ideal Self Regulating Process (Kc increases →)

Kc = 0.3
no oscillation

= 0.6
modest oscillation

= 1.2
PV oscillates

As Kc increases, the controller becomes more active, and as we have grown to expect, this increases the tendency of the PV to display oscillating (or under-damped) behavior.

For comparison, now consider PI control of an ideal integrating process simulation as shown below. As above, reset time, T_i, is held constant throughout the experiment while controller gain, Kc, is increased across the plot.

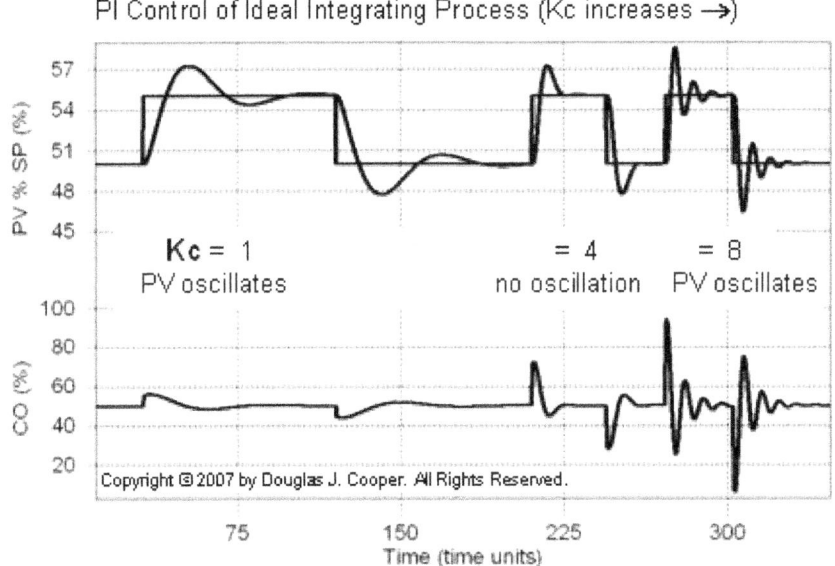

PI Control of Ideal Integrating Process (Kc increases →)

Kc = 1
PV oscillates

= 4
no oscillation

= 8
PV oscillates

A counter-intuitive result is that as Kc becomes small *and* as it becomes large, the PV begins displaying an underdamped (oscillating) response behavior.

While the frequency of the oscillations is clearly different between a small and large Kc when seen together in a single plot as above, it is not always obvious what direction controller gain needs to move to settle the process when looking at such unacceptable performance on a control room display.

Tuning Recipe Required

One of the biggest challenges for practitioners is recognizing that a particular process shows integrating behavior prior to starting a controller design and tuning project. This, like most things, comes with training, experience and practice.

Once in automatic, closed loop behavior of an integrating process can be unintuitive and even confounding. Trial and error tuning can lead us in circles as we try to understand what is causing the problem.

A formal *controller design and tuning recipe* for integrating processes helps us overcome these issues in an orderly and reliable fashion.

A DESIGN AND TUNING RECIPE FOR INTEGRATING PROCESSES

Additionally, a recipe-based approach overcomes many of the concerns that makes control projects challenging in an industrial operating environment. Specifically, a recipe approach causes less disruption to the production schedule, wastes less raw material and utilities, requires less personnel time, and generates less off-spec product.

The Recipe for Integrating Processes

Integrating (or non-self-regulating) processes display counter-intuitive behaviors that make them surprisingly challenging to control. In particular, they do not naturally settle out to a steady operating level if left uncontrolled.

So while the controller design and tuning recipe is generally the same for both self-regulating and integrating processes, there are important differences. Specifically, step 3 of the recipe uses a different dynamic model form and step 4 employs different tuning correlations.

Yet the design and tuning recipe maintains the familiar four step structure :

1. Establish the design level of operation (the normal or expected values for set point and major disturbances).
2. Bump the process and collect controller output (CO) to process variable (PV) dynamic process data around this design level.
3. Approximate the process data behavior with a *first order plus dead time integrating* (FOPDT Integrating) dynamic model.
4. Use the model parameters from step 3 in rules and correlations to complete the controller design and tuning.

It is important to recognize that real processes are more complex than the simple FOPDT Integrating model form used in step 3. In spite of this, the FOPDT Integrating model succeeds in providing an approximation of process behavior that is sufficiently accurate to yield reliable and predictable control performance when used with the rules and correlations in step 4 of the recipe.

The FOPDT Integrating Model

We recall that the familiar first order plus dead time (FOPDT) dynamic model used to approximate self-regulating dynamic process behavior has the form :

$$\text{Tp}\frac{d\text{PV}(t)}{dt} + \text{PV}(t) = \text{Kp} \cdot \text{CO}(t - \theta p) \; \textit{FOPDT Form}$$

Yet this model cannot describe the kind of integrating process behavior shown in these examples. Such behavior is better described with the *FOPDT Integrating model* form :

$$\frac{d\text{PV}(t)}{dt} = \text{Kp}* \cdot \text{CO}(t - \theta p) \; \textit{FOPDT Integrating Form}$$

It is interesting to note when comparing the two models above that the FOPDT Integrating form does not have the lone "+ PV" term found on the left hand side of the FOPDT dynamic model.

Also, individual values for the familiar process gain, Kp, and process time constant, *Tp*, are not separately identified for the FOPDT Integrating model. Instead, an integrator gain, Kp*, is defined that has units of the ratio of the process gain to the process time constant, or :

$$\text{Kp}*[=]\frac{\text{Kp}}{\text{Tp}} \quad \text{or} \quad \text{Kp}* \, [=] \, \text{PV}/(\text{CO} \cdot \text{time})$$

Tuning Correlations for Integrating Processes

Analogous to the FOPDT investigations on this site (*e.g.*, here and here), we will see that the FOPDT Integrating model parameters Kp* and θp of Step 3 can be computed using a graphical analysis of plot data or by automated analysis using commercial software.

Step 4 then provides tuning values for controllers such as the dependent, ideal PI form :

$$\text{CO} = \text{CO}_{\text{bias}} + \text{Kc} \cdot e(t) + \frac{\text{Kc}}{\text{Ti}} \int e(t)dt$$

and the dependent, ideal PID form :

$$\text{CO} = \text{CO}_{\text{bias}} + \text{Kc} \cdot e(t) + \frac{\text{Kc}}{\text{Ti}} \int e(t)dt - \text{Kc} \cdot \text{Td}\frac{d\text{PV}}{dt}$$

One important difference about integrating processes is that since there is no identifiable process time constant in the FOPDT Integrating model, we use dead time, θp, as the baseline marker of time in the design and tuning rules.

Specifically, θp is used as the basis for computing sample time, T, and the closed loop time constant, Tc. Following the procedures widely discussed on this site for self-regulating processes (*e.g.* here and here), we employ a rule to compute the closed time constant, Tc, as :

$Tc = 3\theta p$

The controller tuning correlations for integrating processes use this Tc, as well as the Kp^* and θp from the FOPDT integrating model fit, as :

	Controller Gain Kc	Reset Time Ti	Derive Time Td
PI	$\dfrac{1}{Kp^*}\dfrac{2Tc+\theta p}{(Tc+\theta p)^2}$	$2Tc+\theta p$	
PID	$\dfrac{1}{Kp^*}\dfrac{2Tc+\theta p}{(Tc+0.5\theta p)^2}$	$2Tc+\theta p$	$\dfrac{0.25\theta p^2+Tc\theta p}{2Tc+\theta p}$

Loop Sample Time, T

Determining a proper sample time, T, for integrating processes is somewhat more challenging than for self-regulating processes.

There are two sample times, T, used in process controller design and tuning. One is the control loop sample time that specifies how often the controller samples the measured process variable (PV) and computes and transmits a new controller output (CO) signal. The other is the rate at which CO and PV data are sampled and recorded during a bump test.

All controllers measure, act, then wait until next sample time before repeating the loop. This "measure, act, wait" procedure has a delay (or dead time) of one sample time built naturally into its structure. Thus, the minimum dead time (θp, min) in any control loop is the loop sample time, T.

With this information, we recognize a somewhat circular argument in defining sample time for integrating processes :

- Our time basis for controller design is θp, and as such, then loop sample time, T, should be small relative to dead time, or :

  $T \le 0.1\theta p$

- But the minimum that dead time can be is one sample time, T, or

  θp, min $= T$

Thus, T is based on θp, and if the process is sampled too slowly during a bump test, then θp can be based on T.

To avoid this issue, it is best practice to sample the process as fast as reasonably possible during bump tests so accurate model parameters can be determined during analysis. Loop sample time, T, can then be computed from dead time for controller implementation.

If there is concern about a particular analysis, an alternative and generally conservative way to compute sample time is :

$$T = 0.1 \frac{(PV_{max} - PV_{min})}{(CO_{max} - CO_{min})} \cdot \frac{1}{|Kp^*|}$$

where the subscripts max and min refer to the maximum and minimum values for CO and PV across the signal span of the instrumentation.

Using the Recipe

The tuning recipe for integrating processes has important differences from that used for self-regulating process. When designing and tuning controllers for such processes, we should :

- Use an FOPDT Integrating model form when approximating dynamic model behavior,
- Note that the closed loop time constant, Tc, and sample time, T, are based on model dead time, θp.
- employ PI and PID tuning correlations specific to integrating processes.

ANALYZING PUMPED TANK DYNAMICS WITH A FOPDT INTEGRATING MODEL

The Pumped Tank Process

To better understand the design and tuning of a PID controller for an integrating process, we explore the pumped tank case study from *Control Station's Loop-Pro* software.

As shown below in manual mode, the process has two liquid streams feeding the top of the tank and a single exit stream pumped out the bottom.

As labelled in the figure, the measured process variable (PV) is liquid level in the tank. To maintain level, the controller output (CO) signal adjusts a throttling valve at the discharge of a constant pressure pump to manipulate flow rate out of the bottom of the tank. This approximates the behavior of a centrifugal pump operating at relatively low throughput.

Unlike the *gravity drained tanks* case study where the exit flow rate increases and decreases as tank level rises and falls, the discharge flow rate here is strictly regulated by a pump. As a consequence, the physics do not naturally work to balance the system when any of the stream flow rates change.

This lack of a natural balancing behavior is why the pumped tank is classified as an integrating process. If the total flow into the tank is greater than the flow pumped out, the liquid level will rise and continue to rise until the tank fills or a stream flow changes. If the total flow into the tank is less than the flow pumped out, the liquid level will fall and continue to fall.

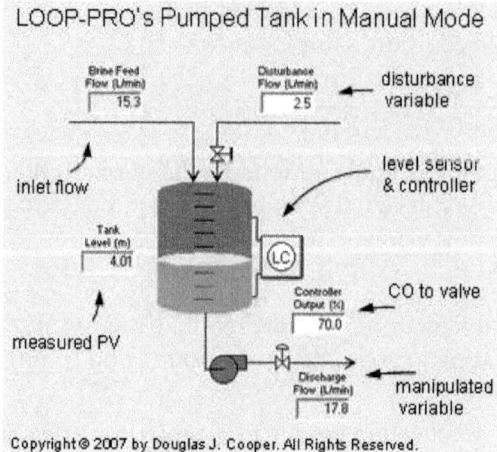

Below is a plot of the pumped tank behavior with the controller in manual mode (open loop). The CO signal is stepped up, increasing the discharge flow rate out of the bottom of the tank. The flow out becomes greater that the total feed into the top of the tank and as shown, liquid level begins to fall.

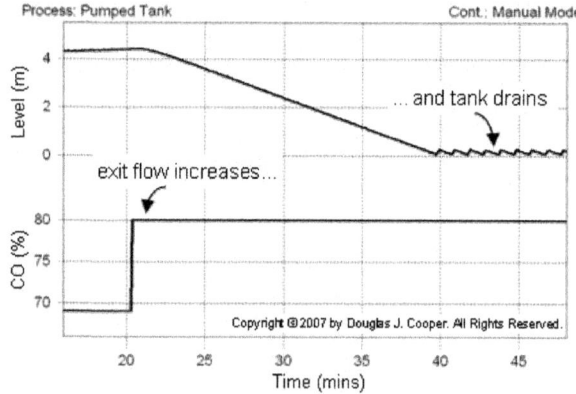

As the situation persists, liquid level continues to fall until the tank is drained. The saw-toothed effect shown when the tank is empty is because the pump briefly surges every time enough liquid accumulates for it to regain suction.

Not shown is that if the controller output were to be decreased enough to cause flow rate out to be less than flow rate in, the tank level would rise until full. If this were a real process, the tank would overflow and spill, creating safety and profitability issues.

The Disturbance Stream

As shown in the process graphic, the disturbance variable is a flow rate of a second-ary feed into the top of the tank. This disturbance flow (D) is controlled independently, as if by another process (which is why it is a disturbance to our process).

When D decreases (or increases), the measured PV level falls (or rises) in response. To illustrate, the plot below shows that the CO is held constant and D is decreased. Characteristic of an integrating process, the PV (tank level) starts falling because total flow into the tank is less than that pumped out. Like the case discussed above, the level continues to fall until the tank is drained.

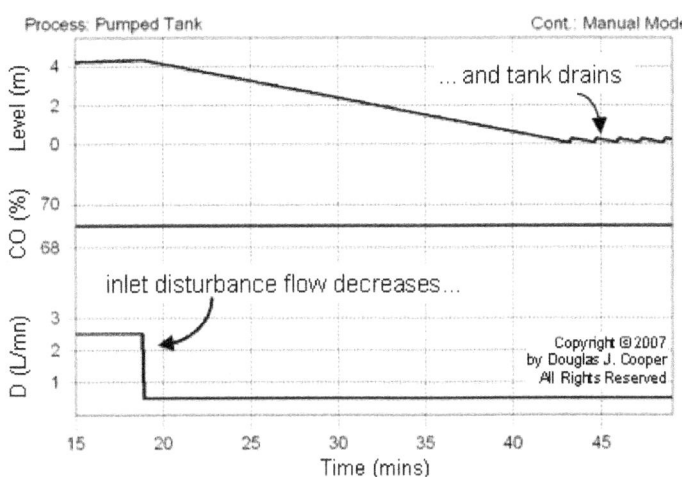

Pumped Tank in Closed Loop

The process graphic below shows the pumped tank in automatic mode (closed loop). The two streams feeding into the top of the tank total 17.8 L/min and level is steady because the controller regulates the discharge flow to this same value.

LOOP-PRO's Pumped Tank in Automatic

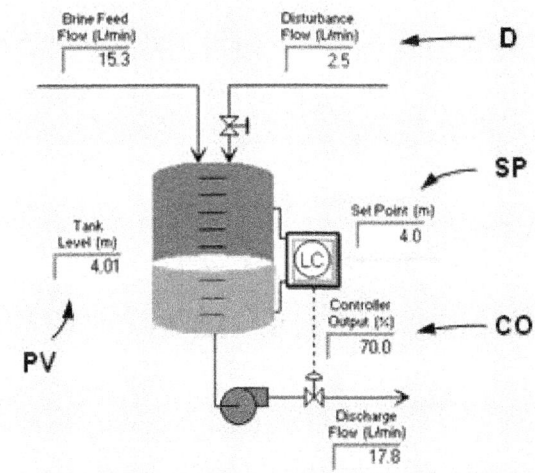

Graphical Modeling of Integrating Process Data

When collecting and analyzing data as detailed in steps 1 and 2 of the *design and tuning recipe for integrating processes*, we begin by specifying a design level of operation (DLO). In this study, we specify :

- Design PV = 4.8 m
- Design D = 2.5 L/min.

Note that while not shown in the plots, D is held constant at 2.5 L/min throughout this study.

Just as with self-regulating processes, step 3 of the recipe uses a simplifying dynamic model approximation to describe the complex behavior of a real process. The FOPDT Integrating model is simple in form yet it provides information sufficiently accurate for controller design and tuning.

The FOPDT Integrating model is :

$$\frac{dPV(t)}{dt} = Kp^* . CO(t - \theta p)$$

where the integrator gain, Kp^*, has units of the ratio of the process gain to the process time constant, or :

$$Kp^*[=] \frac{Kp}{Tp} \text{ or } Kp^*[=] PV / (CO \cdot time)$$

The graphical method of fitting a FOPDT Integrating model to process data requires a data set that includes at least two constant values of controller output, CO_1 and CO_2.

As shown below for the pumped tank, both must be held constant long enough so that a slope trend in the PV response (tank liquid level) can be visually identified.

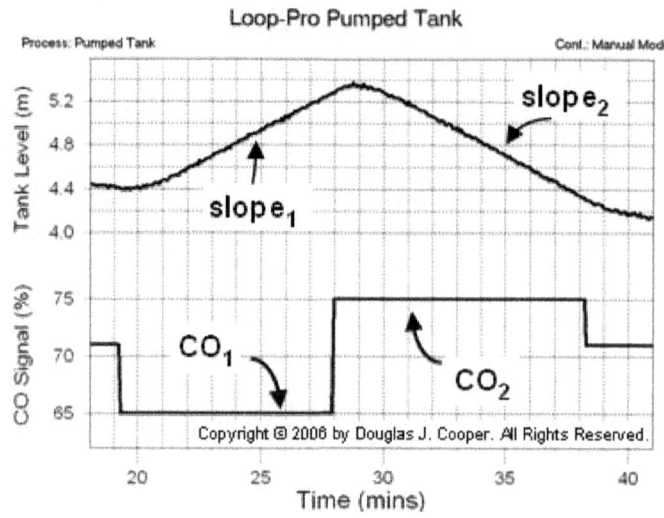

Loop-Pro Pumped Tank

An important difference between the graphical technique for *self-regulating processes* and integrating processes as discussed here is that integrating processes need not start at a steady value (steady-state) before a bump is made to the CO. The graphical technique discussed here is only concerned with the slopes (or rates of change) in PV and the controller output signal that caused each PV slope.

The FOPDT Integrating model describes the PV behavior at each value of constant controller output CO_1 and CO_2 as :

$$\frac{dPV}{dt}\bigg|_1 = Kp*CO_1(t-\theta p)$$

and

$$\frac{dPV}{dt}\bigg|_2 = Kp*CO_2(t-\theta p)$$

Subtracting and solving for $Kp*$ yields :

$$Kp* = \frac{\dfrac{dPV}{dt}\bigg|_2 - \dfrac{dPV}{dt}\bigg|_1}{CO_2 - CO_1} = \frac{slope_2 - slope_1}{CO_2 - CO_1}$$

Graphical Modeling of Pumped Tank Data

- *Computing Integrator Gain :* Below is the same open loop data from the pumped tank simulation as shown above. The CO is stepped from 71% down to 65%, causing the liquid level (the PV) to rise. The controller output is then stepped from 65% up to 75%, causing a downward slope in the liquid level.

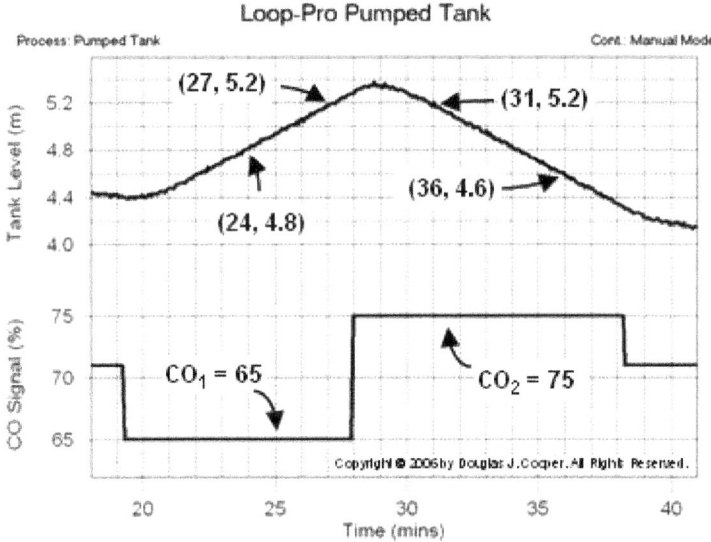

The slope of each segment is calculated as the change in tank liquid level divided by the change in time. From the plot data we compute :

$$\text{slope}_1 = \left.\frac{dPV}{dt}\right|_1 = \frac{\Delta PV_1}{\Delta t_1} = \frac{5.2 - 4.8}{27 - 24} = 0.13\,\frac{m}{min}$$

and

$$\text{slope}_2 = \left.\frac{dPV}{dt}\right|_2 = \frac{\Delta PV_2}{\Delta t_2} = \frac{4.6 - 5.2}{36 - 31} = -0.12\,\frac{m}{min}.$$

Using the two slopes computed above along with their respective CO values from the plot yields the integrator gain, Kp*, for the pumped tank :

$$Kp^* = \frac{\text{slope}_2 - \text{slope}_1}{CO_2 - CO_1} = \frac{(-0.12) - (0.13)}{75 - 65} = -0.025\,\frac{m}{\% \, min}.$$

- *Computing Dead Time :* The dead time is estimated from the plot using the same *method described for the heat exchanger.* That is, dead time, θp, is computed as the difference in time from when the CO signal was stepped and when the measured PV starts a clear response to that change.

As shown above, the pumped tank dead time is estimated from the plot as :

θp = 1.0 min

Thus, the FOPDT Integrator model information needed to proceed with controller tuning using the correlations *presented here* is complete.

Automating the Model Fit

In today's world, there is no need to perform the model fit with graph paper and calculator. Commercial software offers analysis tools that makes fitting an FOPDT Integrating model quite simple.

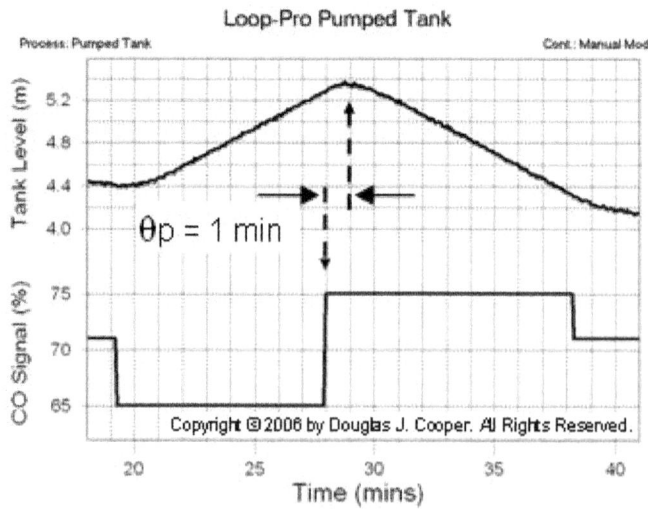

Below, *Control Station's Loop-Pro* software is used to analyze the pumped tank data. As shown, the software displays the data as a plot that includes adjustable nodes and tie lines.

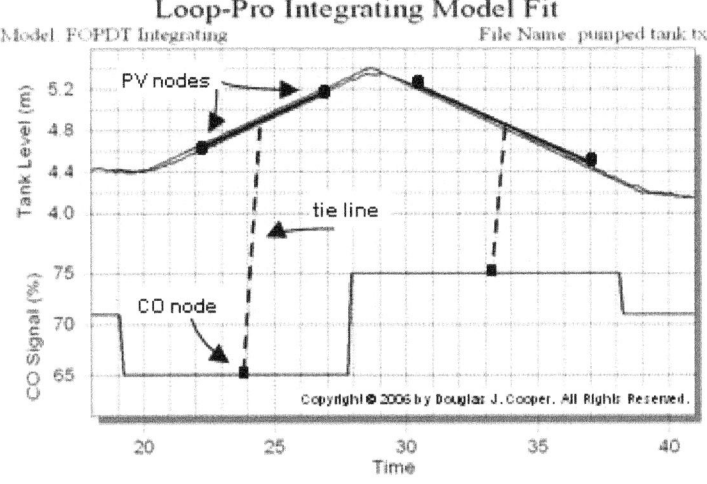

To compute a fit, click on the two CO nodes at the bottom of the plot and drag them to match the two values of constant controller output expected in the data. Both have tie lines to identify their associated PV slope bar. Each slope bar has two end point nodes. Click and drag these so that each bar approximates the sloping segments on the graph.

With the six nodes (two CO and four PV) properly positioned, the FOPDT Integrating model parameters are automatically calculated and the model fit is displayed over the raw data.

We can see that the model PV line matches the measured PV data, so we have confidence that the model fit is good. The image above also shows the FOPDT Integrator values computed by the software :

$Kp^* = -0.023$ m/(% min), $\theta p = 1.0$ min

$Kp^* = -0.025$ m/(% min), $\theta p = 1.0$ min.

In the dynamic modeling world, these values are virtually identical. We note that software offers additional benefits in that it performs the computation quickly, reduces the chance of computational error, and provides a visual confirmation that the model used for controller design and tuning reasonably matches the process data.

PI CONTROL OF THE INTEGRATING PUMPED TANK PROCESS

The process graphic below shows the pumped tank in automatic mode (also called closed loop) :

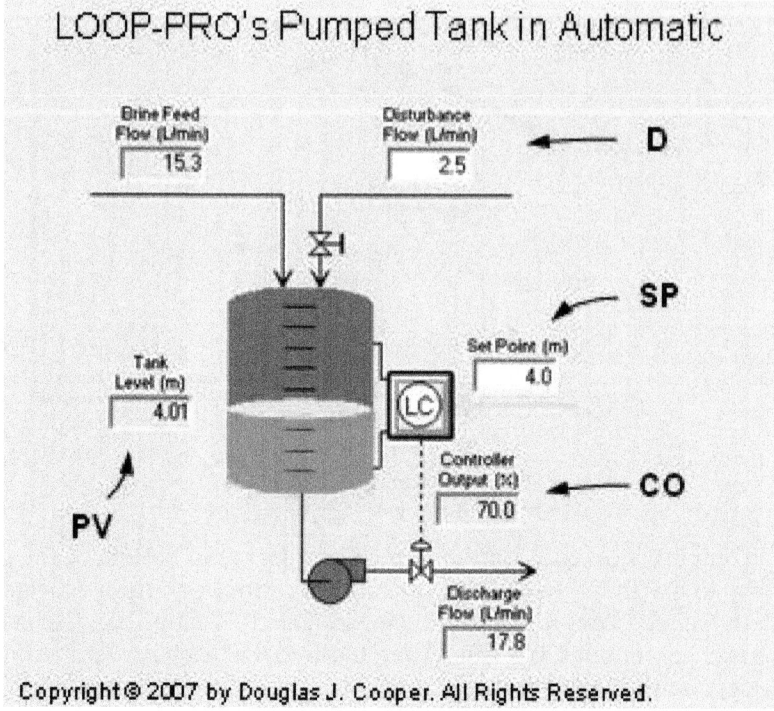

LOOP-PRO's Pumped Tank in Automatic

Brine Feed Flow (L/min) 15.3

Disturbance Flow (L/min) 2.5 ← D

← SP

Tank Level (m) 4.01

Set Point (m) 4.0

Controller Output (%) 70.0 ← CO

PV

Discharge Flow (L/min) 17.8

The important variables for this study are labelled in the above graphic :

CO = signal to valve that adjusts discharge flow rate of liquid (controller output, %)

PV = measured liquid level signal from the tank (measured process variable, m)

SP = desired liquid level in tank (set point, m)

D = flow rate of liquid entering top of tank (major disturbance, L/min)

We follow the controller design and tuning *recipe for integrating processes* in this study as we design and test a PI controller. Please recall that there are subtle yet important differences between this procedure and the *design and tuning recipe* used for the more common *self-regulating process*.

Step 1 : Design Level of Operation (DLO)

- Design value for PV and SP = 4.8 m
- Design value for D = 2.5 L/min

Characteristic of real integrating processes, the pumped tank PV does not naturally settle at a steady operating level if CO is held constant. This lack of a natural "balance point" means we will not specify a CO as part of our DLO.

Step 2 : Collect Process Data Around the DLO

When PV and D are near the design level of operation and D is substantially quiet, we perform a dynamic test and generate CO-to-PV cause and effect dynamic data.

Because the PV of integrating processes tends to drift in manual mode (open loop), one alternative is to perform an open loop dynamic test that does not require bringing the process to steady state. The procedure is to maintain CO at a constant value until a slope trend in the PV can be visually identified. We then move the CO to a different value and hold it until a second PV slope is established.

An alternative approach, presented below, is to use automatic mode data. When in closed loop, dynamic data is generated by bumping the SP. For model fitting purposes, the controller must be tuned such that the CO takes clear and sudden actions in response to the SP changes, and these must force PV movements that dominate the measurement noise.

Because a closed loop approach makes it possible to generate integrating process dynamic test data that begins at steady state, we can use *model fitting software* much like we did in *this closed-loop set point driven study*.

Below we see the pumped tank process under *P-Only control*. As shown in the left half of the plot, while the major disturbance is quiet and at its design level, we are able to obtain good set point tracking performance with P-Only control.

As we *discuss here* and as shown above, a P-Only controller is able to provide good SP tracking performance with no offset as long the major disturbances are quiet and at their design values. Industrial processes can have many disturbances that impact operation. If any one of them changes, as happens in the above plot at roughly 43 minutes, then the simple P-Only controller is incapable of eliminating what becomes a sustained offset (*i.e.*, incapable of making $e(t)$ = SP–PV = 0)

This is why the integral action of a PI controller offers value even though the process itself possesses a naturally integrating behavior.

Note : in a surge tank where exit flow smoothing is more important than maintaining the measured level at SP, offset may not be considered a problem to be solved. Each situation must be considered on its own merits.

The red label in the above plot indicates that the left half contains dynamic response data that begins at steady-state and that is not corrupted by disturbance changes. We isolate this data and model it as described in step 3 below.

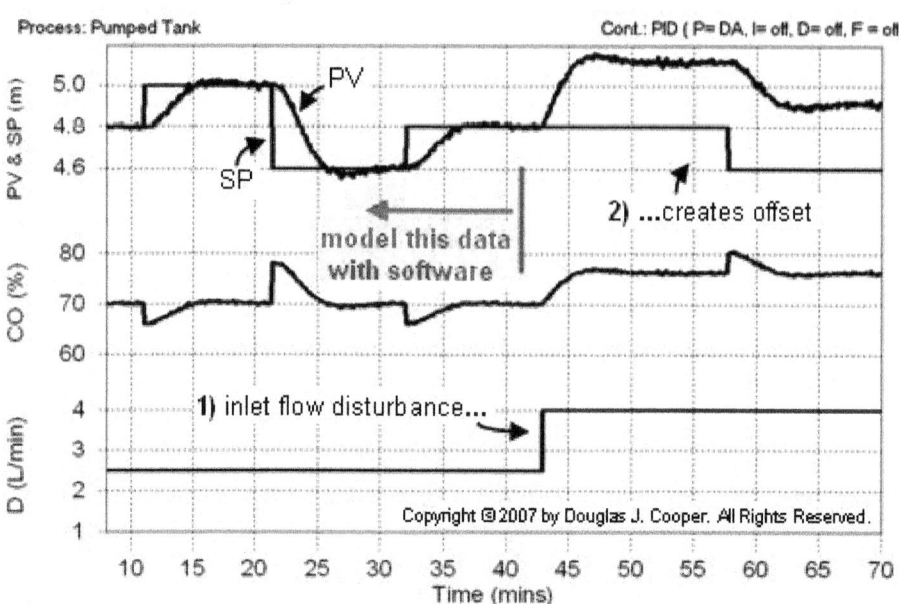

Step 3 : Fit a FOPDT Integrating Model to the Dynamic Process Data

We obtain an approximating description of the closed loop CO to PV dynamic behavior by fitting the process data with a first order plus dead time integrating (*FOPDT Integrating*) model of the form :

$$\frac{dPV(t)}{dt} = Kp^* \cdot CO(t - \theta p)$$

where :

Kp^* = integrator gain, with units [=] PV/(CO ·time)

θp = dead time, with units [=] time

   Cropping the data and fitting the FOPDT Integrator model takes but a few mouse clicks with a *commercial software tool* (recall that an alternative graphical hand calculation method is *described here*).

   The visual similarity between the model and data gives us confidence that we have a meaningful description of the dynamic behavior of this process. The Kp^* and θp for this approximating model are shown at the bottom of the plot.

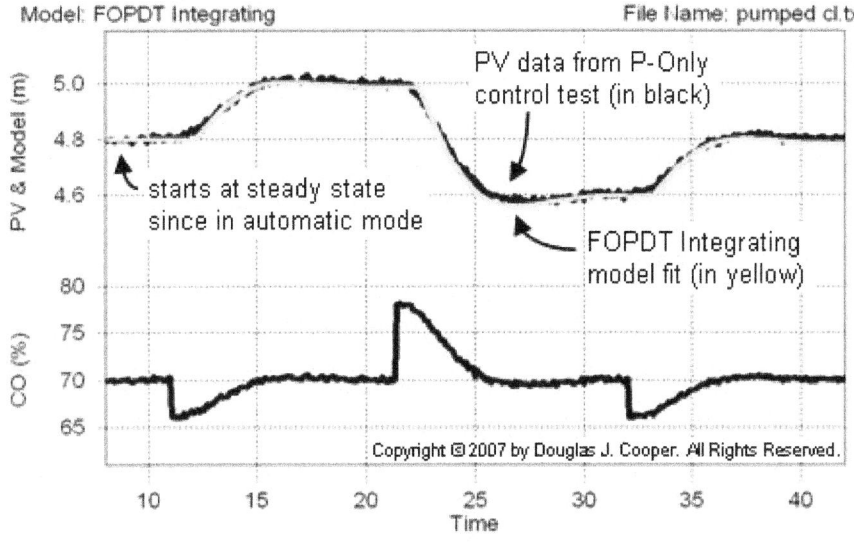

FOPDT Integrating Model Fit of Closed Loop Pumped Tank Data

Integrator Gain, (K*) = -0.0234, Dead Time (TD) = 1.04
Goodness of Fit: R-Squared = 0.9887, SSE = 0.1017

The *model fitting software* performs a systematic search for the combination of model parameters that minimizes the sum of squared errors (SSE), computed as :

$$SSE = \sum_{j=1}^{N}[\text{Measured PV}_j - \text{Model PV}_j]^2$$

The Measured PV is the actual data collected from our process. The Model PV is computed using the model parameters from the search routine and the actual CO data from the file. N is the total number of samples in the file. In general, the smaller the SSE, the better the model describes the data.

The table below summarizes the model parameters from the above closed loop model fit as well as the results from the open loop slope driven analysis.

	-hand fit- Open Loop	-software Open Loop	-software model fits- P-only
Integrator gain, Kp* (m/ % · min)	–0.025	–0.023	–0.023
Dead time, θp (min)	1.0	1.0	1.0

For the control studies in step 4, we will use :

- Kp* =–0.023 m/ % ·min
- θp = 1.0 min.

Step 4 : Use the FOPDT Integrating Parameters to Complete the Design

- *Sample Time, T :* The design and tuning *recipe for integrating processes* suggests setting the loop sample time, T, at one-tenth the process dead time or faster (*i.e.*, $T \le 0.1\theta p$). Faster sampling provides equally good, but not better, performance.

In this study, $T \le 0.1(1.0$ min), so T should be 6 seconds or less. We meet this with the sample time option available from virtually all commercial vendors :

- Sample time, $T = 1$ sec.

- *Control Action (Direct/Reverse)* : The pumped tank has a negative Kp^*, so when CO increases, PV decreases in response. Since a controller must provide negative feedback, if the process is reverse acting, the controller must be direct acting. Thus, if the PV is too high, the controller must increase the CO to correct the error. Since the controller moves in the same direction as the problem, we specify :

  - Controller is direct acting.

- *Computing Controller Error, e(t)* : Set point, SP, is manually entered into a controller. The measured PV comes from the sensor (our *wire in*). Since SP and PV are known values, then at every loop sample time, T, controller error can be directly computed as :

  - Error, $e(t)$ = SP-PV.

- *Determining Bias Value, CO_{bias}* : The lack of a natural balance point with integrating processes makes the determination of a design CO_{bias} problematic. The solution is to use *bumpless transfer*. That is, when switching to automatic, initialize SP to the current value of PV and CO_{bias} to the current value of CO (most commercial controllers are already programmed this way). By choosing our current operation as our design state at switch over, there is no corrective actions needed by the controller and it can smoothly engage, thus :

  - Controller bias, CO_{bias} = CO that exists at switch over.

- *Controller Gain, Kc, and Reset Time, Ti* : We use our FOPDT Integrating model parameters in the industry-proven Internal Model Control (IMC) tuning correlations to compute PI tuning values. Though all PI forms are equally capable, we use the dependent, ideal form of the PI algorithm in this study :

$$CO = CO_{bias} + Kc \cdot e(t) + \frac{Kc}{Ti} \int e(t)dt$$

The first step in using the IMC correlations is to compute Tc, the closed loop time constant. Tc describes how active our controller should be in responding to a set point change or in rejecting a disturbance. For integrating processes, the *design and tuning recipe* suggests :

$Tc = 3\theta p = 3(1.0$ min) = 3 min

With Tc computed, the PI controller gain, Kc, and reset time, Ti, are computed as :

$$Kc = \frac{1}{Kp^*} \frac{2Tc + \theta p}{(Tc + \theta p)^2} \quad Ti = 2Tc + \theta p$$

Substituting the Kp*, θp and Tc identified above into these tuning correlations, we compute :

$$Kc = \frac{1}{-0.023} \frac{2(3)+1}{(3+1)^2} \quad Ti = 2(3)+1$$

or

- Kc = – 19 m/%
- Ti = 7 min

Below is the performance of this PI controller (with Kc = – 19 and Ti = 7) on the pumped tank. The plot includes the same set point tracking and disturbance rejection test conditions as were used in the P-Only controller plot.

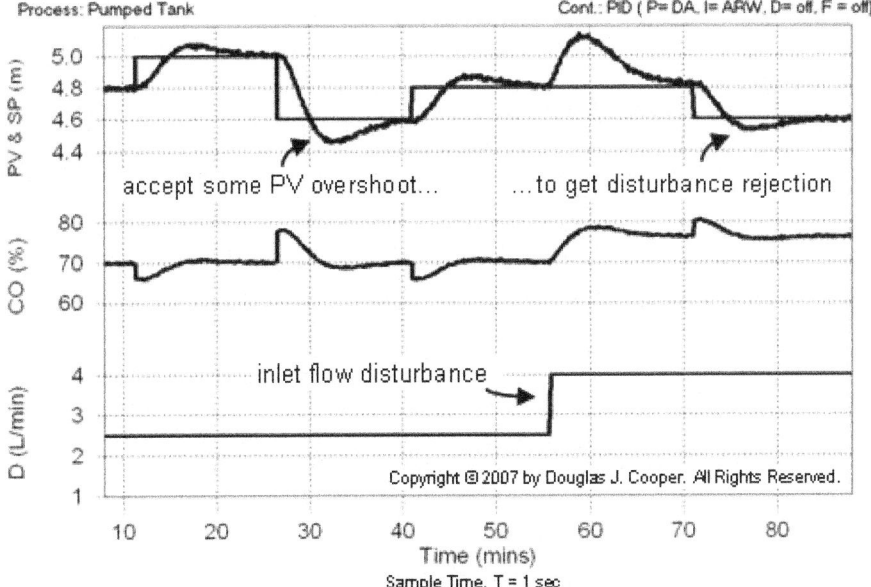

As labelled in the plot, our PI control set point response now includes some overshoot. Recall that the P-Only controller, can provide a rapid set point response with no overshoot, that is, until a disturbance changes the balance point of the process.

The benefit of integral action is that when a disturbance occurs, a PI controller can reject the upset and return the PV to set point. This is because the constant summing of *integral action* continues to move the CO until controller error is driven to zero.

Thus, PI control requires that we accept some overshoot during set point tracking in exchange for the ability to reject disturbances. In many industrial applications, this is considered a fair trade.

Tuning Sensitivity Study

Below is the set point tracking performance of our PI controller on the pumped tank. The controller tuning (Kc = – 19 m/ %, Ti = 7 min) is determined as detailed in the step-by-step recipe above.

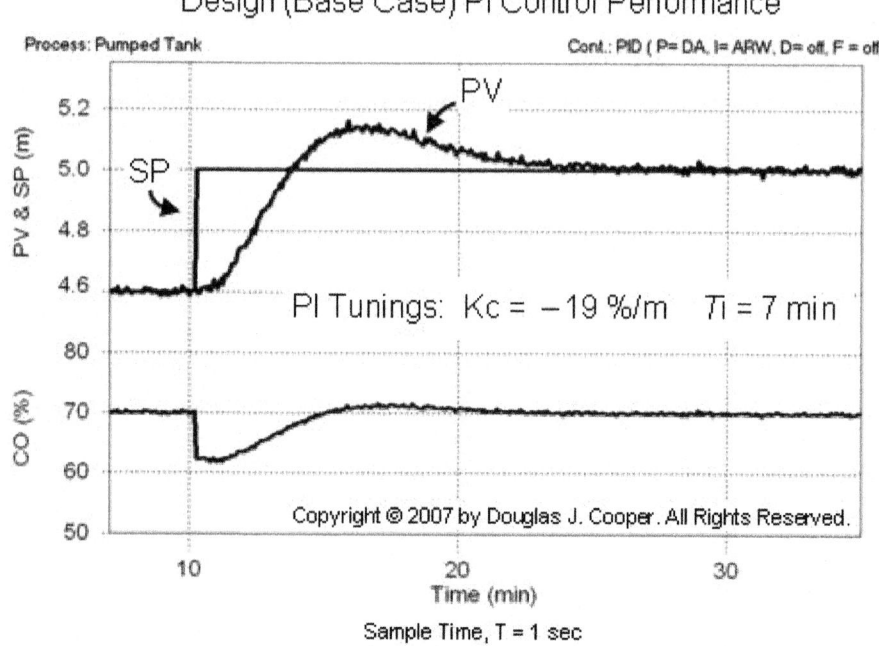

Some questions to consider are :

- How does performance vary as the tuning values change?
- How can we avoid overshoot if we find such behavior undesirable?

Below is a tuning map for a PI controller implemented on the pumped tank integrating process. The center plot is the identical base case performance plot shown above.

The complete map shows set point tracking performance when controller gain (Kc) and reset time (Ti) are individually doubled and halved.

While "good" or "best" performance is a matter best decided by the operations staff, the above map makes it clear that our recipe does an excellent job of meeting the desire for a reasonably rapid rise, a modest overshoot and a quick settling time.

Unfortunately, eliminating overshoot altogether does not appear to be one of our options for PI control of integrating processes.

Impact of Kc and T_i on Integrating Process for PI Controller: $CO = CO_{bias} + Kc\,e(t) + \dfrac{Kc}{T_i}\int e(t)\,dt$

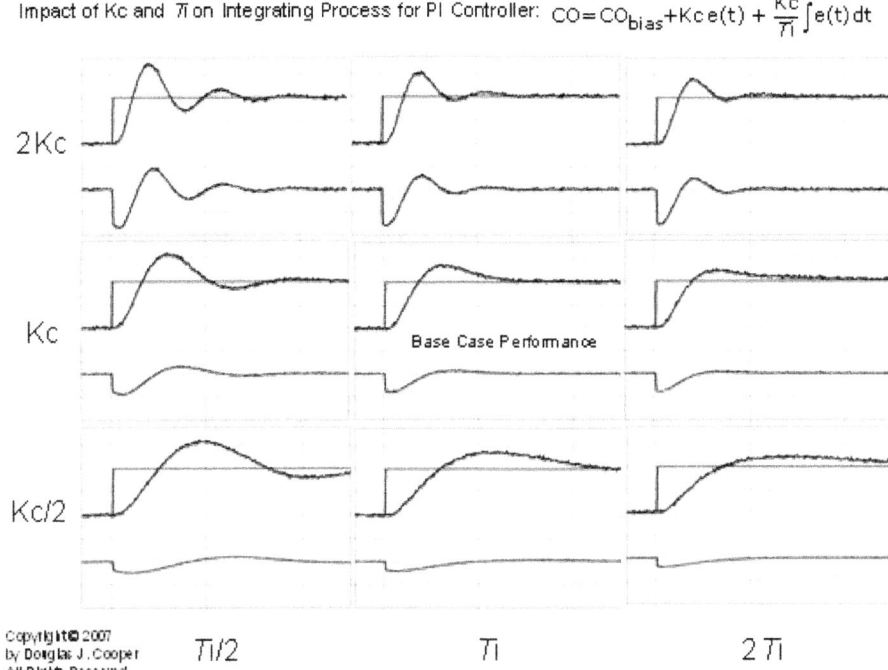

Final Thoughts

The design and tuning recipe for integrating processes provides the above base case performance with minimal testing on our process. In a manufacturing environment where we need a fast solution with minimal disruption, this recipe is certainly one to have in our tool box.

Chapter 10

CASCADE CONTROL FOR IMPROVED DISTURBANCE REJECTION

THE CASCADE CONTROL ARCHITECTURE

Two popular control strategies for improved disturbance rejection performance are cascade control and feed forward with feedback trim.

Improved performance comes at a price. Both strategies require that additional instrumentation be purchased, installed and maintained. Both also require additional engineering time for strategy design, tuning and implementation.

The cascade architecture offers alluring additional benefits such as the ability to address multiple disturbances to our process and to improve set point response performance.

In contrast, the feed forward with feedback trim architecture is designed to address a single measured disturbance and does not impact set point response performance in any fashion.

The Inner Secondary Loop

The dashed line in the block diagram below circles a *feedback control loop*. The only difference is that the words "inner secondary" have been added to the block descriptions. The variable labels also have a "2" after them.

So,

SP2 = Inner secondary set point

CO2 = Inner secondary controller output signal

PV2 = Inner secondary measured process variable signal

And

D2 = Inner disturbance variable (often not measured or available as a signal)

FCE = Final control element such as a valve, variable speed pump or compressor, etc.

Traditional Feedback Control Loop is in the Dashed Circle

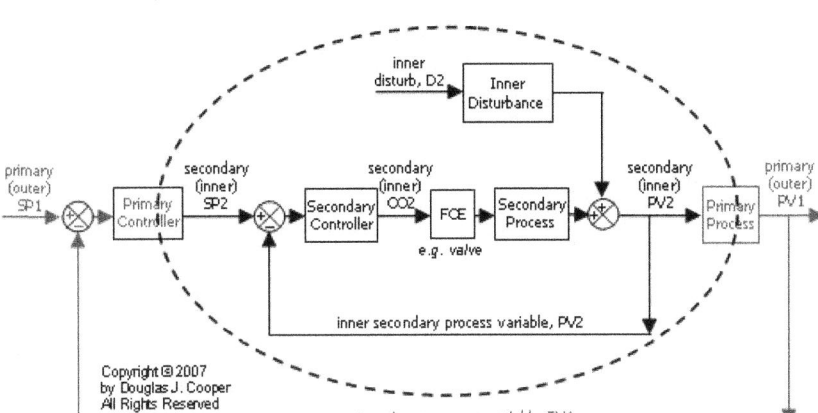

The Nested Cascade Architecture

To construct a cascade architecture, we literally nest the secondary control loop inside a primary loop as shown in the block diagram below.

Note that outer primary PV1 is our process variable of interest in this implementation. PV1 is the variable we would be measuring and controlling if we had chosen a traditional single loop architecture instead of a cascade.

Cascade Structure is a Control Loop within a Control Loop

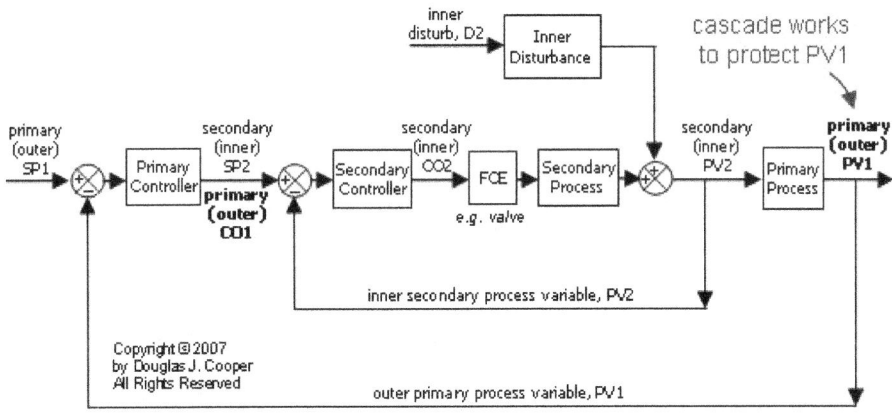

Because we are willing to invest the additional effort and expense to improve the performance response of PV1, it is reasonable to assume that it is a variable important to process safety and/or profitability. Otherwise, it does not make sense to add the complexity of a cascade structure.

Naming Conventions

Like many things in the PID control world, vendor documentation is not consistent. The most common naming conventions we see for cascade (also called nested) loops are :

- Secondary and primary
- Inner and outer
- Slave and master.

Two PVs, Two Controllers, One Valve

Notice from the block diagrams that the cascade architecture has :

- Two controllers (an inner secondary and outer primary controller)
- Two measured process variable sensors (an inner PV2 and outer PV1)
- Only *one* final control element (FCE) such as a valve, pump or compressor.

How can we have two controllers but only one FCE? Because as shown in the diagram above, the controller output signal from the outer primary controller, CO1, becomes the set point of the inner secondary controller, SP2.

The outer loop literally commands the inner loop by adjusting its set point. Functionally, the controllers are wired such that SP2 = CO1 (thus, the master and slave terminology referenced above).

This is actually good news from an implementation viewpoint. If we can install and maintain an inner secondary sensor at reasonable cost, and if we are using a *PLC* or *DCS* where adding a controller is largely a software selection, then the task of constructing a cascade control structure may be reasonably straightforward.

Early Warning is Basis for Success

As shown below, an essential element for success in a cascade design is the measurement and control of an "early warning" process variable.

In the cascade architecture, inner secondary PV2 serves as this early warning process variable. Given this, essential design characteristics for selecting PV2 include that :

- It be measurable with a sensor,
- The same FCE (*e.g.*, valve) used to manipulate PV1 also manipulates PV2,
- The same disturbances that are of concern for PV1 also disrupt PV2, and
- PV2 responds *before* PV1 to disturbances of concern and to FCE manipulations.

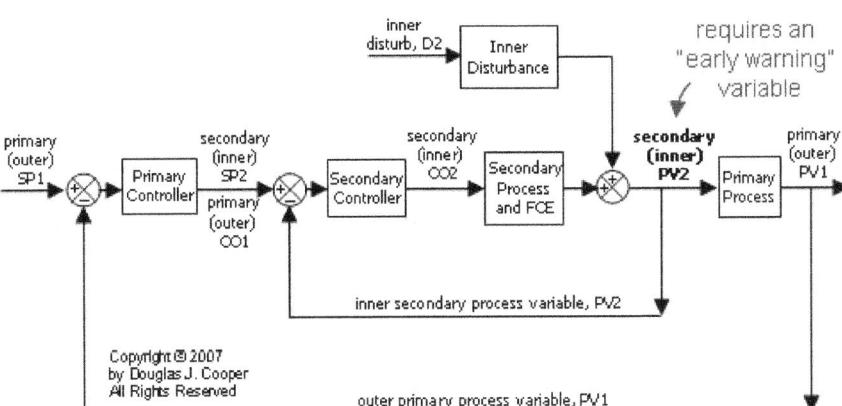

Cascade Control Depends on an Inner "Early Warning" Variable

Since PV2 sees the disruption first, it provides our "early warning" that a disturbance has occurred and is heading toward PV1. The inner secondary controller can begin corrective action immediately. And since PV2 responds first to final control element (*e.g.*, valve) manipulations, disturbance rejection can be well underway even before primary variable PV1 has been substantially impacted by the disturbance.

With such a cascade architecture, the control of the outer primary process variable PV1 benefits from the corrective actions applied to the upstream early warning measurement PV2.

Disturbance Must Impact Early Warning Variable PV2

As shown below, even with a cascade structure, there will likely be disturbances that impact PV1 but do not impact early warning variable PV2.

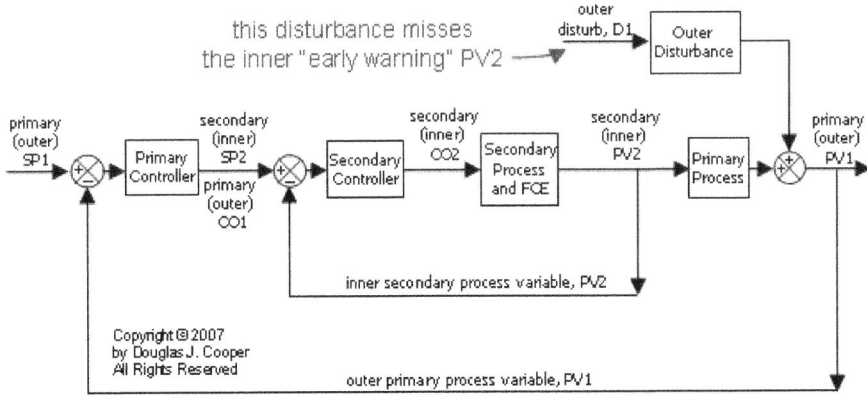

Disturbance Must Hit the Inner PV for Cascade to Provide Benefit

The inner secondary controller offers no "early action" benefit for these outer disturbances. They are ultimately addressed by the outer primary controller as the disturbance moves PV1 from set point.

On a positive note, a proper cascade can improve rejection performance for any of a host of disturbances that directly impact PV2 before disrupting PV1.

An Illustrative Example

To illustrate the construction and value of a cascade architecture, consider the liquid level control process shown below. This is a variation on our *gravity drained tanks*, so hopefully, the behavior of the process below follows intuitively from our previous investigations.

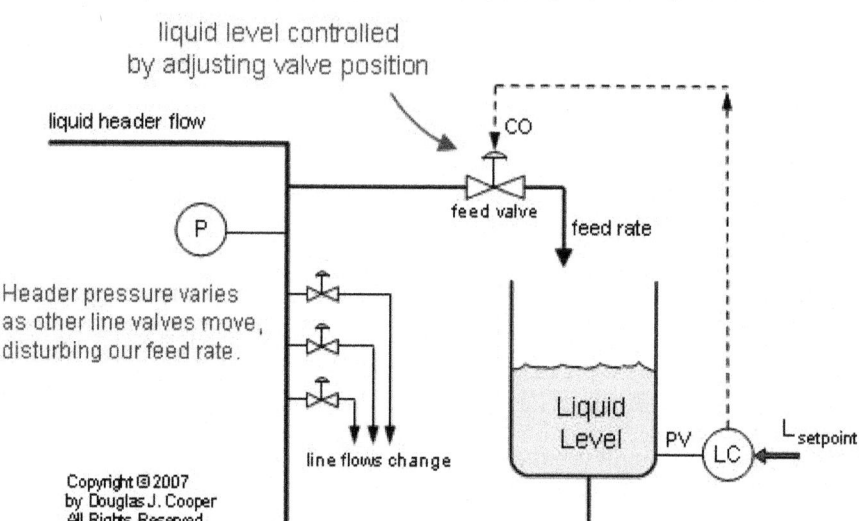

Controlling Liquid Level by Adjusting Feed Flow Rate

As shown above, the tank is essentially a barrel with a hole punched in the bottom. Liquid enters through a feed valve at the top of the tank. The exit flow is liquid draining freely by the force of gravity out through the hole in the tank bottom.

The control objective is to maintain liquid level at set point (SP) in spite of unmeasured disturbances. Given this objective, our measured process variable (PV) is liquid level in the tank. We measure level with a sensor and transmit the signal to a level controller (the LC inside the circle in the diagram).

After comparing set point to measurement, the level controller (LC) computes and transmits a controller output (CO) signal to the feed valve. As the feed valve opens and closes, the liquid feed rate entering the top of the tank increases and decreases to raise and lower the liquid level in the tank.

This "measure, compute and act" procedure repeats every loop sample time, T, as the controller works to maintain tank level at set point.

The Disturbance

The disturbance of concern is the pressure in the main liquid header. As shown in the diagram above, the header supplies the liquid that feeds our tank. It also supplies liquid to several other lines flowing to different process units in the plant.

Whenever the flow rate of one of these other lines changes, the header pressure can be impacted. If several line valves from the main header open at about the same time, for example, the header pressure will drop until its own control system corrects the imbalance. If one of the line valves shuts in an emergency action, the header pressure will momentarily spike.

As the plant moves through the cycles and fluctuations of daily production, the header pressure rises and falls in an unpredictable fashion. And every time the header pressure changes, the feed rate to our tank is impacted.

Problem with Single Loop Control

The single loop architecture in the diagram above attempts to achieve our control objective by adjusting valve position in the liquid feed line. If the measured level is higher than set point, the controller signals the valve to close by an appropriate percentage with the expectation that this will decrease feed flow rate accordingly.

But feed flow rate is a function of two variables :

* Feed valve position, and
* The header pressure pushing the liquid through the valve (a disturbance).

To explore this, we conduct some thought experiments :

Thought Experiment #1 : Assume that the main header pressure is perfectly constant over time. As the feed valve opens and closes, the feed flow rate and thus tank level increases and decreases in a predictable fashion. In this case, a single loop structure provides acceptable level control performance.

Thought Experiment #2 : Assume that our feed valve is set in a fixed position and the header pressure starts rising. Just like squeezing harder on a spray bottle, the valve position can remain constant yet the rising pressure will cause the flow rate through the fixed valve opening to increase.

Thought Experiment #3 : Now assume that the header pressure starts to rise at the same moment that the controller determines that the liquid level in our tank is too high. The controller can be **closing** the feed valve, but because header pressure is rising, the flow rate through the valve can actually be **increasing**.

As presented in Thought Experiment #3, the changing header pressure (a disturbance) can cause a contradictory outcome that can confound the controller and degrade control performance.

A Cascade Control Solution

For high performance disturbance rejection, it is not valve position, but rather, feed flow rate that must be adjusted to control liquid level.

Because header pressure changes, increasing feed flow rate by a precise amount can sometimes mean opening the valve a lot, opening it a little, and because of the changing header pressure, perhaps even closing the valve a bit.

Below is a classic level-to-flow cascade architecture. As shown, an inner secondary sensor measures the feed flow rate. An inner secondary controller receives this flow measurement and adjusts the feed flow valve.

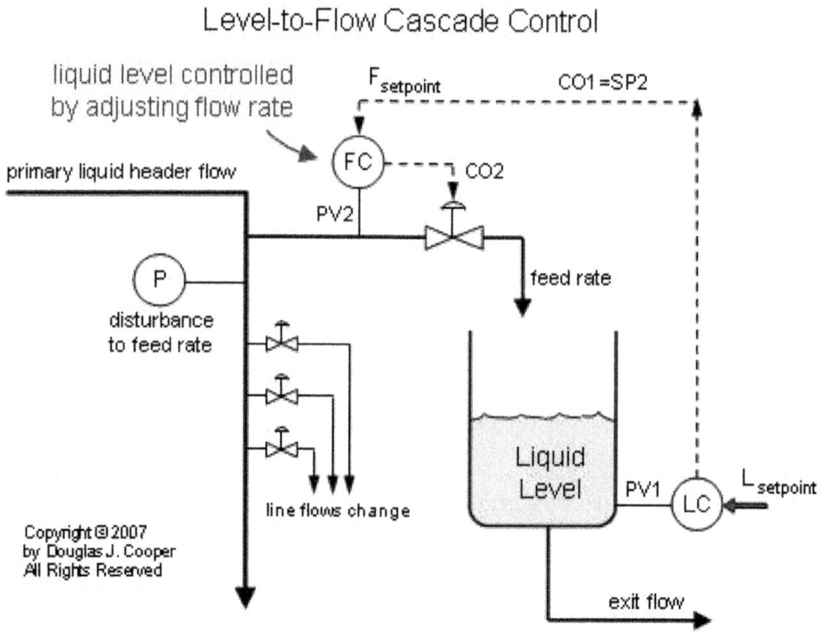

Level-to-Flow Cascade Control

With this cascade structure, if liquid level is too high, the primary level controller now calls for a decreased liquid feed flow rate rather than simply a decrease in valve opening. The flow controller then decides whether this means opening or closing the valve and by how much.

Note in the diagram that, true to a cascade, the level controller output signal (CO1) becomes the set point for the flow controller (SP2).

Header pressure disturbances are quickly detected and addressed by the secondary flow controller. This minimizes any disruption caused by changing header pressure to the benefit of our primary level control process.

The Level-to-Flow Cascade Block Diagram

As shown in the block diagram below, our level-to-flow cascade fits into our block diagram structure. As required, there are :

- Two controllers-the outer primary level controller (LC) and inner secondary feed flow controller (FC)
- Two measured process variable sensors — the outer primary liquid level (PV1) and inner secondary feed flow rate (PV2)
- One final control element (FCE)-the valve in the liquid feed stream.

Formal Level-to-Flow Cascade Structure

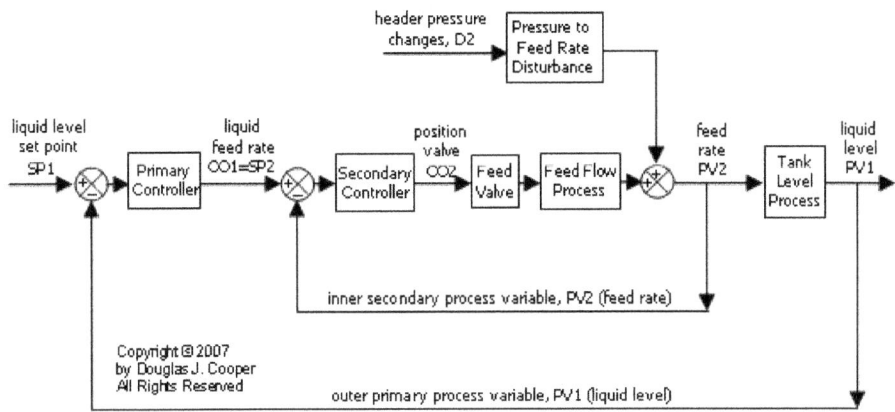

As required for a successful design, the inner secondary flow control loop is nested inside the primary outer level control loop. That is :

- The feed flow rate (PV2) responds *before* the tank level (PV1) when the header pressure disturbs the process or when the feed valve moves.
- The output of the primary controller, CO1, is wired such that it becomes the set point of the secondary controller, SP2.
- Ultimately, level measurement, PV1, is our process variable of primary concern. Protecting PV1 from header pressure disturbances is the goal of the cascade.

Design and Tuning

The inner secondary and outer primary controllers are from the PID family of algorithms. Implementing a cascade builds on many familiar tasks.

There are a number of issues to consider when selecting and tuning the controllers for a cascade. We explore next an *implementation recipe for cascade control.*

AN IMPLEMENTATION RECIPE FOR CASCADE CONTROL

When improved disturbance rejection performance is our goal, one benefit of a *cascade control* (nested loops) architecture over a feed forward strategy is that implementing a cascade builds upon our existing skills.

As shown, a cascade has two controllers. Implementation is a familiar task because the procedure is essentially to employ our *controller design and tuning recipe* twice in sequence.

The Cascade Structure is a Secondary Loop Inside a Primary Loop

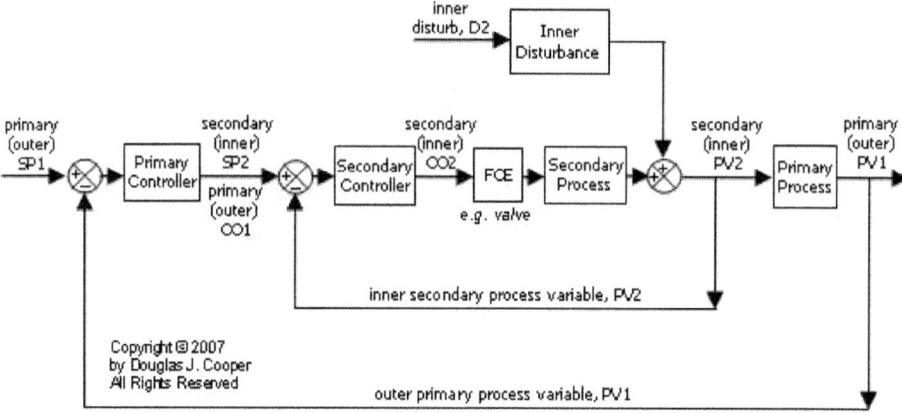

The cascade architecture variables listed in the block diagram include :

$CO2$ = inner secondary controller output signal to the FCE

$PV2$ = the *early warning* inner secondary process variable

$SP2 = CO1$ = inner secondary set point equals the outer primary controller output

$PV1$ = outer primary process variable

$SP1$ = outer primary set point

$D2$ = disturbances that impact the early warning PV2 before they impact PV1

FCE = final control element (*e.g.*, a valve) is continuously adjustable between on/open and off/closed, and that impacts PV2 before it impacts PV1

Two Bump Tests Required

Two bump tests are required to generate the dynamic process response data needed to design and tune the two controllers in a cascade implementation.

- *First the Inner Secondary Controller :* A reliable procedure begins with the outer primary controller in manual mode (open loop) as we apply the design and tuning recipe to the inner secondary controller.

Thus, with our process steady at (or as near as practical to) its *design level of operation (DLO)*, we generate dynamic CO2 to PV2 process response data with either a *manual mode (open loop)* bump test or a more sophisticated *SP driven (closed loop)* bump test.

The objective for the inner secondary controller is timely rejection of disturbances D2 based on the measurement of an "early warning" secondary process variable PV2. Good disturbance rejection performance is therefore of fundamental importance for the inner secondary controller.

Yet as shown in the block diagram, the output signal of the outer primary controller becomes a continually updated set point for the inner secondary controller (SP2 = CO1). Since we expect the inner secondary controller to respond crisply to these rapidly changing set point commands, it must also be tuned to provide good SP tracking performance.

In the perfect world, we would balance disturbance rejection and set point tracking capability for the inner secondary controller. But we cannot shift our attention to the outer primary controller until we have tested and approved the inner secondary controller performance.

In production processes with streams comprised of gases, liquids, powders, slurries and melts, disturbances are often unmeasured and beyond our ability to manipulate at will. So while we desire to balance disturbance rejection and set point tracking performance for the inner secondary controller, in practice, SP tracking tests tend to provide the most direct route to validating inner secondary controller performance.

- *Then the Outer Primary Controller :* Once implemented, the inner secondary controller literally becomes part of the "process" from the outer primary controller's view.

As a result, *any* alteration to the inner secondary controller (*e.g.*, tuning parameter adjustments, algorithm modifications, sample time changes) can change the process gain, Kp, time constant, Tp, and/or dead time, θp, of the outer loop CO1 to PV1 dynamic response behavior. This, in turn, impacts the design and tuning of the outer primary controller.

Thus, we must design, tune, test, accept and then "lock down" the inner secondary controller, leaving it in automatic mode with a fixed configuration. Only then, with our process steady and at (or very near) its DLO, can we proceed with the second bump test to complete the design and tuning of the outer primary controller.

Software Provides Benefit

Given that the outer primary controller design and tuning is based on the specifics of the inner secondary loop, a guess and test approach to a cascaded implementation can prove remarkably wasteful, time consuming and expensive.

In production operations, a *commercial software package* that automates the controller design and tuning tasks will pay for itself as early as a first cascade tuning project.

Minimum Criteria for Success

A successful *cascade implementation requires* that early warning process variable PV2 respond *before* outer primary PV1 both to disturbances of concern (D2) *and* to final control element (FCE) manipulations (*e.g.*, a valve).

Responding first to disturbances means that the inner secondary D2 to PV2 *dead time*, θp, must be smaller than the overall D2 to PV1 dead time, or :

$$\theta p \, (D2 \rightarrow PV2) < \theta p \, (D2 \rightarrow PV1)$$

Responding first to FCE manipulations means that the inner secondary CO2 to PV2 dead time must be smaller than the overall CO2 to PV1 dead time, or :

$$\theta p \, (CO2 \rightarrow PV2) < \theta p \, (CO2 \rightarrow PV1)$$

If these minimum criteria are met, then a cascade control architecture can show benefit in improving disturbance rejection.

P-Only vs. PI for Inner Secondary Controller

A subtle design issue relates to the choice of control algorithm for the inner secondary controller. While perhaps not intuitive, an inner secondary P-Only controller will provide better performance than a PI controller in many cascade implementations.

- *Defining Performance* : We focus all assessment of control performance on outer primary process variable PV1. Performance is "improved" if the cascade structure can more quickly and efficiently minimize the impact of disturbances D2 on PV1. Given the nature of the cascade structure, it is assumed that D2 first disrupts PV2 as it travels to PV1.

  Since PV2 was selected because of its value as an early warning variable, our interest in PV2 control performance extends only to its ability to provide protection to outer primary process variable PV1. We are otherwise unconcerned if PV2 displays offset, shows a large response overshoot, or any other *performance characteristic* that might be considered undesirable in a traditional measured PV.

- *Is the Inner Loop "Fast" Relative to the Overall Process?* : A cascade architecture with a P-Only controller on the inner secondary loop will provide improved disturbance rejection performance over that achievable with a traditional single loop controller if the minimum criteria for success as discussed above are met.

  A PI controller on the inner loop *may* provide even better performance than P-Only, but only if the dynamic character of the inner secondary loop is "fast" relative to that of the overall process.

  If the inner secondary loop dynamic character is not sufficiently fast, then a PI controller on the inner loop, even if properly designed and tuned, will not perform as well as P-Only. It is even possible that a PI controller could degrade performance to an extent that the cascade architecture performs worse than a traditional (non-cascade) single loop controller.

- *Why PI controllers Need a "Fast" Inner Loop* : At every *sample time T*, the outer primary controller computes a controller output signal that is fed as a new set point to the inner secondary controller (CO1 = SP2). The inner secondary controller continually makes CO2 moves as it works to keep PV2 equal to the ever-changing SP2.

The cascade will fail if the inner loop cannot keep pace with the rapid-fire stream of SP2 commands. If the inner secondary controller "falls behind" (or more specifically, if the CO2 actions induce dynamics in the inner loop that do not settle quickly relative to the dynamic behavior of the overall process), the benefit of the early warning PV2 measurement is lost.

A *P-Only controller* can provide energetic control action when tracking set points and rejecting disturbances. Its very simplicity can be a useful attribute in a cascade implementation because a P-Only controller quickly completes its response actions to any control error (E2 = SP2–PV2). While P-Only controllers display *offset* when operation moves from the DLO, this is not considered a performance problem for inner secondary PV2.

With two tuning parameters, a *PI controller* has a greater ability to track set points and reject disturbances. However, this added sophistication yields a controller with a greater tendency to "roll" or oscillate. And the ability to eliminate offset, normally considered a positive attribute for PI controllers, can require a longer series of control actions that extends how quickly a loop settles.

Thus, a PI controller generally needs more time (a faster inner loop) to exploit its enhanced capability relative to that of a P-Only controller. If the dynamic nature of a particular cascade does not provide this time, then an inner-loop P-Only controller is the proper choice.

The inner secondary process dynamics are contained within the overall process dynamics. Therefore, the physics of a cascade implies that the time constant and dead time values for the inner process will not be greater than those of the overall process, or :

$$Tp (CO2 \rightarrow PV2) \leq Tp (CO2 \rightarrow PV1) \text{ and } \theta p (CO2 \rightarrow PV2) \leq \theta p (CO2 \rightarrow PV1)$$

2. Use the time constants to decide whether the inner secondary dynamics are fast enough for a PI controller on the inner loop.

Case 1 : If the inner secondary process is not at least 3 times faster than the overall process, it is not fast enough for a PI controller.

- If $3 \times Tp (CO2 \rightarrow PV2) > Tp (CO2 \rightarrow PV1)$ => Use P-Only controller.

Case 2 : If the inner process is 3 to 5 times faster than the overall process, then P-Only will perform similar to PI control. Use our own preference.

- If $3.Tp (CO2 \rightarrow PV2) \leq Tp (CO2 \rightarrow PV1) \leq 5.Tp (CO2 \rightarrow PV2)$ => Use either P-Only or PI controller.

Case 3 : If the inner process is more than 5 times faster than the overall process, it is "fast" and a PI controller will provide improved performance.

- If $5.Tp (CO2 \rightarrow PV2) < Tp (CO2 \rightarrow PV1)$ => Use PI controller.

4. When we have determine whether the inner secondary controller should be a P-Only or PI algorithm, we tune it and test it. Once acceptable performance has been achieved, leave the inner secondary controller in automatic; it now literally becomes part of the outer primary process.

5. Select an algorithm with integral action for the outer primary controller (*PI, PID* or *PID with CO filter*) to ensure that offset is eliminated. Tune the primary controller using the *design and tuning recipe*. Note that bumping the outer primary process requires stepping, pulse or otherwise perturbing the set point (SP2) of the inner secondary controller.

6. With both controllers in automatic, tuning of the cascade is complete.

Cascade Control of the Jacketed Stirred Reactor

Once a cascade control architecture is put in service, we must remember that every time the inner secondary controller is changed in any way, the outer primary controller should be reevaluated for performance and retuned as necessary.

We should also be aware that the recipe presented above is a general procedure intended for broad application. Thus, there will be occasional exceptions to the rules.

A CASCADE CONTROL ARCHITECTURE FOR THE JACKETED STIRRED REACTOR

Our control objective for the jacketed stirred reactor process is to minimize the impact on reactor operation when the temperature of the liquid entering the cooling jacket changes. We have previously explored *the modes of operation* and *dynamic CO-to-PV behavior* of the reactor. We also have established the performance of a single loop *PI controller* and a *PID with CO Filter controller* in this disturbance rejection application.

Here we consider a *cascade architecture* as a means for improving the disturbance rejection performance in the jacketed stirred reactor.

The Single Loop Jacketed Stirred Reactor

As shown in the process graphic below, the reactor exit stream temperature is controlled by adjusting the flow rate of cooling liquid through an outer shell (or cooling jacket) surrounding the main vessel.

As labelled above for the single loop case :

CO = signal to valve that adjusts cooling jacket liquid flow rate (controller output,%)

PV = reactor exit stream temperature (measured process variable, °C)

SP = desired reactor exit stream temperature (set point, °C)

D = temperature of cooling liquid entering the jacket (major disturbance, °C)

Jacketed Stirred Reactor in Automatic (Closed Loop)

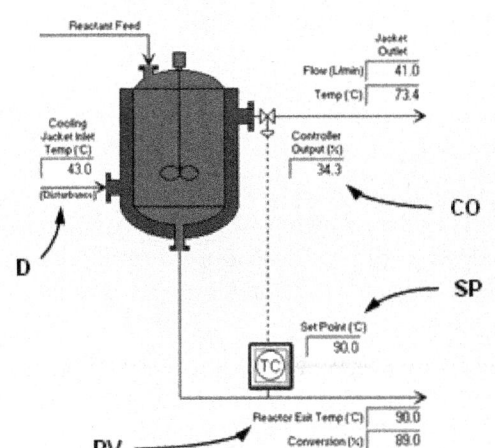

The control objective is to maintain the reactor exit stream temperature (PV) at set point (SP) in spite of unmeasured changes in the temperature of cooling liquid entering the jacket (D).

We measure exit stream temperature with a sensor and transmit the signal to a temperature controller (the TC inside the circle in the diagram). After comparing SP to PV, the temperature controller computes and transmits a CO signal to the cooling jacket liquid flow valve.

As the valve opens and closes, the flow rate of liquid through the jacket increases and decreases. Like holding a hot frying pan under a water faucet, higher flow rates of cooling liquid remove more heat. Thus, a higher flow rate of cooling liquid through the jacket cools the reactor vessel, lowering the reactor exit stream temperature.

Problems with Single Loop Control

The single loop architecture in the diagram above attempts to achieve our control objective by adjusting the flow rate of cooling liquid through the jacket.

If the measured temperature is higher than set point, the controller signals the valve to increase cooling liquid flow by an appropriate percentage with the expectation that this will decrease reactor exit stream temperature accordingly.

The temperature of the cooling liquid entering the jacket (D) can change, sometimes rather quickly. This can disrupt reactor operation as reflected in the measured reactor exit stream temperature PV.

So reactor exit stream temperature PV is a function of two variables :
- Cooling liquid flow rate, and
- The temperature of the cooling liquid entering the cooling jacket (D).

To explore this, we conduct some thought experiments :

Thought Experiment #1 : Assume that the temperature of the cooling liquid entering the jacket (D) is constant over time. If the cooling liquid *flow rate increases* by a certain amount, the reactor exit stream *temperature will decrease* in a predictable fashion (and *vice versa*). Thus, a single loop structure should provide good temperature control performance.

Thought Experiment #2 : Assume that the temperature of cooling liquid entering the jacket (D) starts rising over time. A warmer cooling liquid can carry away less heat from the vessel. If the cooling liquid *flow rate is constant* through the jacket, the reactor will experience less cooling and the exit stream *temperature will increase.*

Thought Experiment #3 : Now assume that the temperature of cooling liquid entering the jacket (D) starts to rise at the same moment that the reactor exit stream temperature moves above set point. The controller will signal for a cooling liquid *flow rate increase*, yet because the cooling liquid temperature is rising, the heat removed from the reactor vessel can actually decrease. Until further corrective action is taken, the reactor exit stream *temperature can increase.*

As presented in Thought Experiment #3, the changing temperature of cooling liquid entering the jacket (a disturbance) can cause a contradictory outcome that can confound the controller and degrade control performance.

Cascade Control Improves Disturbance Rejection

As we established in our study of the *cascade control architecture*, an essential element for success in a cascade (nested loops) design is the measurement and control of an "early warning" process variable, PV2, as illustrated in the block diagram below.

Cascade Control Requires an Inner Secondary "Early Warning" Variable

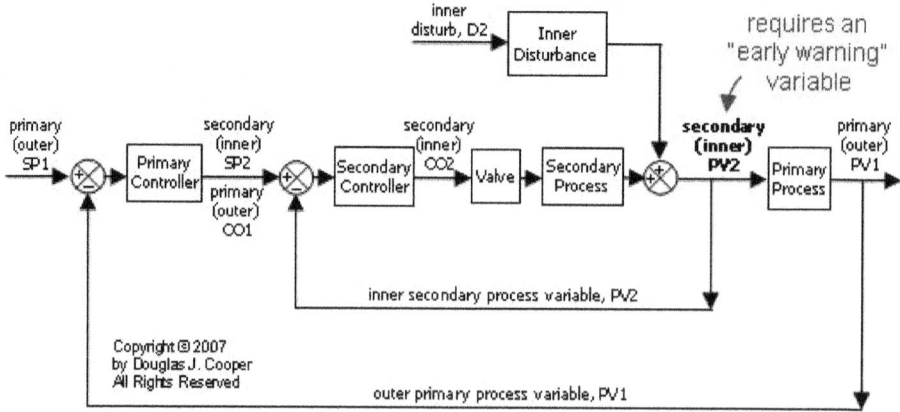

Since disruptions impact PV2 first, it provides our "early warning" that a disturbance is heading toward our outer primary process variable, PV1. The inner secondary controller can begin corrective action immediately. And since PV2 responds first to valve manipulations, disturbance rejection can begin before PV1 has been visibly impacted.

A Reactor Cascade Control Architecture

The thought experiments above highlight that it is problematic to control exit stream temperature by adjusting the cooling liquid flow rate.

An approach with potential for "tighter" control is to adjust the temperature of the cooling jacket itself. This provides a clear process relationship in that, if we seek a higher reactor exit stream temperature, we know we want a higher cooling jacket temperature. If we seek a lower reactor exit stream temperature, we want a lower cooling jacket temperature.

Because the temperature of cooling liquid entering the jacket changes, increasing cooling jacket temperature by a precise amount may mean decreasing the flow rate of cooling liquid a lot, decreasing it a little, and perhaps even increasing the flow rate a bit.

A "*cheap and easy*" proxy for the cooling jacket temperature is the temperature of cooling liquid exiting at the jacket outlet. Hence, we choose this as our inner secondary process variable, PV2, as we work toward the construction of a nested cascade control architecture.

Adding a temperature sensor that measures PV2 provides us the early warning that changes in D, the temperature of cooling liquid entering the jacket, are about to impact the reactor exit stream temperature, PV1.

The addition of a second temperature controller (TC2) completes construction of a jacketed reactor control cascade as shown in the graphic below.

Jacketed Stirred Reactor Cascade Architecture

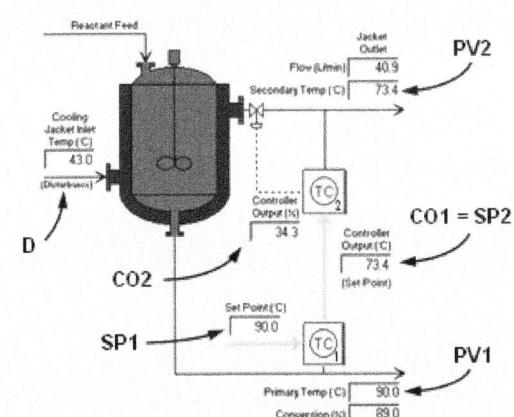

Now, our inner secondary control loop measures the temperature of cooling liquid exiting at the jacket outlet (PV2) and sends a signal (CO2) to the valve adjusting cooling jacket flow rate. The valve increases or decreases the flow rate of cooling liquid if the jacket temperature needs to fall or rise, respectively.

Our outer loop maintains reactor exit stream temperature (our process variable of primary interest and concern) as PV1. Note in the graphic above that the controller output of our primary controller, CO1, becomes the set point of our inner secondary controller, SP2.

If PV1 needs to rise, the primary controller signals a higher set point for the jacket temperature (CO1 = SP2). The inner secondary controller then decides if this means opening or closing the valve and by how much.

Thus, variations in the temperature of cooling liquid entering the jacket (D) are addressed quickly and directly by the inner secondary loop to the benefit of PV1.

The cascade architecture variables are identified on the above graphic and listed below :

PV2 = cooling jacket outlet temperature is our "early warning" process variable (°C)

CO2 = controller output to valve that adjusts cooling jacket liquid flow rate (%)

SP2 = CO1 = desired cooling jacket outlet temperature (°C)

PV1 = reactor exit stream temperature (°C)

SP1 = desired reactor exit stream temperature (°C)

D = temperature of cooling liquid entering the jacket (°C).

The inner secondary PV2 (cooling jacket outlet temperature) is a proper early warning process variable because :

• PV2 is measurable with a temperature sensor.
• The same valve used to manipulate PV1 also manipulates PV2.
• The same disturbance that is of concern for PV1 also disrupts PV2.
• PV2 responds before PV1 to the disturbance of concern and to valve manipulations.

Reactor Cascade Block Diagram

The jacketed stirred reactor block diagram for this nested cascade architecture is shown below :

As expected for a nested cascade, this architecture has :

▪ Two controllers (an inner secondary and outer primary controller)
▪ Two measured process variable sensors (an inner PV2 and outer PV1)
▪ Only one valve (to adjust cooling liquid flow rate)

Jacketed Stirred Reactor Cascade Block Diagram

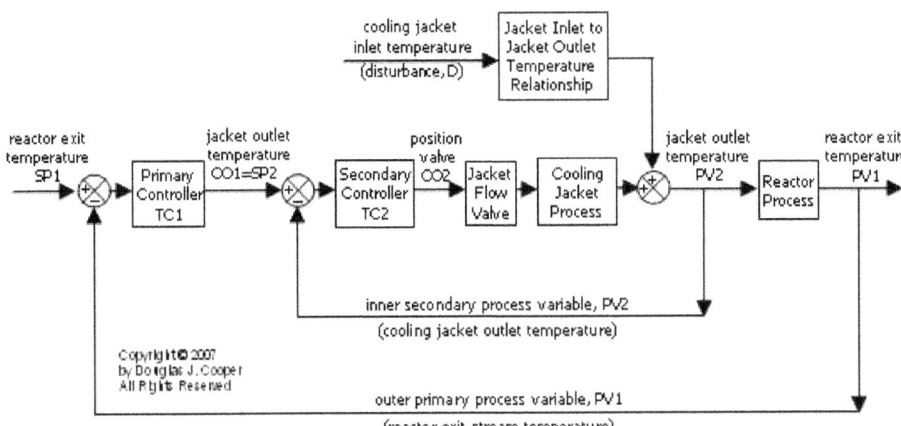

Tuning a Cascade

With a cascade architecture established, we apply our *implementation recipe for cascade control* and *explore the disturbance rejection capabilities* of this structure.

CASCADE DISTURBANCE REJECTION IN THE JACKETED STIRRED REACTOR

Our control objective for the jacketed stirred reactor is to maintain reactor exit stream temperature at set point in spite of disturbances caused by a changing cooling liquid temperature entering the *vessel jacket*.

We also have proposed a *cascade control architecture* for the reactor that offers potential for improving disturbance rejection performance. We now apply our proposed architecture following the *implementation recipe for cascade control*. Our goal is to demonstrate the implementation procedure, understand the benefits and drawbacks of the method, and explore cascade disturbance rejection performance for this process.

The Reactor Cascade Control Architecture

The reactor cascade architecture has been shown in the graphic below :

where :

$CO2$ = controller output to valve that adjusts cooling jacket liquid flow rate (%)

$PV2$ = cooling jacket outlet temperature is our "early warning" process variable (°C)

$SP2 = CO1$ = desired cooling jacket outlet temperature (°C)

$PV1$ = reactor exit stream temperature (°C)

SP1 = desired reactor exit stream temperature (°C)

D = temperature of cooling liquid entering the jacket (°C)

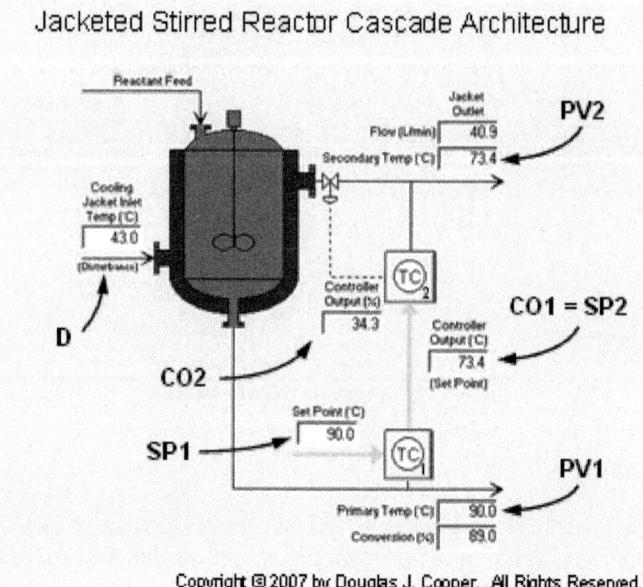

Jacketed Stirred Reactor Cascade Architecture

Design Level of Operation (DLO)

The details and discussion for DLO are *presented* and are summarized :

- Design PV1 and SP1 = 90°C with approval for brief dynamic testing of ±2°C
- Design D = 43°C with occasional spikes up to 50°C.

Minimum Criteria for Success

A *successful cascade implementation* requires that early warning process variable PV2 respond *before* outer primary PV1 both to changes in the jacket cooling liquid temperature disturbance (D) *and* to changes in inner secondary controller output signal CO2. The plots below verify that both of these criteria are met with the architecture shown in the graphic above.

Expressed concisely, the plots show that the delay in response (or *dead time*, θp) follows the rule :

$$\theta p \ (D \rightarrow PV2) < \theta p \ (D \rightarrow PV1) \ \text{and} \ \theta p \ (CO2 \rightarrow PV2) < \theta p \ (CO2 \rightarrow PV1)$$

Thus, a cascade control architecture should improve disturbance rejection performance relative to a single loop architecture.

Verify That PV2 Responds Before PV1 To:

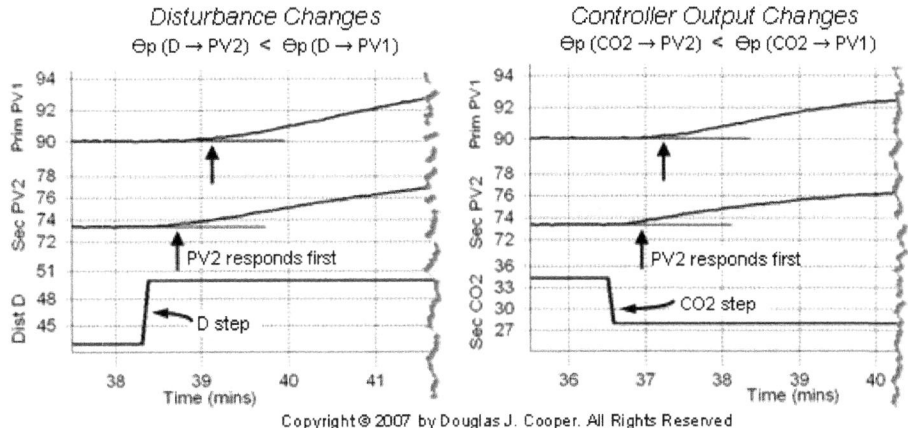

P-Only vs. PI for Inner Secondary Controller

The cascade implementation recipe first helps us decide if the inner secondary controller is fast enough for a PI algorithm or if it is better suited for a P-Only algorithm. The decision is based on the size of the inner secondary time constant relative to that of the overall process time constant.

To compute these time constants, we need to bump the process and analyze the dynamic process response data. In this example we choose to place both inner secondary and outer primary controllers in manual mode (open loop) during the bump test. An alternative not explored here is to have the inner secondary controller in automatic mode with tuning *sufficiently active* to force a clear dynamic response in PV2.

Following the steps of the *cascade implementation recipe* :

1. With the process steady at the *design level of operation* (DLO), we perform a *doublet test*, choosing here to move CO2 from 34.3% → 39.3% → 29.3% → 34.3%. We record both PV2 and PV1 dynamic data as the process responds. We use this "step 1 data set" as we proceed with the implementation recipe.

2. As shown in the plots below, we use *commercial software* to fit a first order plus dead time (FOPDT) model to the inner secondary process (CO2 → PV2) dynamic data and another to the overall process (CO2 → PV1) dynamic data :

   The FOPDT model parameters from these bump tests are summarized :

	Inner Secondary *(CO2 → PV2)*	*Overall Process* *(CO2 → PV1)*
Process gain, Kp =	−0.57 °C/%	−0.51 °C/%
Time constant, T_p =	2.3 min	2.2 min
Dead time, θ_p =	0.25 min	0.81 min

Inner Secondary and Overall Process Bumps Tests

CO2 to PV2 Bump Test Dynamics
Model: First Order Plus Dead Time (FOPDT)   File Name: doublet test.txt

Gain (K) = -0.57, Time Constant (min) = 2.3, Dead Time (min) = 0.25
Goodness of Fit: R-Squared = 0.996, SSE = 13.95

CO2 to PV1 Bump Test Dynamics
Model: First Order Plus Dead Time (FOPDT)   File Name: doublet test.txt

Gain (K) = -0.51, Time Constant (min) = 2.2, Dead Time (min) = 0.81
Goodness of Fit: R-Squared = 0.9972, SSE = 8.022

The *cascade block diagram* implies that the time constant and dead time values for the inner process are contained within and contribute to the dynamics of the overall process. Thus, we can surmise that :

$$Tp\,(CO2 \rightarrow PV2) \leq Tp\,(CO2 \rightarrow PV1) \text{ and } \theta p\,(CO2 \rightarrow PV2) \leq \theta p\,(CO2 \rightarrow PV1)$$

Yet the values in the table above indicate that this seemingly fundamental relationship does not hold true for the jacketed stirred reactor. In fact, when using a FOPDT model approximation of the dynamic response data, the time constant of the inner secondary process is slightly larger (longer) than that of the overall process.

The process graphic shows a main reactor vessel with a large volume of liquid relative to that of the cooling jacket. The function of the cooling jacket is to remove heat energy to regulate reactor vessel temperature. It seems logical that because of the volume differences, as long as the liquid temperature in the large reactor is changing, the temperature of the liquid in the small cooling jacket must follow. And the liquid in the reactor acts as a heat source because the *chemical reaction inside the vessel generates heat energy faster* as reactor temperature rises (and *vice versa*).

This intertwined relationship of heat generation and removal combined with relative sizes of the reactor and jacket offers one physical rationalization as to why the jacket (inner secondary) Tp might reasonably be longer than that of the vessel (overall process) Tp.

A simpler explanation is that the sensor used to measure temperature in the cooling jacket outlet flow was improperly specified at the time of purchase, and unfortunately, it responds slowly to actual cooling liquid changes. This additional response time alone could account for the observed behavior.

In any case, the dead time of the overall process is three times that of the inner secondary controller, and thus, PV2 provides a clear early warning that we can exploit for a cascade design. So while the simple cascade recipe has limitations that require our judgment, it provides valuable insights that enable us to proceed.

3. The *cascade implementation recipe* uses a time constant comparison to decide whether the inner secondary loop is fast enough for a PI controller. We reach a true statement with case 1 of the decision tree in step 3 of the recipe. That decision is :

if $3 \times Tp$ (CO2 → PV2) > Tp (CO2 → PV1) => Use P-Only controller

or using the parameters in the table above :

if 3(2.3 min) > 2.2 min => Use a P-Only controller

Inner Secondary Controller

4. Our cascade implementation recipe states that a P-Only algorithm is the best choice for the inner secondary controller. To explore this decision, we run four trials and compare P-Only and PI algorithms side-by-side. We are able to use the same "step 1 data set" for the design and tuning of all four of these inner secondary controllers :

TRIAL 1 : moderate P-Only

TRIAL 2 : aggressive P-Only

TRIAL 3 : moderate PI

TRIAL 4 : aggressive PI.

As listed in the table in step 2 above, a first order plus dead time (FOPDT) model fit of the "step 1 data set" collected around our design level of operation (DLO) yields inner secondary FOPDT model parameters :

CO2 → PV2 model : Kp = –0.57 °C/%; Tp = 2.3 min; θp = 0.25 min.

Following our *controller design and tuning recipe*, we use these FOPDT model parameters in rules and correlations to complete the secondary controller design.

P-Only Controller design and tuning, including the use of the ITAE tuning correlation for computing controller gain, Kc, from FOPDT model parameters is *summarized in the example*. Following those details :

P-Only algorithm : $CO = CO_{bias} + Kc \cdot e(t)$

TRIAL 1 : Moderate P-Only

$$Kc = \frac{0.2}{Kp}\left(\frac{Tp}{\theta p}\right)^{1.22} = \frac{0.2}{-0.57}\left(\frac{2.3}{0.25}\right)^{1.22} = -5.3\% / °C$$

TRIAL 2 : Aggressive P-Only

Kc = 2.5 (Moderate Kc) = 2.5 (–5.3) = –13.3%/°C

PI Controller design and tuning, including computing controller gain, Kc, and reset time, Ti, from FOPDT model parameters is *summarized in the example*. Following those details :

PI algorithm : $CO = CO_{bias} + Kc \cdot e(t) + \frac{Kc}{Ti}\int e(t)dt$

TRIAL 3 : Moderate PI

Moderate Tc = the larger of $1 \cdot Tp$ or $8 \cdot \theta p$

$\qquad\qquad$ = larger of 1(2.3 min) or 8(0.25 min)

$\qquad\qquad$ = 2.3 min.

$$Kc = \frac{1}{Kp} \frac{Tp}{(\theta p + Tc)} = \frac{1}{-0.57} \frac{2.3}{(0.25 + 2.3)} = -1.6\% / °C$$

Ti = Tp = 2.3 min

TRIAL 4 : Aggressive PI

Aggressive Tc = the larger of 0.1·Tp or 0.8·θp

$\qquad\qquad$ = larger of 0.1(2.3 min) or 0.8(0.25 min)

$\qquad\qquad$ = 0.23 min

$$Kc = \frac{1}{Kp} \frac{Tp}{(\theta p + Tc)} = \frac{1}{-0.57} \frac{2.3}{(0.25 + 0.23)} = -8.4\% / °C$$

Ti = Tp = 2.3 min.

The "step 1 data set" model parameters and the four inner secondary controllers designed from this data are summarized in the upper half of the table below :

Inner Secondary CO2 Bump Test : from 34.39% → 39.3% → 29.3% → 34.3%				
FOPDT CO2 → PV2 Model : Kp = -0.57°C / %; Tp = 2.3 min; θp = 0.25 min				
	Trial 1	Trial 2	Trial 3	Trial 4
Kc [=]%/°C Ti [=] min	Moderate P-Only Kc = - 5.3	Aggressive P-Only Kc = -13.3	Moderate PI Kc = -1.6; Ti = 2.3	Aggressive PI Kc = -8.4; Ti = 2.3
Outer Primary FOPDT CO1 = SP2 → PV1 Models using above inner controllers				
Kp [=]°C/°C	0.70	0.82	0.90	0.93
Tp (min)	0.61	0.43	2.2	0.50
θp (min)	0.63	0.48	0.82	0.62
Outer Primary PI Controllers with moderate tuning using above FOPDT models				
Kc[=]°C/°C Ti[=] min	Moderate PI Kc = 0.15 Ti = 0.61	Moderate PI Kc = 0.12 Ti = 0.43	Moderate PI Kc = 0.33 Ti = 2.2	Moderate PI Kc = 0.10 Ti = 0.50

Outer Primary Controller

5. We normally would implement one inner secondary controller and test it for
    acceptable performance. Here, we "accept" each of the four trial controllers
    and turn our attention to the outer primary loop.

    The four inner secondary controllers are thus implemented one after the other
in a series of studies. When in automatic mode, each literally becomes part of the
overall process. And because each is different, we must perform four separate
bump tests and compute four sets of FOPDT model parameters to describe the
four different outer primary (overall) dynamic process behaviors.

    Recall that the output of the outer primary controller output, CO1, becomes
the set point of the inner secondary controller, SP2. Thus, bumping CO1 is the
same as bumping SP2 (*i.e.*, CO1=SP2).

With each of the four inner secondary trial controllers in automatic, we choose to bump CO1=SP2 from 73.4 °C → 76.4 °C → 70.4 °C → 73.4 °C. We again use *commercial software* to fit a FOPDT dynamic model to the CO1=SP2 → PV2 dynamic response data sets as shown in the plots below. The FOPDT model parameters (Kp, Tp, θp) from each bump test fit are summarized in the table above.

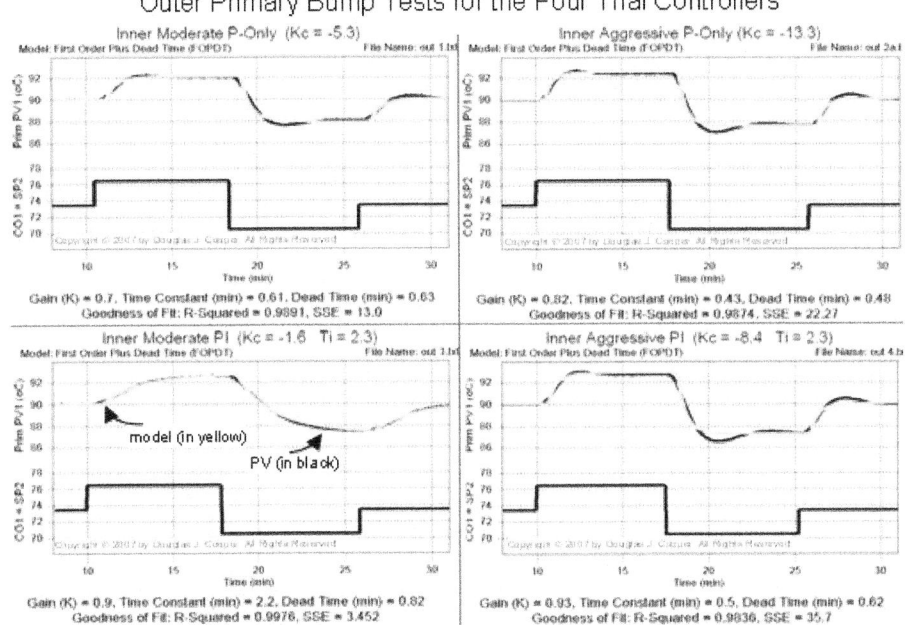

Outer Primary Bump Tests for the Four Trial Controllers

We find that industry practitioners, when designing a strategy for an application as challenging as reactor temperature control, generally seek a moderate response performance. Integral action will always be included to eliminate offset issues. Thus, we pair each inner secondary trial controller with a moderately tuned outer primary PI controller.

Following the identical procedure detailed in trial 3 of step 4 above, we compute four sets of moderate PI controller tuning parameters (one for each inner secondary controller). These are listed in the last row of the above table.

Compare Disturbance Rejection Performance

6. With both controllers in automatic, design and implementation of the cascade is complete. The objective of this cascade is to minimize the disruption to primary process variable PV1 when disturbance D changes.

    The specific D of concern in this study is that the temperature of cooling liquid entering the jacket, normally at 43°C, is known to spike occasionally to 50°C.

    The lower trace in the plot below shows disturbance steps (temperature changes) from 43°C up to 50°C and back. There are four trials shown, one for each of the inner secondary and outer primary controller pairs listed in the table above.

The middle portion of the plot shows the constantly moving inner secondary SP2=C01 in gold and the ability of early warning PV2 to track this ever-changing set point in black.

Recall that offset for early warning variable PV2 is not a concern in many cascade implementations. Here, for example, PV2 is cooling liquid temperature and cooling liquid is not a product destined for market. So our central focus is on how control actions based on this early warning variable help us minimize disturbance disruptions to PV1 and not on how precisely PV2 tracks set point.

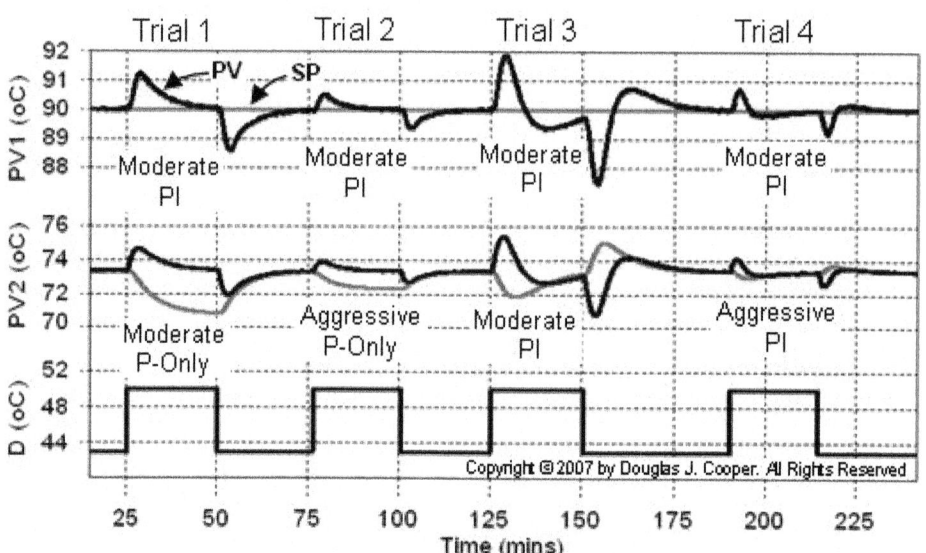

Our success is shown on the upper portion of the plot. The outer primary set point, SP1 (in gold) remains constant throughout the study. Our interest is in the ability of each of the cascade implementations to maintain reactor exit temperature PV1 at SP1 in spite of the abrupt disturbance changes.

Some Observations

In general, a more successful (or "better") cascade performance is one where :

- There is a smaller maximum deviation from set point during the disturbance, and
- PV1 most rapidly returns to and settles out at set point after a disturbance.
- Trials 2 and 4 both have aggressively tuned inner secondary controllers, and these two implementations both have the smallest deviations from set point and settle most rapidly back to SP. This supports the notion that *inner secondary controllers should energetically attack early warning PV2 disruptions* for best cascade performance.

• While an aggressively tuned inner secondary P-Only and PI controller (trials 2 and 4) performed with similar success, the moderately tuned inner secondary PI controller (trial 3) displayed markedly degraded performance. This high sensitivity to inner loop PI tuning strengthens the "use P-Only" conclusion made in step 3 of our cascade implementation recipe.

Comparing Cascade to Single Loop

The central question is whether the extra effort associated with cascade control provides sufficient payoff in the form of improved disturbance rejection performance.

To the left in the plot below is the performance of our trial 2 cascade implementation. The performance of an aggressively tuned single loop PI controller in rejecting the same disturbance is shown to the right.

Cascade vs Single Loop Disturbance Rejection Performance

The cascade architecture reduces the maximum deviation from SP during the disturbance from ±2°C for the single loop controller down to ±0.5°C for the cascade. Settling time is shortened from about 10 minutes for the single loop controller down to about 8 minutes for the cascade.

If the financial return from such improved performance is greater than the cost to install and maintain the cascade, then choose cascade control.

Set Point Tracking Performance

While not our design objective, presented below is the set point tracking performance of the four cascade implementations :

Cascade control is best suited for improved disturbance rejection. As shown above, its impact on set point tracking performance is minimal. While one might argue that our "best" cascade design for disturbance rejection (trial 2) also provides the most rapid SP tracking response, this same improvement can be obtained with more aggressive tuning of a single loop PI controller.

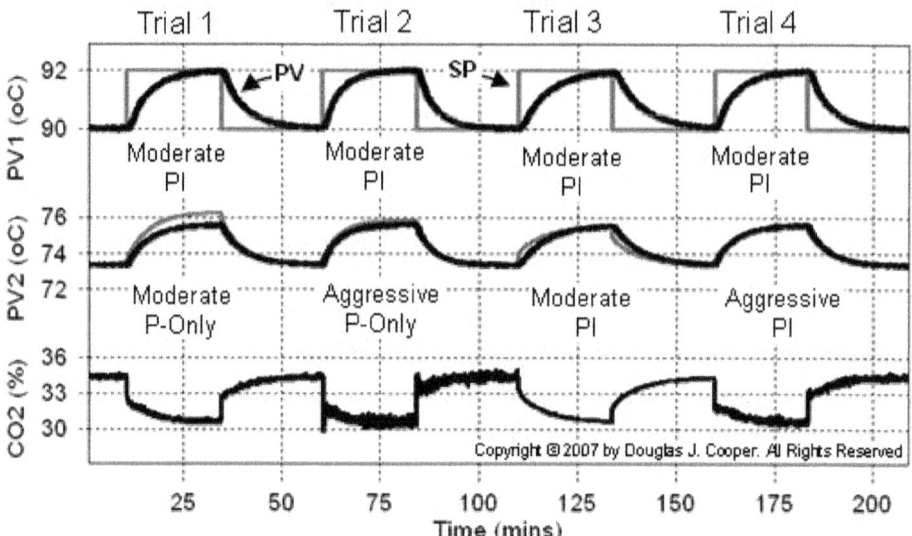

Chapter 11

Feed Forward with Feedback Trim for Improved Disturbance Rejection

THE FEED FORWARD CONTROLLER

The most popular architectures for improved disturbance rejection performance are *cascade control* and the "feed forward with feedback trim" architecture introduced below.

Like cascade, feed forward requires that additional instrumentation be purchased, installed and maintained. Both architectures also require additional engineering time for strategy design, implementation and tuning.

Cascade control will have a small impact on set point tracking performance when compared to a traditional single-loop feedback design and this may or may not be considered beneficial depending on the process application. The feed forward element of a "feed forward with feedback trim" architecture does not impact set point tracking performance in any way.

Feed Forward Involves a Measurement, Prediction and Action

Consider that a process change can occur in another part of our plant and an identifiable series of events then leads that "distant" change to disturb or disrupt our measured process variable, PV.

The *traditional PID controller* takes action only when the PV has been moved from set point, SP, to produce a controller error, $e(t) = SP–PV$. Thus, disruption to stable operation is already in progress before a feedback controller first begins to respond. From this view, a feedback strategy simply starts too late and at best can only work to minimize the upset as events unfold.

In contrast, a feed forward controller measures the disturbance, D, while it is still distant. As shown below, a feed forward element receives the measured

D, uses it to predict an impact on PV, and then computes preemptive control actions, $CO_{feed forward}$, that counteract the predicted impact as the disturbance arrives. The goal is to maintain the process variable at set point (PV = SP) throughout the disturbance event.

Feed Forward With Feedback Trim Architecture

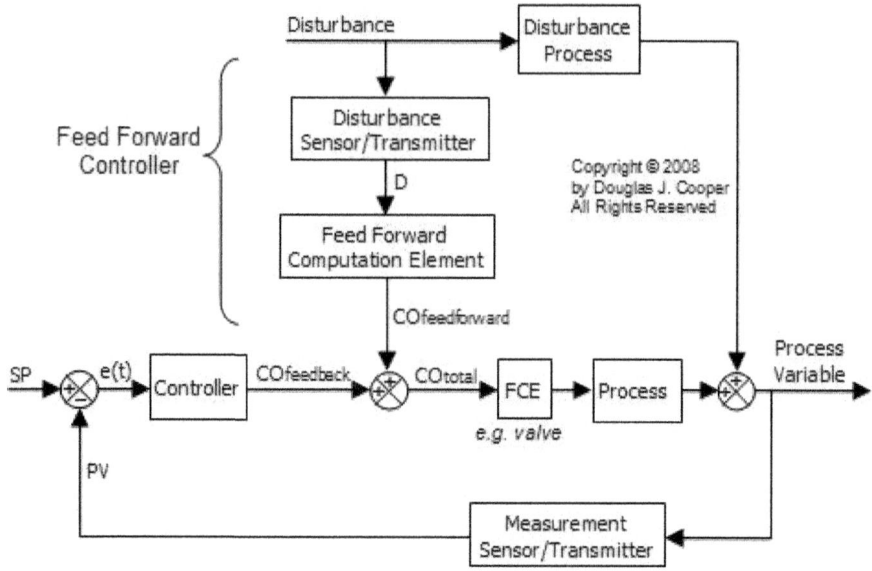

where :

CO = controller output signal

D = measured disturbance variable

$e(t)$ = controller error, SP–PV

FCE = final control element (e.g., valve, variable speed pump or compressor)

PV = measured process variable

SP = set point.

To appreciate the additional components associated with a feed forward controller, we can compare the above to the previously discussed traditional *feed back control loop block\diagram*.

When to Consider Cascade Control

The *cascade architecture requires* that an "early warning" secondary measured process variable, PV2, be identified that is inside (responds before) the primary measured process variable, PV1. Essential elements for success include that :

- PV2 is measurable with a sensor.
- The same final control element (FCE) used to manipulate PV1 also manipulates PV2.

- The same disturbances that are of concern for PV1 also disrupt PV2.
- PV2 responds *before* PV1 to disturbances of concern and to FCE manipulations.

One benefit of a cascade architecture is that it uses two traditional controllers from the PID family, so implementation is a familiar task that builds upon our existing skills. Also, cascade control will help improve the rejection of *any* disturbance that first disrupts the early warning variable, PV2, prior to impacting the primary process variable, PV1.

When to Consider Feed Forward with Feedback Trim

Feed forward anticipates the impact of a measured disturbance on the PV and deploys control actions to counteract the impending disruption in a timely fashion. This can significantly improve disturbance rejection performance, but only for the particular disturbance variable being measured.

Feed forward with feedback trim offers a solution for improved disturbance rejection if no practical secondary process variable, PV2, can be established (*i.e.*, a process variable cannot be located that is measureable, provides an early warning of impending disruption, and responds first to FCE manipulations).

Feed forward also has value if our concern is focused on one specific disturbance that is responsible for repeated, costly disruptions to stable operation. To provide benefit, the additional measurement must reveal process disturbances *before* they arrive at our PV so we have time to compute and deploy preemptive control actions.

The Feed-forward-only Controller

Pure feed-forward-only controllers are rarely found in industrial applications where the process flow streams are composed of gases, liquids, powders, slurries or melts.

The architecture of a feed-forward-only controller for the heat exchanger is illustrated below :

The PV to be controlled is the exit temperature on the tube side of the exchanger. To regulate this exit temperature, the CO signal adjusts a valve to manipulate the flow rate of cooling liquid on the shell side. A side stream of warm liquid combines with the hot liquid entering the exchanger and acts as a measured disturbance, D, to our process.

Because there is no feedback of a PV measurement in our controller architecture, feed-forward-only presents the interesting notion of open loop control. As such, it does not have a tendency to induce oscillations in the PV as can a poorly tuned feedback controller.

If we could mathematically describe how each change in D impacts PV (D → PV) and how each change in CO impacts PV (CO → PV), then we could develop a math model that predicts what manipulations to make in CO to maintain PV at set point whenever D changes.

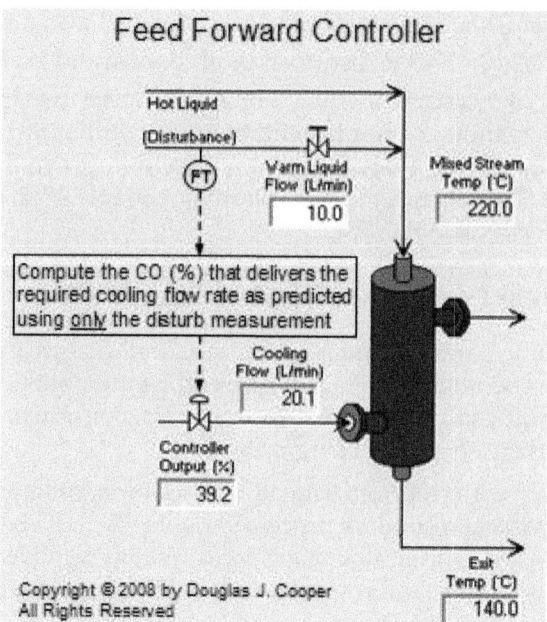

But this would only be true if :

- We have perfect understanding of the D → PV and CO → PV dynamic relationships,
- We can describe these perfect dynamic relationships mathematically,
- These relationships never change,
- There are no other unmeasured disturbances impacting PV, and
- Set point, SP, is always held constant.

The reality, however, is that with only a single measured D, a predictive model cannot account for many phenomena that impact the D → PV and CO → PV behavior. These may include changes in :

- The temperature and flow rate of the hot liquid feed that mixes with our warm disturbance stream on the tube side,
- The temperature of the cooling liquid on the shell side,
- The ambient temperature surrounding the exchanger that drives heat loss to the environment,
- The shell/tube heat transfer coefficient due to corrosion or fouling, and
- Valve performance and capability due to wear and component failure.

Since all of the above are unmeasured, a fixed or stationary model cannot account for them when it computes control action predictions. Installing additional sensors and enhancing the feed forward model to account for each would improve performance but would lead to an expensive and complex architecture. And since there are more potential disturbances and external influences then those listed above, that still would not be sufficient.

This highlights that feed-forward-only control is problematic and should only be considered in rare instances. One situation where it may offer value is if a PV critical to process operation simply cannot be measured or inferred using currently available technology. Feed-forward-only control, in spite of its weaknesses and pitfalls, then offers some potential for improved operation.

Feed Forward with Feedback Trim

The "feed forward with feedback trim" control architecture is the solution widely employed in industrial practice. It balances the capability of a feed forward element to take preemptive control actions for *one* particularly disruptive disturbance while permitting a traditional feedback control loop to :

- Reject all other disturbances and external influences that are not measured,
- Provide set point tracking capability, and
- Correct for the inevitable simplifying approximations in the predictive model of the feed forward element that make preemptive disturbance rejection imperfect.

A feed forward with feedback trim control architecture for the heat exchanger process is shown below :

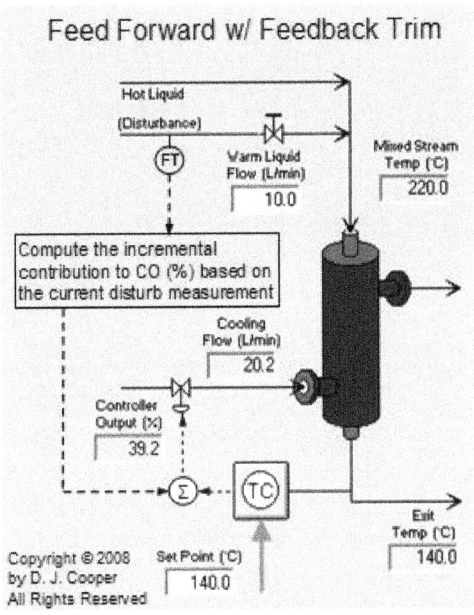

Feed Forward w/ Feedback Trim

To construct the architecture, a feedback controller is first implemented and tested following our controller *design and tuning recipe* as if it were a stand-alone entity. The feed forward controller is then designed based on our *understanding of the relationship* between the D → PV and CO → PV variables. This is generally expressed as a math function that can range in complexity from a simple multiplier to complex differential equations.

With the architecture completed, the disturbance flow is measured and passed to a feed forward element that is essentially a combination disturbance/ process model. The model uses changes in D to predict an impact on PV, and then computes control actions, $CO_{feed\ forward}$ to compensate for the predicted impact.

Conceptual Feed Forward with Feedback Trim Diagram

As shown in the first block diagram above, the $CO_{feed\ forward}$ control actions are combined with $CO_{feed\ back}$ to create an overall control action, CO_{total}, to send to the final control element (FCE).

To illustrate the control strategy in a more tangible fashion, we present the feed forward with feedback trim architecture in a conceptual diagram below :

Feed Forward with Feedback Trim Conceptual Diagram

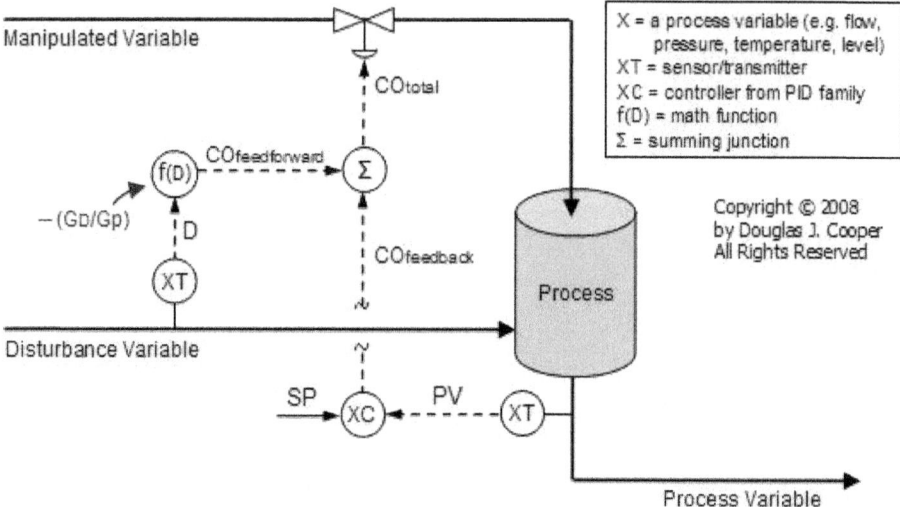

The more accurate the feed forward math function is in computing control actions that will counteract changes in the measured disturbance in a timely fashion, the less impact those disturbances will have on our measured process variable.

Practitioner's note : a potential application of feed forward control exists if we hear an operator say something like, "Every time event *A* happens, process variable *X* is upset. I can usually help the *X* controller by switching to manual mode and moving the controller output." If the variable associated with event *A* is already being measured and logged in the process control system, sufficient data is likely available to allow the implementation of a feed forward element to our feedback controller.

Improved disturbance rejection performance comes at a price in terms of process engineering time for model development and testing, and instrument engineering time for control logic programming. Like all projects, such investment decisions are made on the basis of cost and benefit.

FEED FORWARD USES MODELS WITHIN THE
CONTROLLER ARCHITECTURE

Both "feed forward with feedback trim" and *cascade control* can provide improved disturbance rejection performance. They have different architectures, however, and *choosing between the two* depends on our specific control objective and the ability to obtain certain process measurements.

Feed Forward and Feedback Trim are Largely Independent

As illustrated below, the feed forward with feedback trim architecture is constructed by coupling a *feed-forward-only* controller to a traditional feedback controller.

The feed forward controller seeks to reject the impact of one specific disturbance, D, that is measured *before* it reaches our primary process variable, PV, and starts its disruption to stable operation. Typically, this D is one that has been identified as causing repeated and costly upsets, thus justifying the expense of both installing a sensor to measure it, and developing and implementing the feed forward computation element to counteract it.

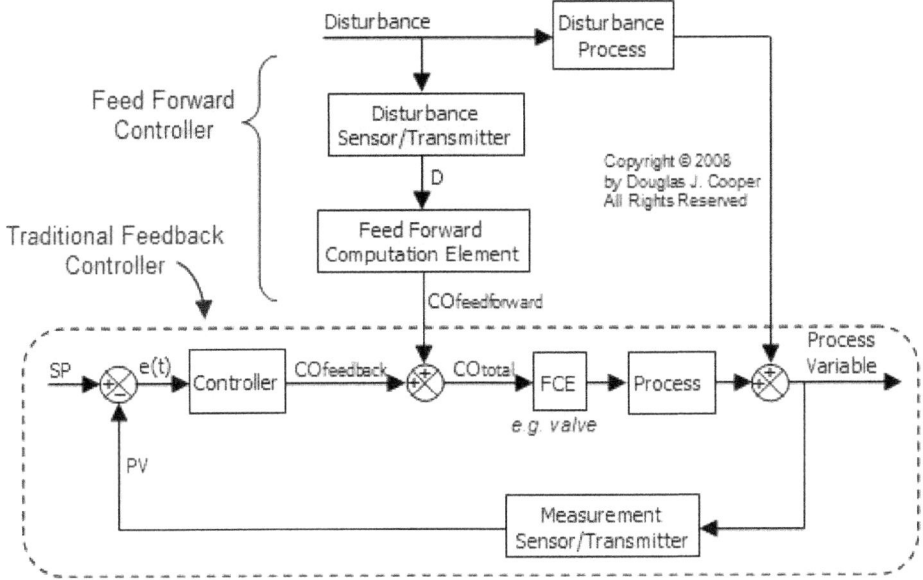

Feed Forward With Feedback Trim Architecture

where :

CO = controller output signal

D = measured disturbance variable

$e(t)$ = controller error, SP–PV

FCE = final control element (*e.g.*, valve, variable speed pump or compressor)

PV = measured process variable

SP = set point

The feedback controller is *designed and tuned* like any stand-alone controller from the PID family of algorithms. The only difference is that it must allow for its controller output signal, $CO_{feedback}$, to be combined with a feed forward controller output signal, $CO_{feedforward}$, to arrive at a total control action, CO_{total}.

Feed Forward Element Uses a Process and Disturbance Model

The diagram below shows that the function block element that computes the feed forward controller output signal, $CO_{feedforward}$, is constructed by combining a process model and disturbance model.

The blocks circled with dotted lines below show where data is collected when developing the process and disturbance models. The feed forward controller *does not* include these circled blocks as separate elements in its architecture.

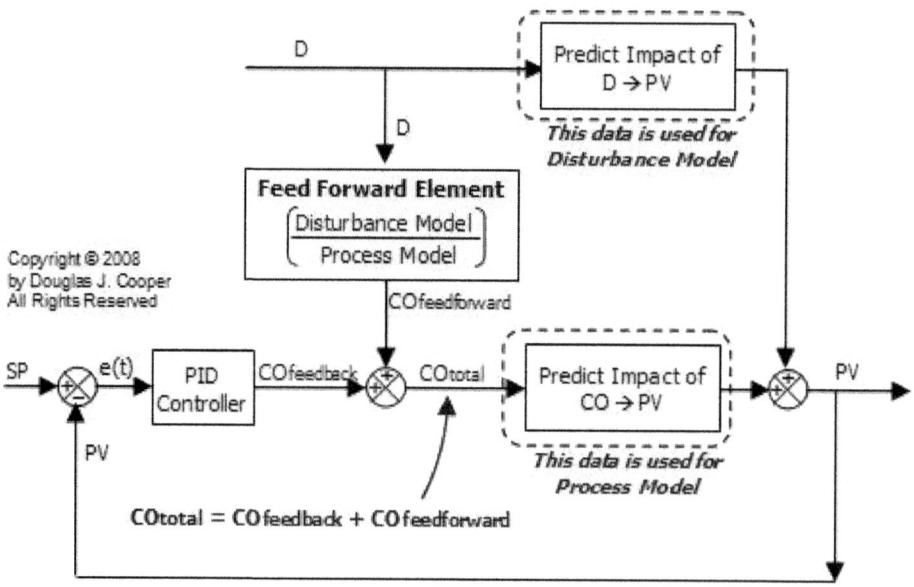

Feed Forward Element is Constructed from
Models of Process and Disturbance

The process model (CO → PV) in the feed forward element describes or predicts how each change in CO will impact PV. The disturbance model (D → PV) describes or predicts how each change in D will impact PV.

In practice, these models can range from simple scaling multipliers (*static* feed forward) through sophisticated differential equations (*dynamic* feed forward). Sophisticated dynamic models can better describe actual process and disturbance behaviors, often resulting in improved disturbance rejection performance. Such models can also be challenging to derive and implement, increasing the time and expense of a project.

Dynamic Feed Forward Based on the FOPDT Model

We first develop a general feed forward element using dynamic models (differential equations). Later, we will explore how we can simplify this general construction into a static feed forward element. Static feed forward is widely employed in industrial practice, in part because it can be implemented with an ordinary multiplying relay that scales the disturbance signal.

A dynamic feed forward element accounts for the *"how far" gain*, the *"how fast" time constant* and the *"how much delay" dead time* behavior of both the process (CO → PV) and disturbance (D → PV) relationships.

The simplest differential equation that describes such "how far, how fast, and with how much delay" behavior for either the process or disturbance dynamics is the familiar *first order plus dead time (FOPDT)* model.

- *The* CO → PV *Process Model* : Describing the CO → PV process behavior with a FOPDT model is not a new challenge. For example, we *presented all details* as we developed the FOPDT dynamic CO → PV model for the gravity drained tanks process from step test data as :

$$1.4\frac{dPV(t)}{dt} + PV(t) = 0.09 \cdot CO(t-0.5)$$

Which matches the general FOPDT (first order plus dead time) dynamic model form :

$$Tp\frac{dPV(t)}{dt} + PV(t) = Kp \cdot CO(t-\theta p)$$

where for a change in CO, the FOPDT model parameters are :

- Kp = *process gain* (the direction and how far PV will travel)
- Tp = *process time constant* (how fast PV moves after it begins its response)
- θp = *process dead time* (how much delay before PV first begins to respond).

This equation describes how each change in CO causes PV to respond (CO → PV) as time, t, passes. Our past modeling experience also includes developing and documenting FOPDT CO → PV models for the *heat exchanger* and *jacketed reactor* processes.

- *The* D → PV *Disturbance Model* : The procedure used to develop the FOPDT (first order plus dead time) process models referenced above can be used in an identical fashion to develop a dynamic D → PV disturbance model. While we do not show the graphical calculations at this point, consider the plot below from the *gravity drained tanks* process.

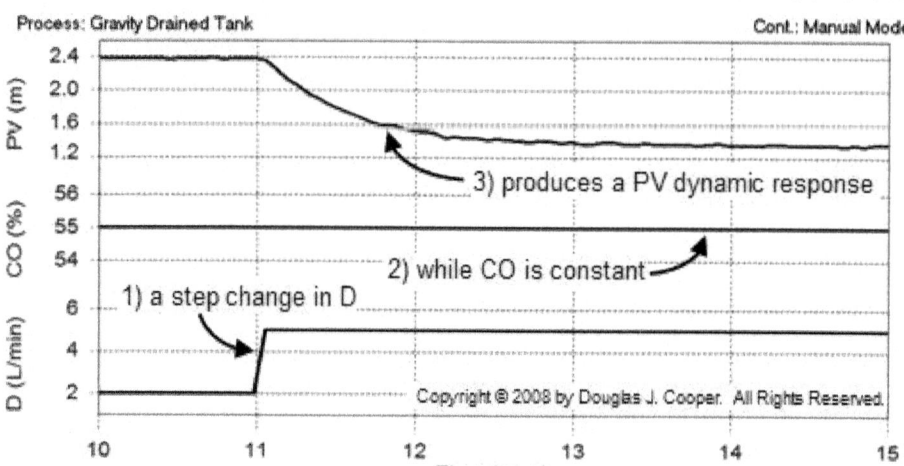

Disturbance Step With Constant CO Reveals D → PV Dynamic Behavior

Instead of analyzing a plot where a CO step forces a PV response while D remains constant, here we would analyze a D step forcing a PV response while CO remains constant.

We presume that an analogous *graphical modeling procedure* can be followed to determine the "how far, how fast, and with how much delay" dynamic D → PV disturbance model :

$$TD\frac{dPV(t)}{dt} + PV(t) = KD \cdot D(t - \theta D)$$

where for a step change in D :

- KD = disturbance gain (the direction and how far PV will travel)
- TD = disturbance time constant (how fast PV moves after it begins its response)
- θD = disturbance dead time (how much delay before PV first begins to respond)

Dynamic Feed Forward as a Thought Experiment

A feed forward element typically performs a model computation every sample time, T, to address any changes in measured disturbance, D. To help us visualize events, we discuss this as if it occurs as a two step "prediction and corrective action" procedure for a single disturbance :

1. The D → PV disturbance model receives a change in the measured value of D and predicts an open-loop or uncontrolled "impact profile" on PV. This includes a prediction of how much delay will pass before the disruption first arrives at PV, the direction PV will travel for this particular D once it begins to respond, and how fast and how far PV will travel before it settles out at a predicted new steady state.

2. The CO $\rightarrow$ PV process model then uses this PV impact profile to back-calculate a series of corrective control actions, CO$_{\text{feedforward}}$. These are CO moves sent to the final control element (FCE) to cause an "equal but opposite" response in PV such that it remains at set point, SP, throughout the event. Thus, the CO model seeks to exactly counteract the disruption profile predicted in step 1. The first CO actions are delayed as needed so they meet D upon arrival at PV. A series of CO actions are then deployed to counteract the predicted disruption over the life of the event.

Even sophisticated dynamic models are too simple to precisely describe the behavior of real processes. So although a feed forward element can dramatically reduce the impact of a disturbance on our PV, it will not provide a perfect "prediction and corrective action" disturbance rejection.

To account for model inaccuracies, the feed forward signal is combined with traditional feedback control action, CO$_{\text{feedback}}$, to create a total controller output, CO$_{\text{total}}$. Whether it be a *P-Only*, *PI*, *PID* or *PID w/ CO Filter* algorithm, the feedback controller plays the important role of :

- Minimizing the impact of disturbance variables other than D that can disrupt the PV,

- Providing set point tracking capability to the overall strategy, and

- Correcting for the simplifying approximations used in constructing the feed forward computation element that ultimately makes it imperfect in its actions.

The Sign of CO$_{\text{feedforward}}$ Requires Careful Attention

Notice in the block diagram above that CO$_{\text{feedforward}}$ is added as :

$$CO_{\text{total}} = CO_{\text{feedback}} + CO_{\text{feedforward}}$$

We write the equation this way because it is consistent with typical vendor documentation and standard piping/process & instrumentation diagrams (P&IDs).

The "plus" sign in the equation above requires our careful attention. To understand the caution, consider a case where the D $\rightarrow$ PV disturbance model predicts that D will cause the PV to *move up* in a certain fashion or pattern over a period of time. According to our thought experiment above, the CO $\rightarrow$ PV process model must compute feed forward CO actions that cause the PV to *move down* in an identical pattern.

But the sign of both the process gain, Kp, and disturbance gain, KD, together determine whether we need to send an increasing or decreasing signal to the FCE to compensate for a particular D. If Kp and K$_D$ are both positive or both negative, for example, then as a disturbance D moves up, the CO$_{\text{feedforward}}$ signal must move down to compensate. If Kp and K$_D$ are of opposite sign, then D and CO$_{\text{feedforward}}$ move in the same direction to counteract each other.

We show a standard "plus" sign in the equation above. But the computed feed forward signal, $CO_{feedforward}$, will be positive or negative depending on the signs of the process and disturbance gains as just described. This ensures that the impact of the disturbance and the compensation from the feed forward element move in opposite directions to provide improved disturbance rejection performance.

Dynamic Feed Forward in Math

Suppose we define a generic process model, Gp, and generic disturbance model, GD, as :

Gp = generic CO → PV process model (*describing how a CO change will impact PV*)

GD = generic D → PV disturbance model (*describing how a D change will impact PV*)

We allow Gp and GD to have forms that can range from simple to the sophisticated.

For example, they can both be pure gain values Kp and K_D and nothing more. They can be full FOPDT differential equations that include Kp, Tp and θp, and K_D, T_D and $θ_D$. One or the other (or both) can be *non-self regulating (integrating)*, or perhaps self-regulating but second or *third order*.

Leaving the exact form of the models undefined for now, we develop our feed forward element with the following steps :

1. Our generic CO → PV process model, Gp, allows us to compute a PV response to changes in CO as :

    $$PV = Gp \cdot CO$$

    With the generic model approach, we can rearrange the above to compute controller output actions that would reproduce a known or specified PV as :

    $$CO = (1/Gp) \cdot PV$$

2. Our generic D → PV disturbance model, G_D, lets us compute a PV response to changes in D as :

    $$PV = GD \cdot D$$

3. Following the logic in the above thought experiment, we use the D → PV model of step 2 to predict an impact profile on PV for any measured disturbance D :

    $$PV_{impact} = GD \cdot D$$

4. We then use our rearranged equation of step 1 to back-calculate a series of corrective feed forward control signals that will move PV in a pattern that is opposite (and thus negative in sign) to the predicted PV impact profile from Step 3 :

    $$CO_{feedforward} = -(1/Gp) \cdot PV_{impact}$$

5. We finish by substituting the "$PV_{impact} = GD \cdot D$" equation of step 3 into the $CO_{feedforward}$ equation of step 4 :

$$CO_{feedforward} = -(1/Gp) \cdot (G_D \cdot D)$$

and rearrange to arrive at our final feed forward computational element composed of a disturbance model divided by a process model :

$$CO_{feedforward} = -(G_D/Gp) \cdot D$$

Note : Above is a math argument that we hope seems reasonable and easy to follow.

But please be aware that for such manipulations to be proper, all variables and equations must first be mapped into the Laplace domain using *Laplace transforms*. The ease with which complex equations can be manipulated in the Laplace domain is a major reason control theorists use it for their derivations.

So, to be mathematically correct, we must first cast all variables and models into the Laplace domain as :

$PV(s) = Gp(s) \cdot CO(s)$ when the disturbance is constant

and

$PV(s) = G_D(s) \cdot D(s)$ when the controller output signal is constant

where the Laplace domain models $Gp(s)$ and $G_D(s)$ are called *transfer functions*.

At the end of our derivation, our final feed forward computational element should be expressed as :

$$CO_{feedforward}(s) = -[G_D(s)/Gp(s)] \cdot D(s)$$

Thus, while we had omitted important details, our feed forward equation is indeed correct and we will use it going forward. We will continue to downplay the complexities of the math for now as we focus on methods of use to industry practitioners.

Feed Forward with Feedback Trim Conceptual Diagram

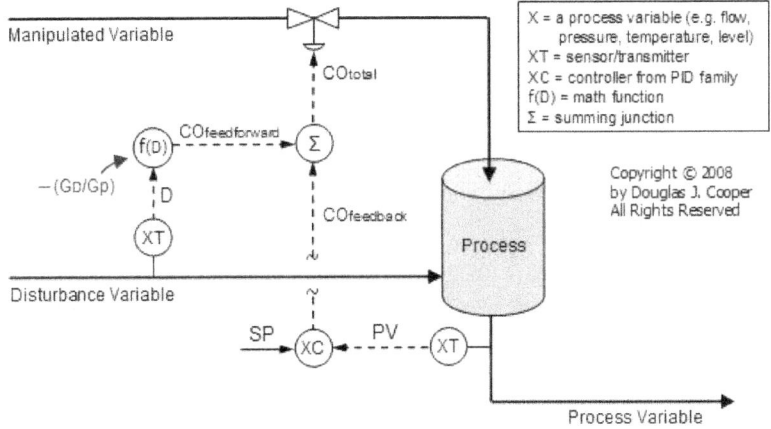

Conceptual Feed Forward With Feedback Trim Diagram

We now have the basis for why we express the feed forward element math function, $f(D) = -(G_D/Gp)$ as shown in our generalize feed forward with feedback trim conceptual diagram below :

Implementation and Testing of Feed Forward with Feedback Trim

We next explore the widely used and surprisingly powerful static feed forward controller. We will discover the ease with which we can develop such an architecture, and also explore some of the limitations of this simplified approach.

STATIC FEED FORWARD AND DISTURBANCE REJECTION IN THE JACKETED REACTOR

The purpose of the *feed forward controller* of the *feed forward with feedback trim* architecture is to reduce the impact of one specific disturbance, D, on our primary process variable, PV. An additional sensor must be located upstream in our process so we have a disturbance measurement that provides warning of impending disruption. The feed forward element uses this D measurement signal to compute and implement corrective control actions so the disturbance has minimal impact on stable operation.

Here we build on the mathematical foundation of this previous material as we explore the popular and surprisingly powerful **static** feed forward computational element for this disturbance rejection architecture.

Static Feed Forward Uses the Simplest Model Form

If we define a generic process model, Gp, and generic disturbance model, G_D, as :

Gp = generic CO → PV process model (*describing how a CO change will impact PV*)

G_D = generic D → PV disturbance model (*describing how a D change will impact PV*)

then we can *show details to derive* a general feed forward computational element as a ratio of the disturbance model divided by the process model :

$$CO_{feedforward} = -(G_D/Gp) \cdot D$$

Models Gp and G_D can range from the simple to the sophisticated. With static feed forward, we limit Gp and G_D to their constant "which direction and how far" gain values :

Gp = Kp (the CO → PV process gain)

$G_D = K_D$ (the D → PV disturbance gain)

And the static feed forward element is thus a simple gain ratio multiplier :

$$CO_{feedforward} = -(K_D/Kp) \cdot D \text{ (static feed forward element)}$$

The static feed forward controller does not consider how the controller output to process variable (CO → PV) dynamic behavior differs from the disturbance to process variable (D → PV) dynamic behavior.

- We do not account for the size of the process time constant, Tp, relative to the disturbance time constant, T_D. As a consequence, we cannot compute and deploy a series of corrective control actions over time to match how fast the disturbance event is causing the PV to move up or down.

- We do not consider the size of the process dead time, θp, relative to the disturbance dead time, θ_D. Thus, we cannot delay the implementation of corrective actions to coordinate their arrival with the start of the disturbance disruption on PV.

For processes where the CO → PV dynamic behavior is very similar to the D → PV dynamic behavior, like many liquid level processes for example, static feed forward will perform virtually the same as a fully dynamic feed forward controller in rejecting our measured disturbance.

Visualizing the action of this static feed forward element as a two step "prediction and corrective action" procedure for a single disturbance :

1. The D → PV disturbance gain, K_D, receives a change in D and predicts the total final impact on PV. The computation can only account for information contained in K_D, which includes the direction and how far PV will ultimately travel in response to the measured D before it settles out at a new steady state.

2. The CO → PV process gain, Kp, then uses this disturbance impact prediction of "which direction and how far" to back-calculate one CO move as a corrective control action, $CO_{feedforward}$. This $CO_{feedforward}$ move is sent immediately to the final control element (FCE) to cause an "equal but opposite" response in PV.

Static Feed Forward with Feedback Trim Conceptual Diagram

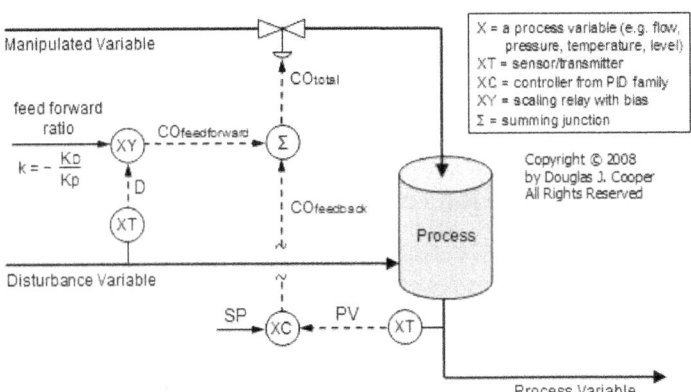

Limited in Capability but (Reasonably) Easy to Implement

The static feed forward element makes one complete and final corrective action for every measured change in D. It does not delay the feed forward signal so it will meet the D impact when it arrives at PV. It does not compute and deploy a series of CO actions to try and counteract a predicted disruption pattern over an event life.

Even with this limited performance capability, the benefit of the static form that makes it popular with industry practitioners is that it is reasonably straight-forward to implement in a production environment. As shown below, we can construct a static feed forward element with :

- A sensor/transmitter to measure disturbance D
- A scaling relay that multiplies signal D by our static feed forward ratio, $(-K_D/Kp)$
- A summing junction that adds $CO_{feedforward}$ to COfeedback to produce CO_{total}

$CO_{feedforward}$ is Normally Zero

An important implementation issue is that $CO_{feedforward}$ should equal zero when D is at its *design level of operation (DLO)* value. Thus, the D used in our calculations is actually the disturbance signal from the sensor/transmitter ($D_{measured}$) that has been shifted or biased by the design level of operation disturbance value (DDLO), or :

$D = D_{measured} - D_{DLO}$

With this definition, both D and $CO_{feedforward}$ will be zero when the disturbance is at its normal or expected value. Such a biasing capability is included with most all commercial scaling relay function blocks.

Static Feed Forward and the Jacketed Reactor Process

We have previously explored *the modes of operation* and *dynamic CO → PV behavior* of the jacketed stirred reactor process. We also have established the performance of a single loop *PI controller*, a *PID with CO Filter controller* and a *cascade control im-plementation* when our control objective is to minimize the impact of a disturbance caused by a change in the temperature of the liquid entering the cooling jacket.

Here we explore the design, implementation and performance of a static *feed forward with feedback trim architecture* for this same disturbance rejection objective.

Limitations of the single loop architecture

The control objective is to maintain the reactor exit stream temperature (PV) at set point (SP) in spite of changes in the temperature of cooling liquid entering the jacket (D) by adjusting controller output (CO) signals to the cooling liquid flow valve.

where :

CO = signal to valve that adjusts cooling jacket liquid flow rate (controller output, %)

PV = reactor exit stream temperature (measured process variable, °C)

SP = desired reactor exit stream temperature (set point, °C)

D = temperature of cooling liquid entering the jacket (major disturbance, °C)

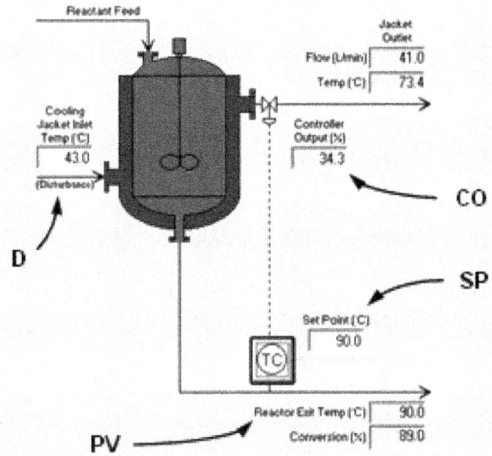

A Feed Forward with Feedback Trim Reactor Architecture

Below is the jacketed stirred reactor process with a feed forward with feedback trim controller architecture.

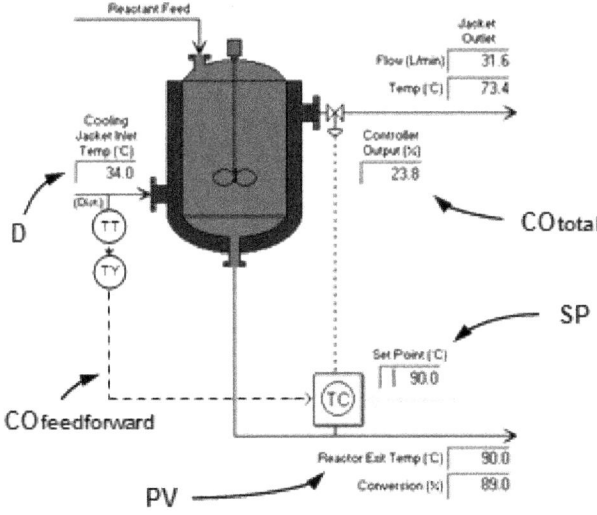

The loop architecture from this *commercial software* simulation shows that D is measured, scaled and transmitted as $CO_{feedforward}$ to the controller. There it is combined with the traditional feedback signal to produce the CO_{total} sent to the cooling jacket flow valve.

Design Level of Operation (DLO)

The details and discussion of the DLO used in our disturbance rejection studies for the jacketed stirred reactor are presented and are summarized :

- Design PV and SP = 90 °C with approval for brief dynamic testing of ±2 °C
- Design D = 43 °C with occasional spikes up to 50 °C

We note that D moves between 43 °C and 50 °C. We seek a single DLO value that lets us conveniently compare results from two different design methods explored below.

We choose here a D_{DLO} as the average value of $(43+50)/2 = 46.5$ °C and acknowledge that other choices (such as simply using 43 °C) are reasonable. As long as we are consistent in our methodology, the conclusions we draw when comparing the two methods will remain unchanged.

When D = 46.5 °C and CO = 40%, then our measured process variable settles at the design value of PV = 90 °C. This relationship between the three variables explains the DLO values indicated on the plots that follow.

Design Method 1 : Compute KD/Kp from Historic Data

Below is a trend from our data historian showing approximately three hours of operation from the jacketed reactor process under PI control. No feed forward controller is active. The set point (SP) is constant and a number of disturbance events force the PV from SP. All variables are near our DLO as described above.

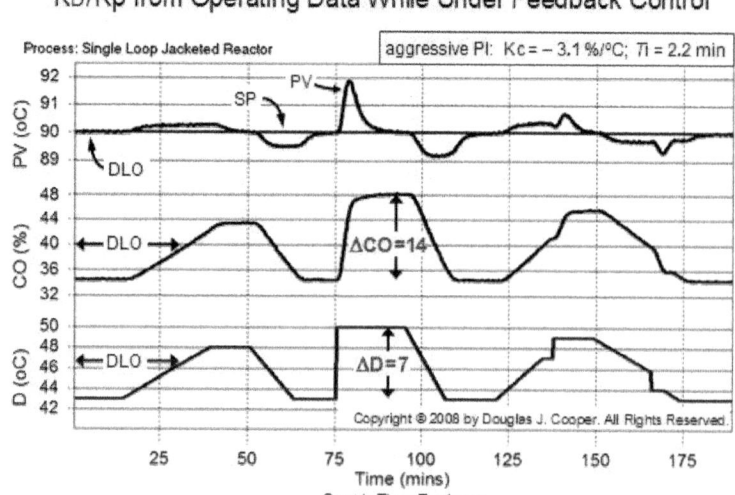

KD/Kp from Operating Data While Under Feedback Control

If *we recall the definition* that Kp = ΔPV/ΔCO and K_D = ΔPV/ΔD, then for our static feed forward design :

$$CO_{feedforward} = - (K_D/Kp) \cdot D$$
$$= - [(\Delta PV/\Delta D)/(\Delta PV/\Delta CO)] \cdot D$$
$$= - [(\Delta CO/\Delta D)] \cdot D$$

With this last equation, our design challenge is reduced to finding a disturbance change that lasts long enough for the controller output response to settle. If this event occurs reasonably close to our DLO, then we can directly compute our gain ratio feed forward element by measuring the disturbance and controller output changes from a plot.

- *About the Feedback Controller :* The plot shows disturbance rejection performance when the process is using a PI controller tuned for an aggressive response action.

Note that with this "measure from a plot" approach of Method 1, the process can be controlled by any algorithm from the PID family, though integral action must be included to return the PV to SP (eliminate *offset*) after a disturbance. Also, our feedback controller tuning can range from conservative (sluggish) through an aggressive (active) response without affecting the applicability of the method.

Our interest is limited to finding a ΔD disturbance event with a corresponding ΔCO controller output signal response that lasts long enough for the PV to be returned to SP. Then as shown above, the ΔD and ΔCO relationship can be measured directly from the plot data.

- *Accounting for Negative Feedback :* If we are using automatic mode (closed loop) data as shown in the plot above, we must account for the *negative feedback* of our controller in our calculations. A controller always takes action that moves the CO signal in a direction that counteracts the developing controller error. Thus, when using automatic mode (closed loop) data as above, we must consider that a negative sign has been introduced into the signal relationship.

On the plot above, we have labelled a ΔD disturbance change with corresponding ΔCO controller output signal response. We introduce the sign change from negative feedback of our controller and compute :

$$CO_{feedforward} = - [(\Delta CO/\Delta D)] \cdot D \cdot (-1 \text{ for negative feedback})$$
$$= [(14\%)/(7 \text{ °C})] \cdot D$$
$$= [2\%/\text{°C}] \cdot D$$

Design Method 2 : Perform two Independent Bump Tests

To validate that a feed forward gain ratio of 2%/°C is a reasonable number as determined from automatic mode (closed loop) data, here we perform two independent step tests, compute individual values for K_D and Kp, and then compare the ratio results to Method 1.

This approach is largely academic because the challenges of steadying a real process and then stepping individual parameters in such a perfect fashion is unrealistic in the chaotic world of most production environments. This exercise holds value, however, because it provides an alternate route that confirms the results presented in Method 1.

We follow the *established procedure* for computing a process gain, Kp, from a manual mode (open loop) step test response plot. That is, we set our disturbance parameter at D_{DLO}, set the controller output at a constant CO value and wait until the PV is steady. We then step CO to force a PV response that is *centered around our DLO.*

Such a step response plot is shown below for the jacketed stirred reactor. We measure and compute $Kp = \Delta PV / \Delta CO = -0.4\ °C/\%$ as indicated on the plot.

Kp from Open Loop Process Step Response at the DLO

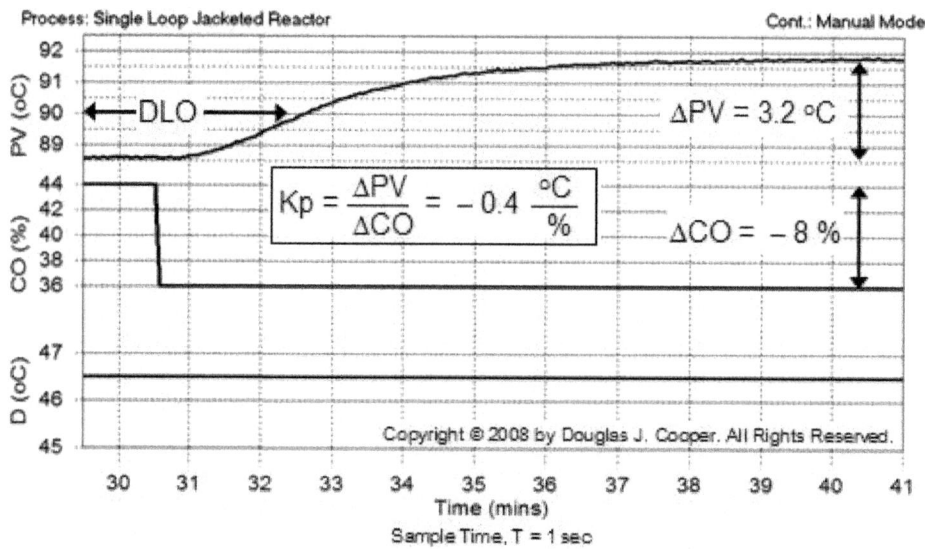

We repeat the procedure to compute a disturbance gain, K_D, from an open loop step response plot. Here, we set our CO signal at the DLO value of 40%, set the disturbance parameter at a constant D value and wait until the PV is steady. We then step D to force a PV response that is again centered around our DLO.

Such a disturbance step response plot is shown below. We measure and compute $KD = \Delta PV / \Delta D = 0.8\ °C/°C$ as labelled on the plot.

With values for Kp and K_D, we compute our gain ratio feed forward multiplier. Since we are in manual mode (open loop), we need not account for any sign change due to negative feedback in our calculation.

$$CO_{feedforward} = -(K_D/Kp) \cdot D$$
$$= -[(0.8\ °C/°C)/(-0.4\ °C/\%)] \cdot D$$
$$= [2\%/°C] \cdot D$$

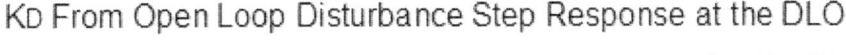

KD From Open Loop Disturbance Step Response at the DLO

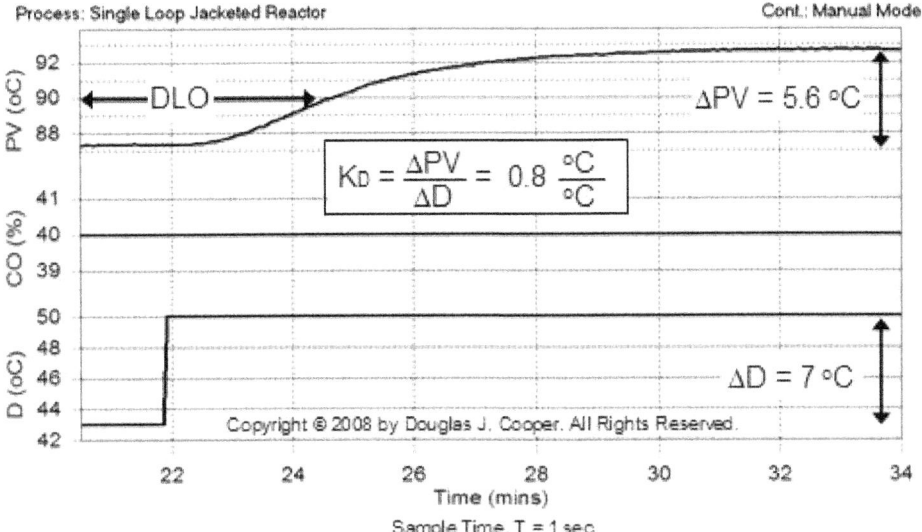

Thus, with careful testing using a consistent and repeatable *commercial process simulation*, we observe that the practical approach of Method 1 provides the same results as the academic approach of Method 2.

Implement and Test

The all-important question we now consider is whether the extra effort associated with designing and implementing a "static feed forward with feedback trim" architecture provides sufficient payoff in the form of improved disturbance rejection performance.

To the left in the plot below is the performance of a dependent, ideal PI controller with aggressive tuning values for controller gain Kc = − 3.1%/°C and reset time Ti = 2.2 min.

To the right in the plot above is the disturbance rejection performance of our static feed forward with feedback trim architecture. The feed forward gain ratio multiplier used is the 2%/°C as determined by two different methods. The feedback controller remains the aggressively tuned PI algorithm as described above.

The static feed forward controller makes one complete preemptive corrective CO action whenever a change in D is detected as noted in the plot above. There is no delay of the feed forward signal based on relative dead time considerations, and there is no series of CO actions computed and deployed based on relative time constant considerations.

Nevertheless, the static feed forward controller is able to reduce the maximum deviation from SP during a disturbance event to half of its original value. The settling time is also reduced, though less dramatically. Like any control project, the operations staff must determine if this represents a sufficient payback for the effort and expenses required.

PI vs 'PI with Static Feed Forward' in Jacketed Stirred Reactor

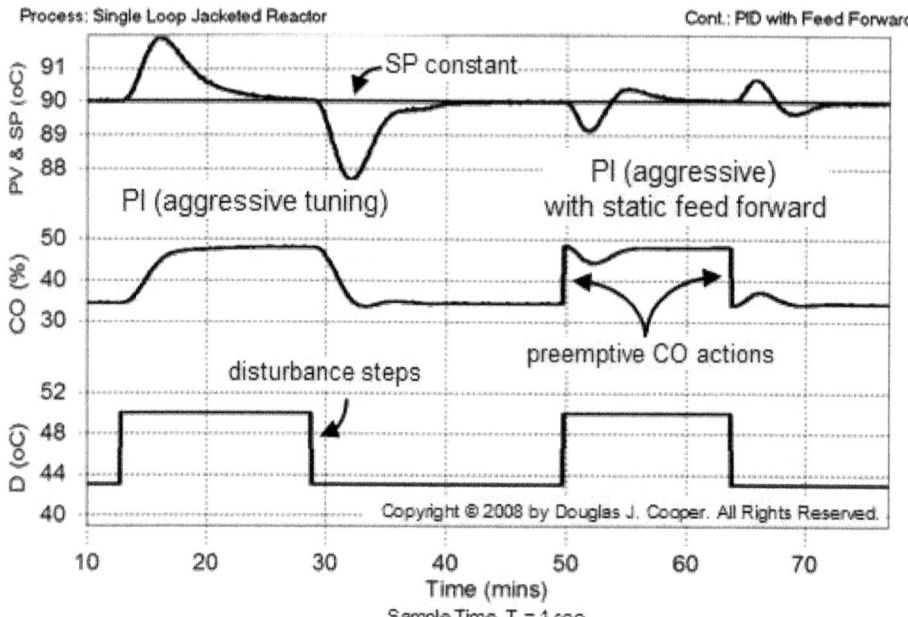

Practitioner's note : Our decision to add feed forward to a feedback control loop is driven by the character of the disturbance and its effect on our PV. If the controller can react more quickly than the D can change, feed forward is not likely to significantly improve control. However, if the disturbance changes rapidly and fairly often, feed forward control can be a powerful tool to stabilize our process.

No Impact on Set Point Tracking Performance

While not our design objective, presented below is the set point tracking performance of the single loop PI controller compared to that of the static feed forward with feedback trim architecture. The same aggressive PI tuning values used above are maintained for this study.

Feed forward with feedback trim is designed for the improved rejection of one measured disturbance. As shown above, a feed forward controller has no impact on set point tracking performance when the disturbance is constant. This makes sense since the computed $CO_{feedforward}$ signal does not change unless D changes. Indeed, with no change in the measured disturbance, both architectures provide identical performance.

Set Point Tracking Performance Unaffected by Feed Forward

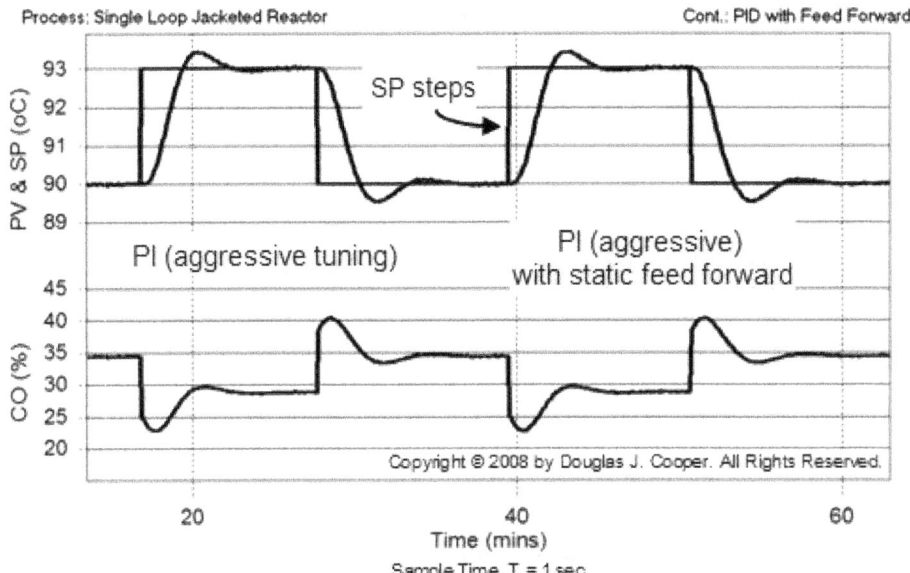

Chapter 12

RATIO, OVERRIDE AND CROSS-LIMITING CONTROL

THE RATIO CONTROL ARCHITECTURE

The ratio control architecture is used to maintain the flow rate of one stream in a process at a defined or specified proportion relative to that of another. A common application for ratio control is to combine or blend two feed streams to produce a mixed flow with a desired composition or physical property.

Ratio Control Conceptual Diagram

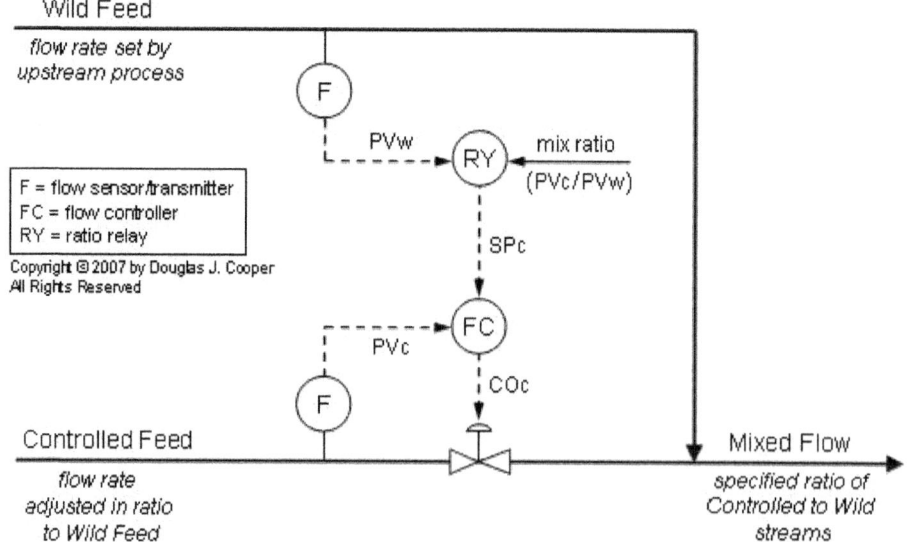

F = flow sensor/transmitter
FC = flow controller
RY = ratio relay

Wild Feed — flow rate set by upstream process

PVw

RY — mix ratio (PVc/PVw)

SPc

FC

PVc

COc

F

Controlled Feed — flow rate adjusted in ratio to Wild Feed

Mixed Flow — specified ratio of Controlled to Wild streams

The conceptual diagram below shows that the flow rate of one of the streams feeding the mixed flow, designated as the *wild feed*, can change freely. Its flow rate might change based on product demand, maintenance limitations, feedstock variations, energy availability, the actions of another controller in the plant, or it may simply be that this is the stream we are least willing to manipulate during normal operation.

The other stream shown feeding the mixed flow is designated as the *controlled feed*. A final control element (FCE) in the controlled feed stream receives and reacts to the controller output signal, COc, from the ratio control architecture.

Relays in the Ratio Architecture

As the above diagram illustrates, we measure the flow rate of the wild feed and pass the signal to a relay, designated as RY in the diagram. The relay is typically one of two types :

- A *ratio relay*, where the mix ratio is entered once during configuration and is generally not available to operations staff during normal operation.
- A *multiplying relay* (shown), where the mix ratio is presented as an adjustable parameter on the operations display and is thus more readily accessible for change.

In either case, the relay multiplies the measured flow rate of the wild feed stream, PVw, by the entered mix ratio to arrive at a desired or set point value, SPc, for the controlled feed stream. A flow controller then regulates the controlled feed flow rate to this SPc, resulting in a mixed flow stream of specified proportions between the controlled and wild streams.

Linear Flow Signals Required

A ratio controller architecture as described above requires that the signal from each flow sensor/transmitter change linearly with flow rate. Thus, the signals from the wild stream process variable, PVw, and the controlled stream process variable, PVc, should increase and decrease in a straight-line fashion as the individual flow rates increase and decrease.

Turbine flow meters and certain other sensors can provide a signal that changes linearly with flow rate. Unfortunately, a host of popular flow sensors, including inferential head flow elements such as orifice meters, do not. Additional computations (function blocks) must then be included between the sensor and the ratio relay to transform the nonlinear signal into the required linear flow-to-signal relationship.

Flow Fraction (Ratio) Controller

A classic example of ratio control is the blending of an additive into a process stream. As shown below, an octane booster is blended with straight-run gasoline stream being produced by an atmospheric distillation column. For any number of reasons, the production rate of straight-run gasoline will vary over time in a refinery. Therefore, the amount of octane booster required to produce the desired octane rating in the mixed product flow must also vary in a coordinated fashion.

Flow Fraction (Ratio) Controller

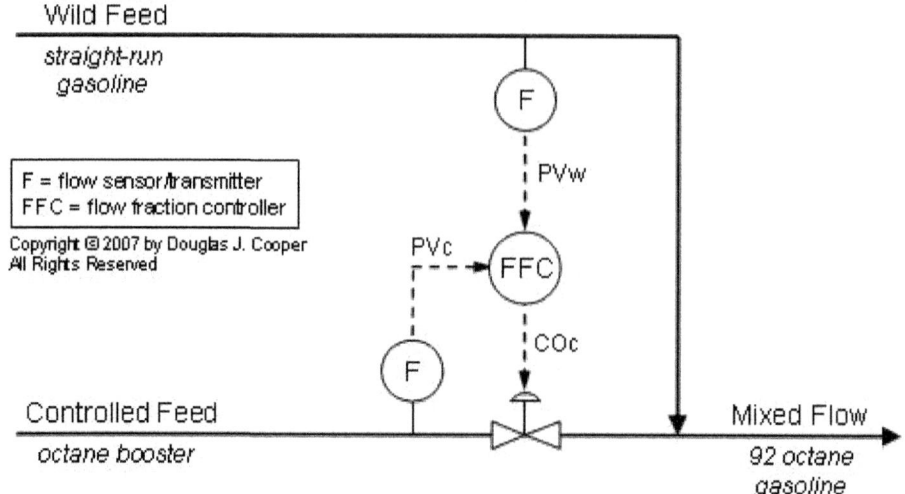

Rather than using a relay, we present an alternative ratio control architecture based on a flow fraction controller (FFC). The FFC is essentially a "pure" ratio controller in that it receives the wild feed and controlled feed signals directly as inputs. A ratio set point value is entered into the FCC, along with tuning parameters and other values required for any controller implementation.

Ratio Relay or Flow Fraction Controller

The flow fraction (ratio) controller is a preconfigured option in many modern computer based *DCS* or advanced *PLC* control systems. It provides exactly the same functionality as the ratio relay combined with a single-input single-output controller as discussed above.

The choice of using a relay or an FFC is a practical matter. The entered ratio multiplier value in a relay is not a readily accessible parameter. It therefore requires a greater level of permission and access to adjust. Consequently, the use of the ratio relay has the advantage (or disadvantage depending on the application) of requiring a higher level of authorization before a change can be made to the ratio multiplier.

Multiplying Relay with Remote Input

The ratio controller shown below presents an additional level of complexity in that, like the *cascade architecture*, our ratio controller is contained within and is thus part of a larger control strategy.

In the example below, an analyzer sensor measures the composition or property we seek to maintain in the mixed flow stream. The measured value is compared to a set point value, SPA, and a mix ratio controller output signal, CO_A, is generated based on the difference. Thus, like a cascade, the outer analyzer controller continually sends mix ratio updates to the inner ratio control architecture.

Ratio Relay with Remote Input

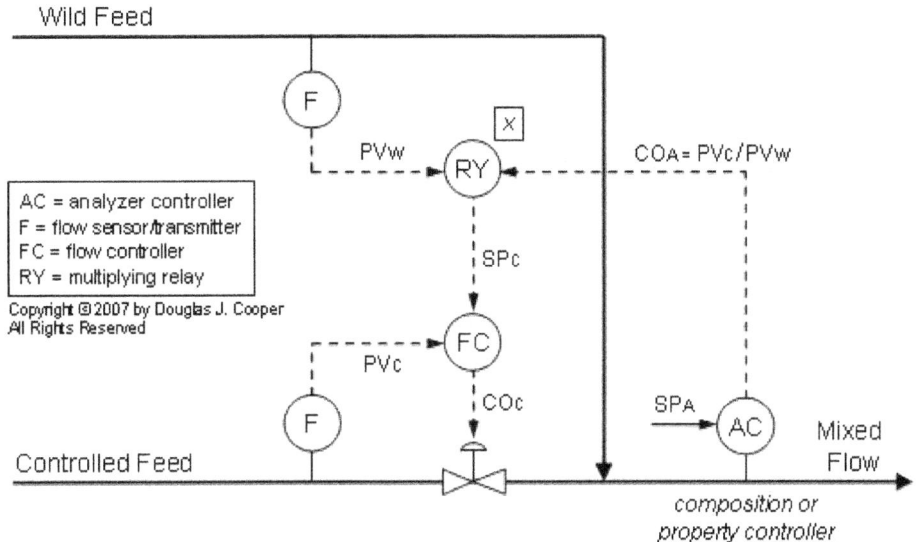

The updated mix ratio CO_A value enters the multiplying relay as an external set point. The objective of this additional complexity is to correct for any unmeasured changes in the wild feed or controlled feed, thus maintaining the mixed flow composition or property at the set point value, SP_A.

The term "analyzer" is used broadly here. Hopefully, we can indentify a fast, inexpensive and reliable sensor that allows us to infer the mixed flow composition or property of interest. Examples might include a capacitance probe, an in-line viscometer, or a pH meter.

If we are required to use a chromatograph, spectrometer or other such instrument, we must allow for the increased maintenance and attention such devices often demand. Perhaps more important, the time to complete a sample and analysis cycle for these devices can introduce a long dead time into our feedback loop. As *dead time increases*, best attainable control performance decreases.

RATIO CONTROL AND METERED-AIR COMBUSTION PROCESSES

A *ratio control strategy* can play a fundamental role in the safe and profitable operation of fired heaters, boilers, furnaces and similar fuel burning processes. This is because the air-to-fuel ratio in the combustion zone of these processes directly impacts fuel combustion efficiency and environmental emissions.

A requirement for ratio control implementation is that both the fuel feed rate and combustion air feed rate are measured and available as process variable (PV) signals. Shown below is a conceptual air/fuel ratio control strategy.

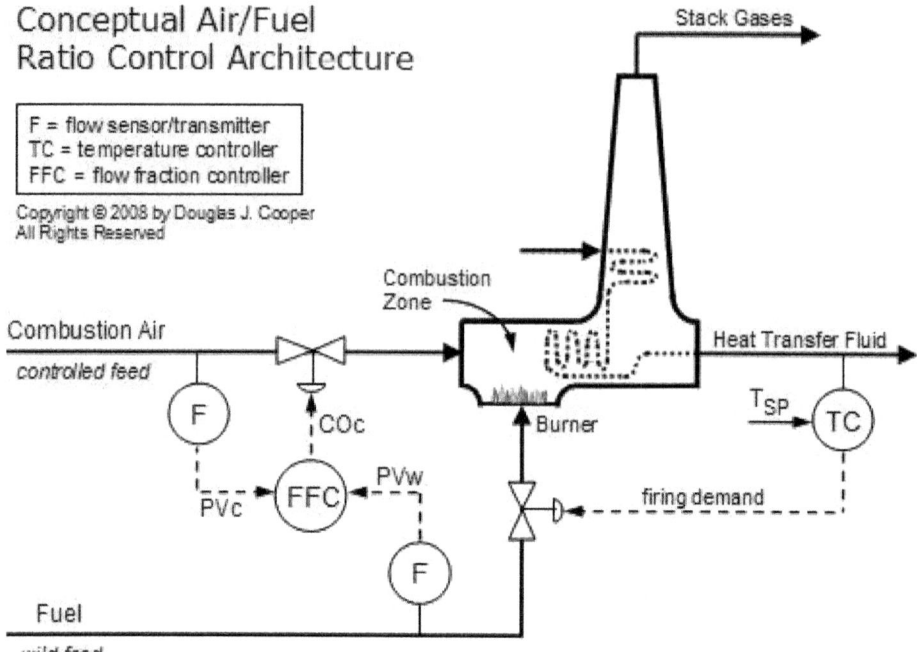

Conceptual Air/Fuel Ratio Control Architecture

F = flow sensor/transmitter
TC = temperature controller
FFC = flow fraction controller

In this representative architecture, the fuel flow rate is adjusted to maintain the temperature of a heat transfer fluid exiting a furnace. On other processes, fuel flow rate might be adjusted to maintain the pressure in a steam header, the duct temperature downstream of the burner, or similar variable that must be regulated for efficient operation.

The combustion air feed rate is then adjusted by a flow fraction (ratio) controller to maintain a desired air/fuel ratio. While a simple sensor and valve is shown above, we will expand and modify this conceptual architecture as we progress in this discussion because :

- The final control element (FCE) for the combustion air stream, rather than being a valve, is more commonly a variable speed blower, perhaps with adjustable dampers or louvers.

- Measuring combustion air flow rate is challenging and can involve measuring a pressure drop across a portion of the combustion gas exhaust flow path.

- In different applications, the air flow rate can be the wild feed while fuel flow rate is the controlled feed.

- Stack gas analyzers add value and sophistication as they monitor the chemistry associated with combustion efficiency and environmental emissions.

Why Air/Fuel Ratio is Important

In combustion processes, air/fuel ratio is normally expressed on a mass basis. We get maximum useful heat energy if we provide air to the combustion zone at a mass flow rate (*e.g.*, lb/min, Kg/hr) that is properly matched to the mass flow rate of fuel to the burner.

Consider this generic equation for fuel combustion chemistry :

$$Fuel + Air = \frac{Useful}{Heat} + CO_2 + H_2O + CO + \underbrace{\frac{Unburned}{Fuel}}_{\substack{Increases\ as \\ combustion\ air \\ Decreases}} + \underbrace{\frac{Waste\ Heat}{Up\ the\ Stack}}_{\substack{Increases\ as \\ combustion\ air \\ Increases}}$$

Where :

CO_2 = carbon dioxide

CO = carbon monoxide

H_2O = water

Air = 21% oxygen (O_2) and 79% nitrogen (N_2)

Fuel = hydrocarbon such as *natural gas* or *liquid fuel oil*.

Air is largely composed of oxygen and nitrogen. It is the oxygen in the air that combines with the carbon in the fuel in a highly energetic reaction called combustion. When burning hydrocarbons, nature strongly prefers the carbon-oxygen double bonds of carbon dioxide and will yield significant heat energy in an *exothermic reaction* to achieve this CO_2 form.

Thus, carbon dioxide is the common green house gas produced from the complete combustion of hydrocarbon fuel. Water vapor (H_2O) is also a normal product of hydrocarbon combustion.

Aside : Nitrogen oxide (NOx) and sulfur oxide (SOx) pollutants are not included in our combustion chemistry equation. They are produced in industrial combustion processes principally from the nitrogen and sulfur originating in the fuel. As the temperature in the combustion zone increases, a portion of the nitrogen in the air can also convert to NOx.

Too Little Air Increases Pollution and Wastes Fuel

The oxygen needed to burn fuel comes from the air we feed to the process. If the air/fuel ratio is too small in our heater, boiler or furnace, there will not be enough oxygen available to completely convert the hydrocarbon fuel to carbon dioxide and water.

A too-small air/fuel ratio leads to *incomplete combustion* of our fuel. As the availability of oxygen decreases, noxious exhaust gases including *carbon monoxide* will form first. As the air/fuel ratio decreases further, partially burned and unburned fuel can appear in the exhaust stack, often revealing itself as smoke and soot. Carbon monoxide, partially burned and unburned fuel are all poisons whose release is regulated by the government (the Environmental Protection Agency in the USA).

Incomplete combustion also means that we are wasting expensive fuel. Fuel that does not burn to provide useful heat energy, including carbon monoxide that could yield energy as it converts to carbon dioxide, literally flows up our exhaust stack as lost profit.

Too Much Air Wastes Fuel

The issue that makes the operation of a combustion process so interesting is that if we feed too much air to the combustion zone (if the air/fuel ratio is too high), we also waste fuel, though in a wholly different manner.

Once we have enough oxygen available in the burn zone to complete combustion of the hydrocarbon fuel to carbon dioxide and water, we have addressed the pollution portion of our combustion chemistry equation. Any air fed to the process above and beyond that amount becomes an additional process load to be heated.

As the air/fuel ratio increases above that needed for complete combustion, the extra nitrogen and unneeded oxygen absorb heat energy, decreasing the temperature of the flame and gases in the combustion zone. As the operating temperature drops, we are less able to extract useful heat energy for our intended application.

So when the air/fuel ratio is too high, we produce a surplus of hot air. And this hot air simply carries its heat energy up and out the exhaust stack as lost profit.

Theoretical (Stoichiometric) Air

The relationship between the air/fuel ratio, pollution formation and wasted heat energy provides a basis for control system design. In a meticulous laboratory experiment with exacting measurements, perfect mixing and unlimited time, we could determine the precise amount of air required to just complete the conversion of a hydrocarbon fuel to carbon dioxide and water. This minimum amount is called the "theoretical" or "stoichiometric" air.

Unfortunately, real combustion processes have imperfect mixing of the air with the fuel. Also, the gases tend to flow so quickly that the air and fuel mix have limited contact time in the combustion zone. As such, if we feed air in the exact theoretical or stoichiometric proportion to the fuel, we will still have incomplete combustion and lost profit.

Real burners generally perform in a manner similar to the graph below. The cost associated with operating at increased air/fuel ratios is the energy wasted in heating extra oxygen and nitrogen. Yet as the air/fuel ratio is decreased, losses due to incomplete combustion and pollution generation increase rapidly.

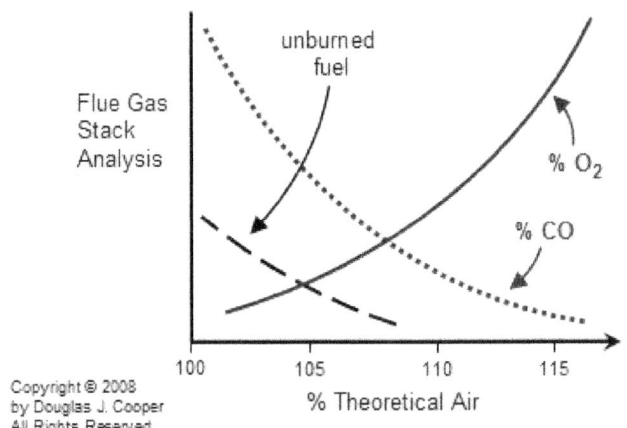

For any particular burner design, there is a target air/fuel ratio that balances the competing effects to minimize the total losses and thus maximize profit. As the graph above suggests (note that there is no scale on the vertical axis), a gas or liquid fuel burner generally balances losses by operating somewhere between 105% to 120% of theoretical air. This is commonly referred to as operating with 5% to 20% excess air.

Sensors Should Be Fast, Cheap and Easy

Fired heaters, boilers and furnaces in processes with streams composed of gases, liquids, powders, slurries and melts are found in a broad range of manufacturing, production and development operations. Knowing that the composition of the fuel, the design of the burners, the configuration of the combustion zone, and the purpose of the process can differ for each implementation hints at a dizzying array of control strategy design and tuning possibilities.

To develop a standard control strategy, we require a flexible method of measuring excess air so we can control to a target air/fuel ratio. We normally seek sensors that are reliable, inexpensive, easy to install and maintain, and quick to respond. If we cannot get these qualities with a direct measurement of the process variable (PV) of interest, then an effective alternative is to measure a related variable if it can be done with a "fast, cheap and easy" sensor option.

Excess air is an example of a PV that is very challenging to directly measure in the combustion zone, yet oxygen and energy content in the stack gases is an appropriate alternative. As it turns out, operating with 5% to 20% excess air equates to having about 1% to 3% oxygen by volume in the stack gases.

Measuring the Stack Gases

By measuring exhaust stack gas composition, we obtain information we need to properly monitor and control air/fuel ratio in the combustion zone. Stack analyzers fall into two broad categories :

- *Dry Basis Extractive Analyzers* pull a gas sample from the stack and cool it to condense the water out of the sample. Analysis is then made on the dry stack gas.
- *Wet Basis In Situ Analyzers* are placed in very close proximity to the stack. The hot sample being measured still contains the water vapor produced by combustion, thus providing a wet stack gas analysis.

A host of stack gas (or flue gas) analyzers can be purchased that measure O_2. The wet basis analyzers yield a lower oxygen value than dry basis analyzers by perhaps 0.3%–0.5% by volume.

Instruments are widely available that also include a carbon monoxide measurement along with the oxygen measurement. A common approach is to pass the stack gas through a catalyst chamber and measure the energy released as the carbon monoxide and unburned fuel converts to carbon dioxide. The analyzer results are expressed as an equivalent percent CO in the sample. The single number, expressed as a CO measurement but representing fuel wasted because of insufficient air, simplifies control strategy design and process operation.

With a measurement of O_2 and CO (representing all lost fuel) in the stack of our combustion process, we have critical PV measurements needed to implement an air/fuel ratio control strategy. Note that it is the responsibility of the burner manufacturer and/or process design staff to specify the target set point values for a particular combustion system prior to controller tuning.

Air Flow Metering

Combustion processes generally have combustion air delivered in one of three ways:

- A forced draft process uses a blower to feed air into the combustion zone.
- An induced draft process has a blower downstream of the burner that pulls or draws air through the combustion zone.
- A natural draft process relies on the void left as hot exhaust gases naturally rise up the stack to draw air into the combustion zone.

For this discussion, we assume *a blower* is being used to either force or induce combustion air feed because natural draft systems are not appropriately designed for active air flow manipulation.

Even with a blower, measuring the air feed rate delivered at low pressure through the twists and turns of irregular ductwork and firebrick is not cheap or easy. A popular alternative is to measure the pressure drop across some part of the exhaust gas stream. The bulk of the exhaust gas is nitrogen that enters with the combustion air. As long as the air/fuel ratio adjustments are modest, the exhaust gas flow rate will track the combustion air feed rate quite closely.

Thus, a properly implemented differential pressure measurement is a "fast, cheap and easy" method for inferring combustion air feed rate. The figure below illustrates such a measurement across a heat transfer section and up the exhaust stack.

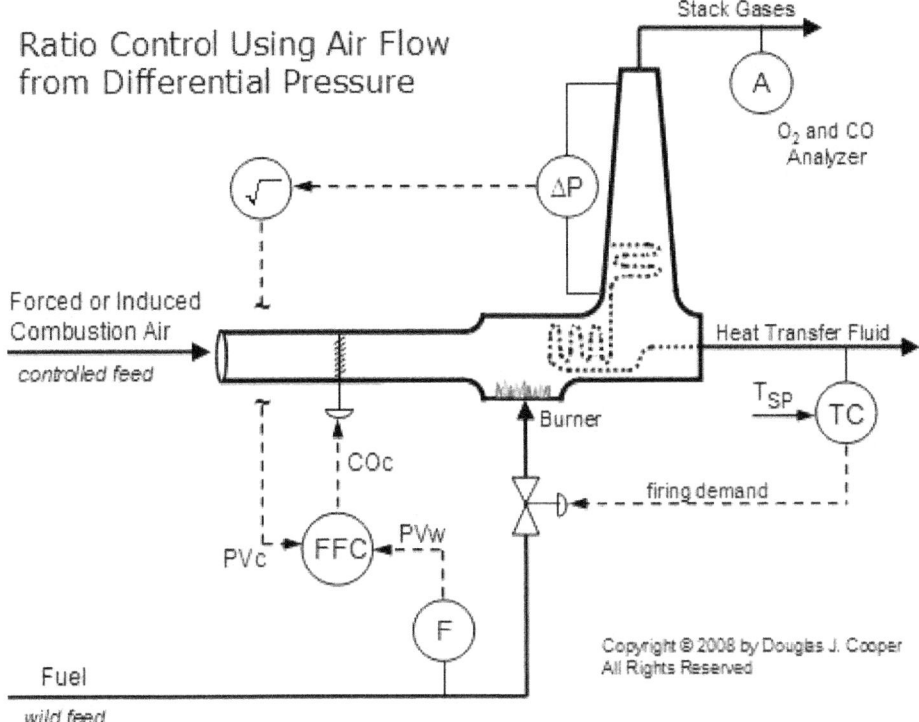

Ratio Control Using Air Flow from Differential Pressure

Also shown is that the controller output signal from the flow fraction (ratio) controller, CO_C, adjusts louvers to modulate the flow through the combustion zone. As the louvers open and close to permit more or less flow, the differential pressure measurement will increase or decrease, respectively.

The signal from the wild and controlled flow sensors must change linearly with flow rate. The differential pressure transmitter connected across a portion of the exhaust gas path becomes a linear gas flow sensor by recognizing that total gas flow, F, is proportional to the square root of the pressure differential (ΔP), or $F = \alpha \sqrt{\Delta P}$. Thus, the controlled feed process variable signal, PV_C, is linear with flow when the square root of the differential pressure signal is extracted as shown in the diagram.

Practitioner's Note : The differential pressure measurement must not be connected across the portion of the gas flow path that includes the adjustable louvers. Each change in louver position changes the F vs. ΔP relationship. Success would require that during calibration, we somehow determine a different coefficient α for each louver position. This unrealistic task is easily avoided by proper location of the differential pressure taps.

Calibrating the differential pressure signal to a particular air feed rate is normally achieved while the fired heater, boiler or furnace is operating with the air/fuel ratio controller in manual mode. The maximum or full scale differential pressure calibration is determined by bringing the fuel flow firing rate to maximum (or as close as practical) and then adjusting the air feed flow rate until the design O_2 level is being measured in the stack gas.

The differential pressure being measured by these sensors is very small and the exhaust gas contains water vapor that can condense in sensing lines. Even one or two inches of condensate in one side of the differential pressure transmitter can dramatically corrupt the measurement signal.

Choosing Air or Fuel for Firing Rate Control

With a means of measuring both the combustion air flow and the fuel flow, and with a stack analyzer to permit calibration and monitoring, we can implement the simple air/fuel ratio control as shown in the diagram above.

Because it can be measured accurately, fuel feed rate is a popular choice for the firing rate control variable. Yet in certain applications it is more desirable to employ the combustion air flow rate in this capacity.

If fuel is the firing rate control variable, a rapid increase in firing rate with air following behind in time will lead to incomplete combustion as discussed above. On the other hand, if air is made the firing rate control variable, a rapid decrease in firing rate will lead to the same situation.

OVERRIDE (SELECT) ELEMENTS AND THEIR USE IN RATIO CONTROL

A select element receives two input signals and forwards one of them onward in the signal path. A *low select*, shown below to the left, passes the lowest of the two signals, while a *high select*, shown to the right, passes the larger value onward.

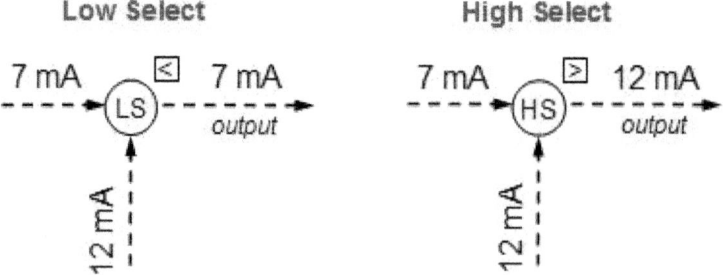

A select element can be implemented as a *DCS* or *PLC* function block, as a few lines of computer code, or as a simple hardware circuit. And while the elements above are using electrical current, they can also be designed to select between high and low voltage or *digital (discrete) counts*.

The above pictures are not meant to imply that the selected output value has anything to do with signal location. If the 12 mA signal shown entering on the lower input were to drop down to 5 mA while the 7 mA input entering from the left side remained constant, then the low select output would be 5 mA while the high select output would be 7 mA.

Logic Permits Increased Sophistication

The simple select element enables decision-making logic to be included in a control strategy, which in turn provides a means for increasing strategy sophistication. One such example is to use a select element to construct an architecture designed to control to a maximum or minimum limit or constraint.

Another popular application, and the one explored here, is to employ a select as an *override element* in a ratio control architecture. In particular, we explore how a select override might be included in an air/fuel ratio combustion control strategy to enhance safety, limit emissions and maximize useful energy from fuel.

Ratio Strategy Without Override

Before demonstrating the use of a select override, we consider a variation on our previously discussed *ratio control of a metered-air combustion process.*

Ratio Control with Remote Set Point to Wild Feed Loop

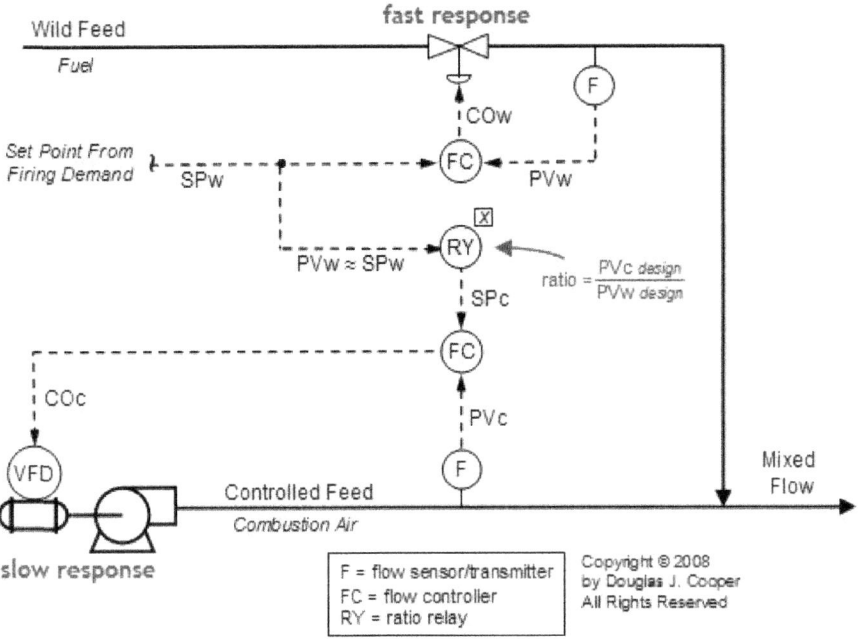

In this design, the fuel mass flow rate is regulated by a flow controller whose set point, SPw, arrives as a firing demand from elsewhere in the plant. SPw might be generated, for example, by a controller adjusting the duct temperature downstream of a burner, the temperature of a heat transfer fluid exiting a furnace or the pressure in a steam header.

There is an implicit assumption in this architecture that the fuel mass flow rate closely tracks the firing demand set point, that is, PVw ≈ SPw. Thus, an integral term (*e.g., PI control*) is required in the wild feed flow control algorithm.

Since SPw is set elsewhere, we are not free to adjust fuel flow rate separately to maintain a desired air/fuel ratio. It is thus appropriately designated as the "wild feed" in this construction.

As SPw (and thus PVw) increases and decreases, a ratio relay shown in the control diagram multiplies the incoming signal by a design air/fuel ratio value (or in the general case, a controlled/wild feed ratio value) to compute the combustion air set point, SPc, for the controlled feed flow loop.

Practitioner's Note : A ratio controller architecture requires that the signal from each mass flow sensor/transmitter change linearly with flow rate. Thus, the signals from the wild stream process variable, PVw, and the controlled stream process variable, PVc, should increase and decrease in a straight-line fashion as the individual mass flow rates increase and decrease. If the flow sensor is not linear, additional computations (function blocks) must be included between the sensor and the ratio relay to transform the nonlinear signal into the required linear flow-to-signal relationship.

If the fuel flow control loop and the combustion air control loop both respond quickly to flow commands COw and COc respectively, then the architecture above should maintain the desired air/fuel ratio even if the demand set point signal, SPw, moves rapidly and often.

Problem if the Combustion Air Loop is Slow

The diagram shows a valve as the final control element (FCE) adjusting the fuel mass flow rate, and a *variable frequency drive (VFD)* and blower assembly as the FCE adjusting the combustion air mass flow rate. Valves generally respond quickly to controller output signal commands, so we expect the fuel mass flow rate to closely track changes in COw.

In contrast, air blower assemblies vary in capability. Here we consider a blower that responds slowly to control commands, COc, relative to the valve (the *time constant* of the blower "process" is much larger than that of the valve).

While we desire that the mass flow rates of the two streams move together to remain in ratio, the different response times of the FCEs means that during a firing demand change (a change in SPw), the feed streams may not be matched at the desired air/fuel ratio for a period of time.

To illustrate, consider a case where the firing demand, SPw, suddenly increases. The fuel flow valve responds quickly, increasing fuel feed to the burner. The ratio relay will receive SPw and raise the set point of the combustion air mass flow rate, SPc, so the two streams can remain in ratio.

If the air blower response is slow, however, a fuel rich environment can temporarily develop. That is, there will be a period of time when we are below the desired 5% to 20% of excess air (below the 105% to 120% of *theoretical or stoichiometric air*) as we wait for the blower to ramp up and deliver more air to the burner.

If there is insufficient air for complete combustion, then carbon monoxide and partially burned fuel will appear in the exhaust stack. As such, we have a situation where we are wasting expensive fuel and violating environmental regulations.

Solution 1 : Detune the Fuel Feed Controller

One solution is to enter conservative or sluggish tuning values into the fuel feed controller. By detuning (slowing down) the wild feed control loop so it moves as slowly as the combustion air blower, the two feed streams will be able to track together and stay in ratio. We thus avoid creating the fuel rich environment as just described.

Unfortunately, however, we also have made the larger firing demand control system less responsive, and this diminishes overall plant performance. In some process applications, a slow or sluggish ratio control performance may be acceptable. In the particular case of combustion control, it likely is not.

Solution 2 : Use a Low Select Override

The addition of an override to our control architecture is shown below. The diagram is the same as that above except a second ratio relay feeding a low select element has been included in the design.

The second ratio relay receives the *actual* measured combustion air mass flow rate, PVc, and computes a matching fuel flow rate based on the design air/fuel ratio. This "fuel flow matched to the actual air flow" value is transmitted to the low select element. As shown below, the low select element also receives the firing demand fuel flow rate, SPw, set elsewhere in the plant.

Ratio with Low Select Override Control

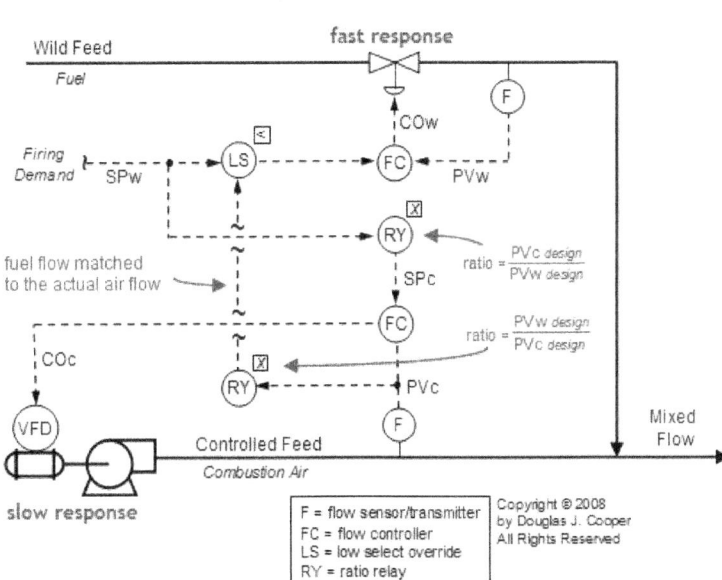

F = flow sensor/transmitter
FC = flow controller
LS = low select override
RY = ratio relay

A low select element passes the lowest of the two input signals forward. In this case, if SPw is a fuel rate that exceeds the availability of combustion air required to burn it, the select element will *override* the demand signal and forward the lower "fuel flow matched to the actual air flow" signal.

The override strategy shown above thus ensures that the feed streams remain in ratio for a rapid increase in firing demand, SPw, but it has no effect when there is a rapid decrease in firing demand.

When SPw rapidly decreases, the fuel flow rate will respond quickly and we will be in a "lean" environment (too much combustion air) until the blower slows to match the decreased fuel rate. When there is more air than that needed for complete combustion, the extra nitrogen and unneeded oxygen absorb heat energy, decreasing the temperature of the flame and gases in the combustion zone.

So while the select override element has eliminated pollution concerns when firing demand rapidly *increases*, we produce a surplus of hot air that simply leaves the exhaust stack as lost profit when firing demand rapidly *decreases*. In effect, we have solved only half the air/fuel balance problem with a single select override element.

A Simulated Furnace Air/Fuel Ratio Challenge

To further our understanding of the select override in an air/fuel ratio strategy, we consider a furnace simulation (images shown below) available in *commercial software*. The furnace burns natural gas to heat a process liquid flowing through tubes in the fire box. Firing demand is determined by a temperature controller located on the process liquid as it exits the furnace.

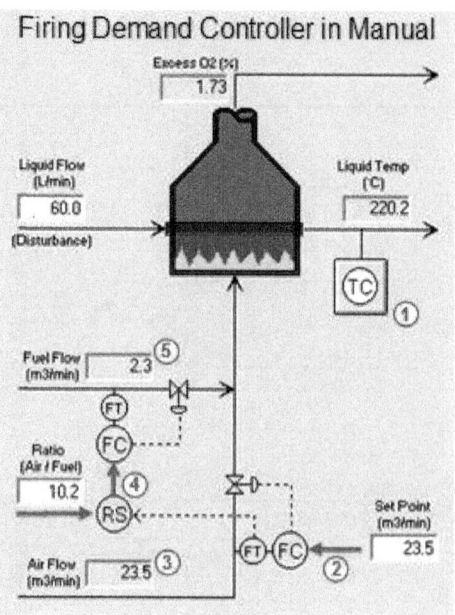

Because the output of the firing demand temperature controller becomes the set point to the wild feed of the air/fuel ratio strategy, it is in fact a primary (or outer) controller in a *cascade control* architecture. If the temperature of the process liquid is too hot (greater than set point), the firing demand controller seeks to reduce energy input. If the temperature is below set point, it seeks to add energy.

Unlike the example above, combustion air is the wild feed in this furnace simulation. Thus, when the firing demand temperature controller is in automatic (the cascade is enabled) set point changes are transmitted to the air flow controller. If the temperature controller is in manual, the set point of the combustion air flow controller must be entered manually by operations staff.

- *Firing Demand in Manual :* We first consider furnace operation when the firing demand temperature controller is in manual mode as shown below.

  Following the number labels on the above diagram :

  1. The firing demand temperature controller on the process liquid exiting the furnace is in manual mode. As such, it takes no control actions.

  2. With the firing demand controller in manual, operations staff enter a set point (SP) into the combustion air flow controller. The controller adjusts the air flow valve to ensure the measured combustion air feed equals the SP value.

  3. The display is in volume units (m3/min), though ratio controllers traditionally employ a mass flow basis.

  4. Operations staff enter a desired air/fuel ratio into the ratio station (RS). The entered value is much like a set point. The ratio station receives the combustion air flow signal and forwards a fuel flow set point to maintain the desired ratio.

  5. The flow controller adjusts the fuel valve to maintain the desired ratio. Here, the air/fuel ratio is : 23.5/2.3 = 10.2.

  Notes :

  - The fuel flow controller has its set point established by the signal from the ratio station (RS), which could be constructed, for example, by inverting the desired air/fuel ratio with a division function and then using a multiplying relay to compute a desired fuel flow rate.

  - The flow transmitters for the combustion air and fuel rate must be linearized.

  - Ratio control traditionally uses a mass flow basis. The use of volumetric flow units implies that the air and fuel are delivered at fixed temperature and pressure, thus making the volume flows proportional to the mass flows. Alternatively, a sophisticated function could translate mass flows to volume flows for display purposes. An air/fuel ratio in mass units (kg for example) would have a different value from the volume ratio because of the difference in molecular weights for the two streams.

- *Firing Demand in Automatic :* With operation steady, we switch the firing demand temperature controller to automatic as shown below :

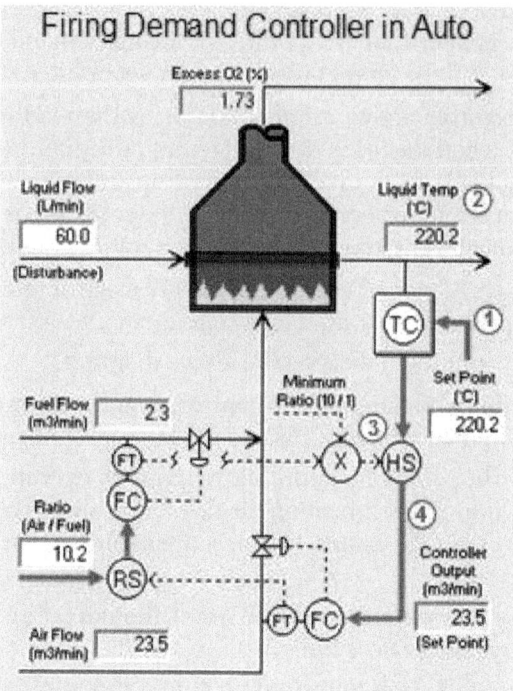

Following the number labels on the above diagram :

1. The firing demand controller measures the temperature of the process liquid exiting the furnace (the measured process variable, PV), compares it to the set point (SP) value, and computes as its controller output signal an air feed rate set point.

2. In the figure above, the measured PV equals the SP of the temperature controller, so the air feed set point from the firing demand controller is the same as when it was in manual mode.

3. A high select element receives the air feed set point from the firing demand controller and a minimum permitted air feed set point based on the current flow of fuel.

4. The larger of the two air feed set points is forwarded by the high select element to the air flow controller, ensuring that there is always sufficient air to completely combust the fuel in the firebox. Because the firing demand controller generates an air feed set point that is above the minimum 10/1 ratio specified by the designers, then the high select element passes it onward to the combustion air flow controller.

- *Firing Demand Override :* A process upset requires that the high select element override the firing demand controller as shown below :

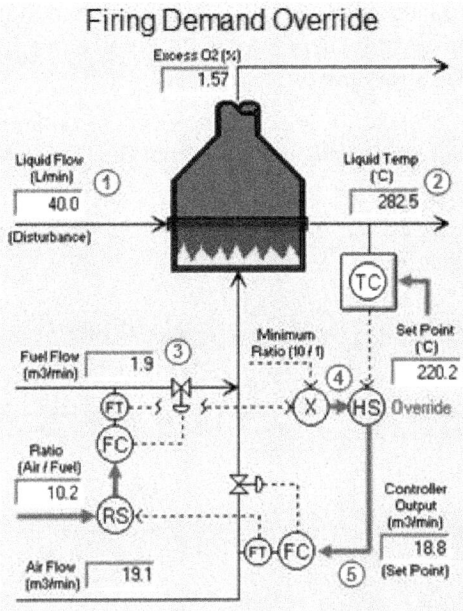

Following the number labels on the above diagram :

1. The flow rate of the process liquid drops from 60 to 40 L/min, reducing the heat energy demand on the furnace.

2. As flow rate drops, the excess energy in the furnace raises the exit tempera-ture of the process liquid. The measured PV temperature moves well above the SP, causing the firing demand controller to decrease energy input by rapidly lowering the set point to the air flow controller.

3. The current flow rate of fuel is 1.9 (actually 1.88 but the display has round off).

4. The minimum ratio that ensures enough air to complete combustion in the firebox of this furnace is a 10/1 ratio of air/fuel, or 1.88 x 10 = 18.8 m3/min. The high select element receives a combustion air feed rate from the firing demand controller that is below this minimum.

5. The high select element overrides the air flow set point from the firing demand controller and forwards the minimum 18.8 m3/min air flow suf-ficient for complete combustion.

   This second furnace example used combustion air as the wild feed. Yet in both cases, an override element was required to implemented a control strategy that enhanced safety, limited emissions and maximized the useful energy from fuel.

Cross-Limiting Ratio Control Strategy

If we were to increase the process liquid flow rate through the furnace, the firing demand controller would quickly ramp up the combustion air feed rate to provide more energy. Temporarily, there would be more air than that needed for complete combustion. That temporary surplus of hot air will carry its heat energy up and out the exhaust stack as lost profit.

So similar to the first example, a single select override element provides only half the solution depending on the direction that the upstream demand is moving.

While fairly complex, the cross-limiting structure offers benefit in that it provides protection in an air/fuel ratio strategy both when firing demand is increasing *and* decreasing.

Ratio with Low Select Override and Carefully Scaled Sensors

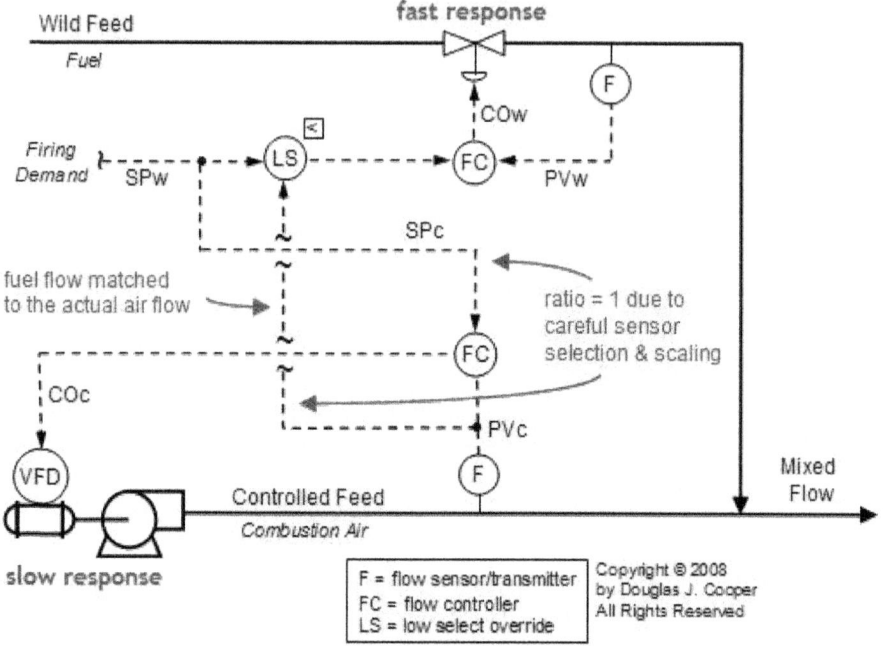

By careful sensor selection and scaling, we can maintain the "ratio with low select override" strategy while eliminating the multiplying relays from our design. As long as we use control algorithms with an integrating term (PI or PID), the upstream demand signal becomes the set point for both controllers and the desired ratio will be maintained.

RATIO WITH CROSS-LIMITING OVERRIDE CONTROL
OF A COMBUSTION PROCESS

Variations on this cross-limiting architecture are widely employed within the air/fuel ratio logic of a broad range of industrial combustion control systems.

Steam Boiler Process Example

To provide a larger context for this topic, we begin by considering a multi-boiler steam generation process as shown below :

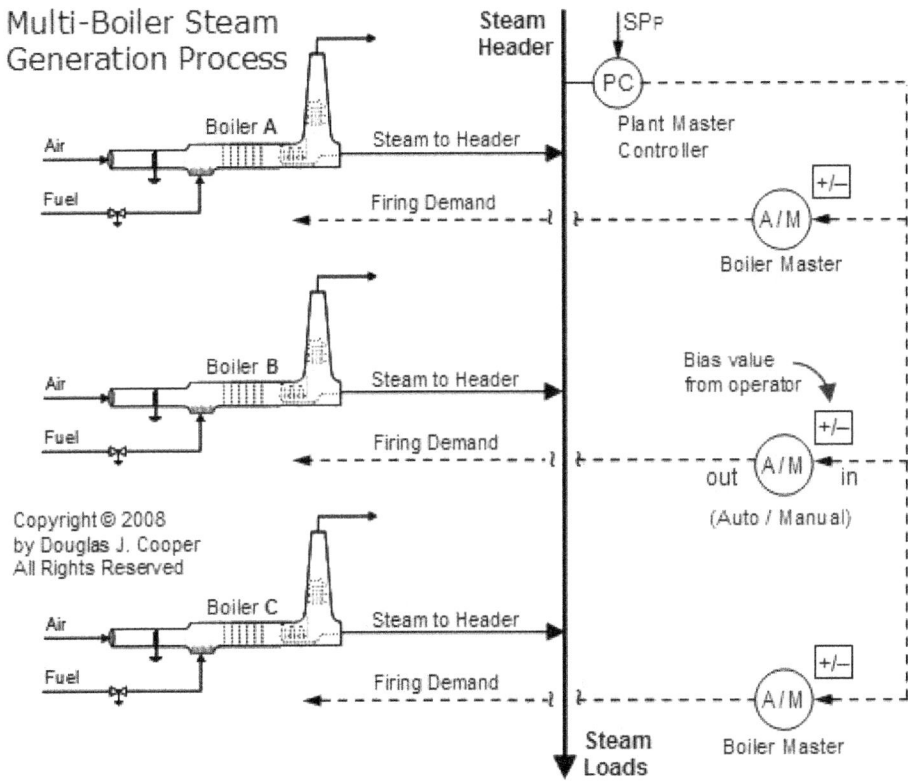

Steam generation processes often have multiple boilers that feed a common steam header. When steam is needed anywhere in the plant, the load is drawn from this common header. Steam turbines, for example, drive generators, pumps and compressors. Steam is widely used for process heating, can be injected into production vessels to serve as a reactant or diluent, and may even be used to draw a vacuum in a vessel via jet ejectors.

With so many uses, steam loads can vary significantly and unpredictably over time in a plant. The individual boilers must generate and feed steam to the common header at a rate that matches these steam load draws. Controlling the steam header to a constant pressure provides an important stabilizing influence to plant-wide operation.

- *Plant Master Controller* : A popular multi-boiler architecture for maintaining header pressure is to use a single pressure controller on the common header that outputs a firing demand signal for *all* of the boilers in the steam plant. This steam header pressure controller is widely referred to as the *Plant Master*.

Based on the difference between the set point (SP_p) and measured pressure in the header, the Plant Master controller computes a firing demand output that signals all of the boilers in the plant to increase or decrease firing, and thus, steam production.

- *Boiler Master Controller* : The Boiler Masters in the above multi-boiler process diagram are Auto/Manual selector stations with biasing (+/-) values. If all three of the Boiler Masters are in automatic, any change in the Plant Master output signal will pass through and create an associated change in the firing demand for the three boilers.

If a Boiler Master is in automatic, that boiler is said to be operating as a *swing boiler*. As such, its firing demand signal will vary (or swing) directly as the Plant Master signal varies. If each of the fuel flow meters are scaled so that 100% of fuel flow produces maximum rated steam output, then each boiler will swing the same amount as the Plant Master calls for variations in steam production.

But suppose Boiler B has cracked *refractory brick* in the fire box or some other mechanical issue that, until repaired, requires that it be operated no higher than, for example, 85% of its design steam production rate. That is, Boiler B has been *derated* and its maximum permissible steam generating capacity has been lowered from the original design rating. Two options we can consider include :

1. When a Boiler Master is in automatic, then : signal out = signal in + bias

   where the bias value is set by the operator. If the bias value of Boiler Master B is set in this example to –15%, then no matter what output is received from the Plant Master (0% to 100%), the firing demand signal will never exceed 85% (100% plus the negative 15% bias). In this mode of operation, Boiler B will still swing with Boiler A and Boiler C in response to the Plant Master, but it will operate at a firing rate 15% below the level of the other two boilers (assuming their bias values are zero).

2. If a boiler is suffering from refractory problems, then allowing the firing rate to swing can accelerate refractory degradation. Thus, Boiler Master B might alternatively be switched to manual mode where the output firing demand signal is set to a constant value. In manual mode, Boiler B is said to provide a base load of steam production. With the firing rate of Boiler B set manually from the Boiler Master, it is unresponsive to firing demand signal variations from the Plant Master. We then would have two swing boilers (Boiler A and Boiler C) and one base loaded boiler (Boiler B).

Combustion Control Process

As shown below, each furnace and steam boiler has its own control system. Of particular interest here is the maintenance of a specified air/fuel mass ratio for efficient combustion at the burners.

As shown above, the air/fuel ratio control strategy receives a firing demand from the Boiler Master. Air mass flow rate may be measured downstream of the combustion zone and is thus shown as an input to the ratio control strategy.

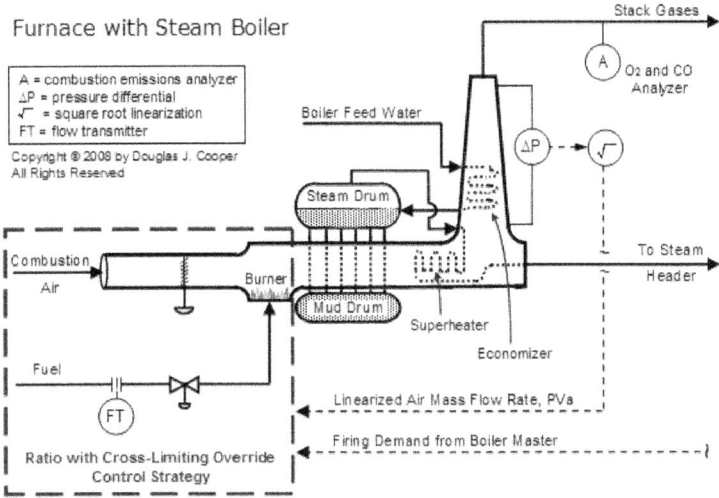

Furnace with Steam Boiler

Ratio with Cross-Limiting Override Control

Certain assumption are used in the presentation that follows :

1. Air/fuel ratio is normally expressed as a mass flow ratio of air to fuel.

2. The air and fuel flow transmitter signals are linear with respect to the mass flow rate and have been scaled to range from 0-100%.

3. The flow transmitters have been carefully calibrated so that both signals at the design air/fuel ratio are one to one. That is, if the fuel flow transmitter signal, PV_f, is 80%, then an air flow signal, PVa, of 80% will produce an air flow rate that meets the design air/fuel mass ratio.

Shown below are the sensors, controllers, final control elements (FCEs) and function blocks that might be included in the above dashed box labelled "ratio with cross-limiting override control strategy."

Ratio with Cross-Limiting Override Control Strategy

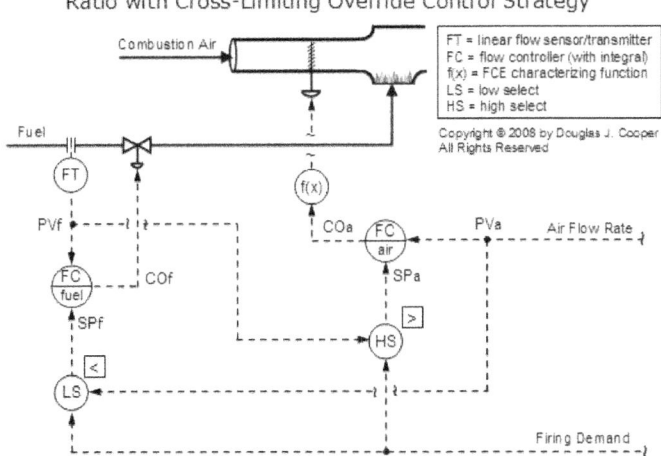

Before discussing the details of the strategy, we rearrange the loop layout to make the symmetry of the design more apparent. Specifically, we reverse the fuel flow direction (fuel now flows from right to left below) and show the air mass flow rate transmitter as a generic measurement within the control architecture. The control diagram above is otherwise identical to that below.

Rearranged Ratio with Cross-Limiting Override Control

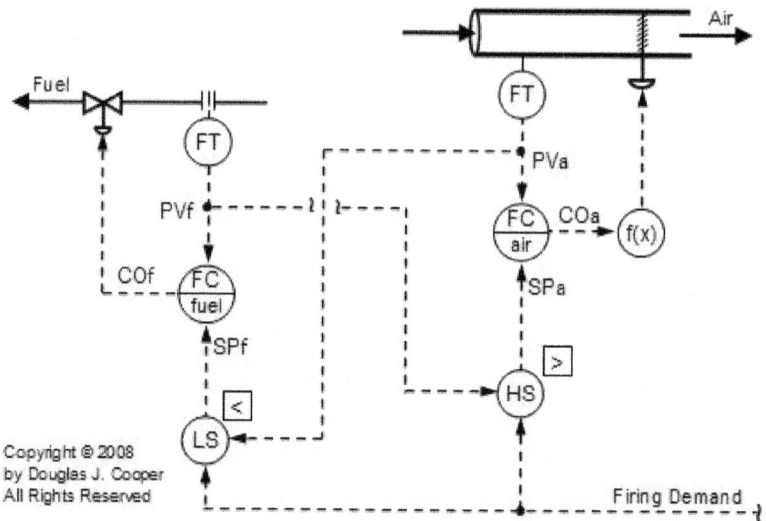

Practitioner's Note : In any real process, different flow loops will have different *process gains* (the same change in controller output signal, CO, will produce a different change in flow rate) and each loop itself will display a *nonlinear behavior* over its range of operation (the process gain, time constant and/or dead time will change as operating level changes). The purpose of the characterizing function block, $f(x)$, is to match the process gain of one loop over the range of operation with that of the other loop. With matching signal-to-flow gains, this optional function block simplifies the tuning of a ratio control strategy with two flow control loops. The characterizing function block, $f(x)$, also simplifies manual operation because the two flow CO signals will be approximately equal at the design air/fuel ratio.

As shown above, the firing demand signal enters the high select override as a candidate for the set point of the air flow controller (SPa). In this cross-limiting strategy, the same firing demand signal enters the low select override as a candidate for the set point of the fuel flow controller (SP$_f$).

As discussed in assumption 3 above, the flow transmitters have been calibrated so that when both signals match, we are at the design air/fuel mass flow ratio. Thus, because of the high select override, SPa is always the greater of the the firing demand signal or the value that matches the current fuel flow signal. And because of the low select override, SP$_f$ is always the lesser of the firing demand signal or the value that matches the current air flow signal.

The result is that if firing demand moves up, the high select will pass the firing demand signal through as SPa, causing the air flow to increase. Because of the low select override, the fuel set point, SP_f, will not match the firing demand signal increase, but rather, will follow the increasing air flow rate as it responds upward.

And if the firing demand moves down, the low select will pass the firing demand signal through as SP_f, causing the fuel flow to decrease. Because of the high select override, the air set point, SPa, will not match the firing demand signal decrease, but rather, will track the decreasing fuel flow rate as it moves downward.

In short, the control system ensures that during sudden operational changes that move us in either direction from the design air/fuel ratio, the burner will temporarily receive extra air until balance is restored (we will be temporarily lean). While a lean air/fuel ratio means we are heating extra air that then goes up and out the stack, it avoids the environmentally harmful emission of carbon monoxide and unburned fuel.

Variable Air/Fuel Ratio

The basic cross-limiting strategy we have described to this point provides no means for adjusting the air/fuel ratio. This may be necessary, for example, if the composition of our fuel changes, if the burner performance changes due to corrosion or fouling, or if the operating characteristics of the burner change as firing level changes.

Shown below is a cross-limiting override control strategy that also automatically adjusts the air/fuel ratio based on the oxygen level measured in the exhaust stack.

Ratio with Cross-Limiting Override & Oxygen Trim (Remote Input)

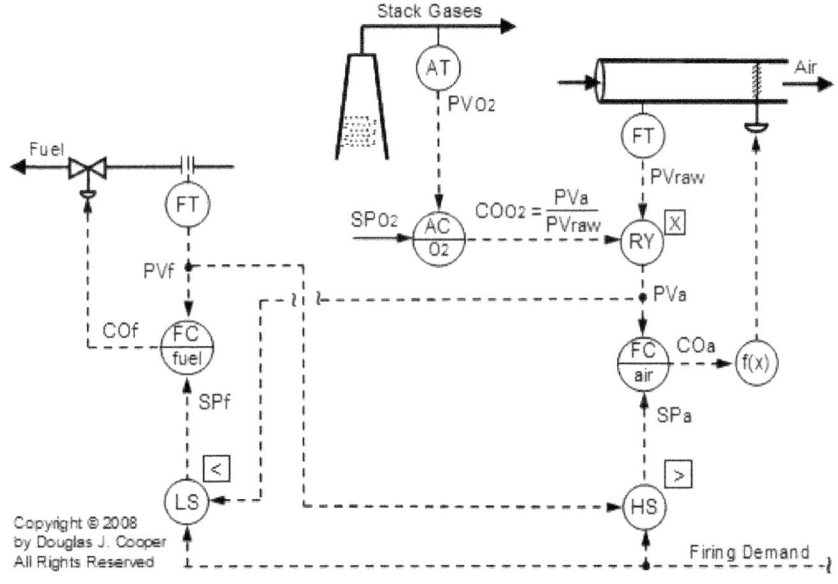

As shown in the diagram, the signal from the air flow transmitter, PV_{raw}, is multiplied by the output of the analyzer controller, COO_2, and the product is forwarded as the measured air flow rate process variable, PV_a. With this construction, if the measured exhaust oxygen, PVO_2, matches the oxygen set point, SPO_2, then the analyzer controller (AC) output, COO_2 will equal one and PVa will equal PV_{raw}.

But if the oxygen level in the stack is too high, COO_2 will become greater than one. By multiplying the raw air flow signal, PV_{raw}, by a number greater than one, PV_a appears to read high. And if the oxygen level in the stack is too low, we multiply PV_{raw} with a number smaller than one so that PV_a appears to read low.

The ratio strategy reacts based on the artificial PV values, adjusting the air/fuel ratio until the measured oxygen level, PVO_2, is at set point SP_{02}.

This manipulation to the air/fuel ratio based on measured exhaust oxygen is commonly called oxygen trim control. By essentially changing the effective calibration of the air flow transmitter to a new range, the signal ratio of the carefully scaled air and fuel transmitters can remain 1 : 1.

Practitioner's Note : Analyzers fail more often than other components in the strategy, so when designing and tuning the analyzer controller, it is important to limit how far COO_2 can move from its baseline value of one. Also, the analyzer controller is effectively the primary (outer) controller in a *cascade loop*. The secondary (inner) loop is the same air flow control loop being driven by the Plant Master. As a result, it is advisable to tune the oxygen (or combustibles) trim controller significantly more conservatively than the Plant Master to minimize loop interactions.

Chapter 13

CASCADE, FEED FORWARD AND THREE-ELEMENT CONTROL

CASCADE, FEED FORWARD AND BOILER LEVEL CONTROL

One common application of cascade control combined with feed forward control is in level control systems for boiler steam drums.

The control strategies now used in modern industrial boiler systems had their beginnings on shipboard steam propulsion boilers. When boilers operated at low pressure, it was reasonably inexpensive to make the steam drum large. In a large drum, liquid level moves relatively slowly in response to disturbances (it has a long time constant). Therefore, manual or automatic adjustment of the feedwater valve in response to liquid level variations was an effective control strategy.

But as boiler operating pressures have increased over the years, the cost of building and installing large steam drums forced the reduction of the drum size for a given steam production capacity.

The consequence of smaller drum size is an attendant reduction in process time constants, or the speed with which important process variables can change. Smaller time constants mean upsets must be addressed more quickly, and this has led to the development of increasingly sophisticated control strategies.

3 Element Strategy

As shown below, most boilers of medium to high pressure today use a "3-element" boiler control strategy. The term "3-element control" refers to the number of process variables (PVs) that are measured to effect control of the boiler feedwater control valve. These measured PVs are :

- Liquid level in the boiler drum,
- Flow of feedwater to the boiler drum, and

- Flow of steam leaving the boiler drum.

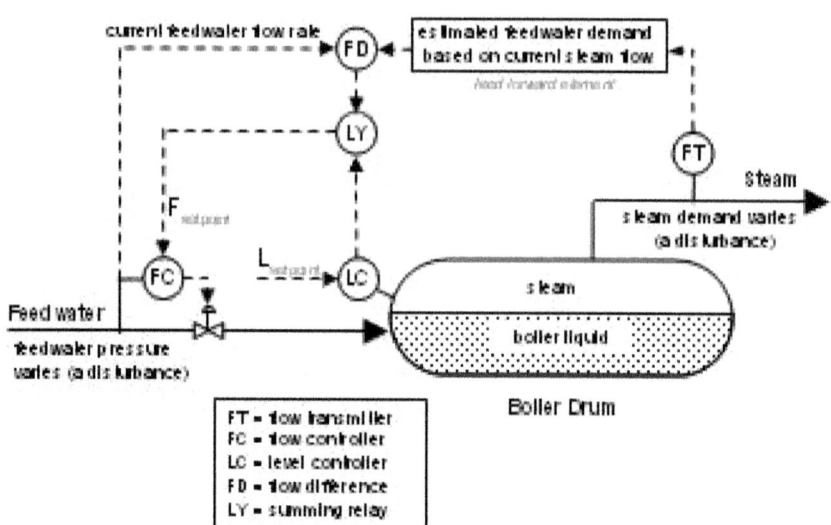

Maintaining liquid level in the boiler steam drum is the highest priority. It is critical that the liquid level remain low enough to guarantee that there is adequate disengaging volume above the liquid, and high enough to assure that there is water present in every steam generating tube in the boiler. These requirements typically result in a narrow range in which the liquid level must be maintained.

The feedwater used to maintain liquid level in industrial boilers often comes from multiple sources and is brought up to steam drum pressure by pumps operating in parallel. With multiple sources and multiple pumps, the supply pressure of the feedwater will change over time. Every time supply pressure changes, the flow rate through the valve, even if it remains fixed in position, is immediately affected.

So, for example, if the boiler drum liquid level is low, the level controller will call for an increase in feedwater flow. But consider that if at this moment, the feedwater supply pressure were to drop. The level controller could be opening the valve, yet the falling supply pressure could actually cause a decreased flow through the valve and into the drum.

Thus, it is not enough for the level controller to directly open or close the valve. Rather, it must decide whether it needs more or less feed flow to the boiler drum. The level controller transmits its target flow as a set point to a flow controller. The flow controller then decides how much to open or close the valve as supply pressure swings to meet the set point target.

This is a "2-element" (boiler liquid level to feedwater flow rate) cascade control strategy. By placing this feedwater flow rate in a fast flow control loop, the flow controller will immediately sense any variations in the supply conditions which produce a change in feedwater flow. The flow controller will adjust the boiler feedwater valve position to restore the flow to its set point before the boiler drum liquid level is even affected. The level controller is the primary controller (sometimes referred to as the master controller) in this cascade, adjusting the set point of the flow controller, which is the secondary controller (sometimes identified as the slave controller).

The third element in a "3-element control" system is the flow of steam leaving the steam drum. The variation in demand from the steam header is the most common disturbance to the boiler level control system in an industrial steam system.

By measuring the steam flow, the magnitude of demand changes can be used as a feed forward signal to the level control system. The feed forward signal can be added into the output of the level controller to adjust the flow control loop set point, or can be added into the output of the flow control loop to directly manipulate the boiler feedwater control valve. The majority of boiler level control systems add the feed forward signal into the level controller output to the secondary (feedwater flow) controller set point. This approach eliminates the need for characterizing the feed forward signal to match the control valve characteristic.

Actual boiler level control schemes do not feed the steam flow signal forward directly. Instead, the difference between the outlet steam flow and the inlet water flow is calculated. The difference value is directly added to the set point signal to the feedwater flow controller. Therefore, if the steam flow out of the boiler is suddenly increased by the start up of a turbine, for example, the set point to the feedwater flow controller is increased by exactly the amount of the measured steam flow increase.

Simple material balance considerations suggest that if the two flow meters are exactly accurate, the flow change produced by the flow control loop will make up exactly enough water to maintain the level without producing a significant upset to the level control loop. Similarly, a sudden drop in steam demand caused by the trip of a significant turbine load will produce an exactly matching drop in feedwater flow to the steam drum without producing any significant disturbance to the boiler steam drum level control.

Of course, there are losses from the boiler that are not measured by the steam production meter. The most common of these are boiler blow down and steam vents (including relief valves) ahead of the steam production meter. In addition, boiler operating conditions that alter the total volume of water in the boiler cannot be corrected by the feed forward control strategy. For example, forced circulation boilers may have steam generating sections that are placed out of service or in service intermittently. The level controller itself must correct for these unmeasured disturbances using the normal feedback control algorithm.

Notes on Firing Control Systems

In general, firing control is accomplished with a Plant Master that monitors the pressure of the main steam header and modulates the firing rate (and hence, the steam production rate) of one or more boilers delivering steam to the steam header. The firing demand signal is sent to all boilers in parallel, but each boiler is provided with a Boiler Master to allow the Plant Master demand signal to be overridden or biased. When the signal is overridden, the steam production rate of the boiler is set manually by the operator, and the boiler is said to be base-loaded. Most boilers on a given header must be allowed to be driven by the Plant Master to maintain pressure control. Boilers that have the Boiler Master set in automatic mode (passing the steam demand from the Plant Master to the boiler firing control system) are said to be swing boilers as opposed to base-loaded boilers.

The presence of heat recovery steam boilers on a steam header raises new control issues because the steam production rate is primarily controlled by the horsepower demand placed on the gas turbine providing the heat to the boiler. If the heat recovery boiler operates at a pressure above the header pressure, a separate pressure control system can be used to blow off excess steam from the heat recovery boiler when production is above the steam header demand. Note that for maximum efficiency, most heat recovery boilers are fitted with duct burners to provide additional heat to the boiler. The duct burner is controlled with a Boiler Master like any other swing boiler. As long as there are other large swing boilers connected to the steam header, the other fired boilers can reduce firing as required when output increases from the heat recovery boiler.

DYNAMIC SHRINK/SWELL AND BOILER LEVEL CONTROL

Boiler Start-up

As high pressure boilers ramp up to operating temperature and pressure, the volume of a given amount of saturated water in the drum can expand by as much as 30%. This natural expansion of the water volume during start-up is not dynamic shrink/swell, though it does provide its own unique control challenges.

The expansion (or more precisely, decrease in density) of water during start-up of the boiler poses a problem if a differential pressure or displacer instrument is used for level measurement. Such a level transmitter calibrated for saturated water service at say, 600 psig, will indicate higher than the true level when the drum is filled with relatively cool boiler feedwater at a low start-up pressure.

If left uncompensated at low pressure conditions, the "higher than true level" indication will cause the controller to maintain a lower than desired liquid level in the drum during the start-up period. If the low level trip device is actually sensitive to the interface (e.g. conductance probes or float switches), troublesome low level trip events become very likely during start-up.

This variation in the sensitivity of the level transmitter with operating conditions can be corrected by using the drum pressure to compensate for the output of the level transmitter. The compensation can be accomplished with great accuracy using steam table data. The compensation has no dynamic significance and can be used independent of boiler load or operating pressure.

Dynamic Shrink/Swell

Dynamic shrink/swell is a phenomenon that produces variations in the level of the liquid surface in the steam drum whenever boiler load (changes in steam demand) occur. This behavior is strongly influenced by the actual arrangement of steam generating tubes in the boiler.

I have significant experience with "water wall" boilers that have radiant tubes on three sides of the firebox. There is a steam drum located above the combustion chamber and a mud drum located below the combustion chamber.

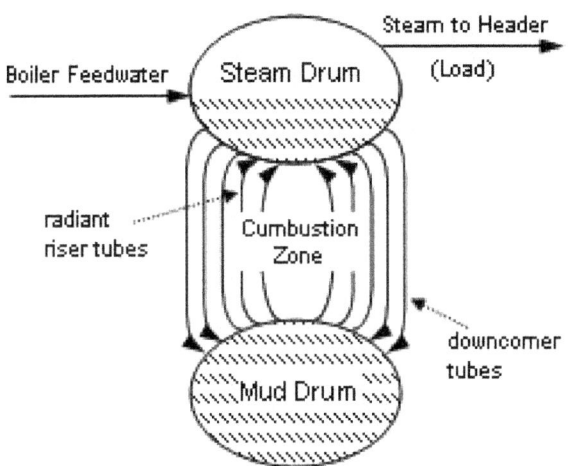

During operation, the tubes exposed to the radiant heat from the flame are always producing steam. As the steam rises in the tubes, boiler water is also carried upward and discharged into the steam drum. Tubes that are not producing significant steam flow have a net downward flow of boiler water from the steam drum to the mud drum.

The tubes producing large quantities of steam are termed risers and those principally carrying water down to the mud drum from the steam drum are termed downcomers. Excluding the tubes subject to radiant heat input from the firebox flame, a given tube will serve as a riser at some firing rates and a downcomer at other firing rates.

The mechanics of the natural convection circulation of boiler water within the steam generator is the origin of the dynamic shrink/swell phenomenon. Consider what happens to a boiler operating at steady state at 600 psig when it is subjected to a sudden increase in load (or steam demand).

A sudden steam load increase will naturally produce a drop in the pressure in the steam drum, because, initially at least, the firing rate cannot increase fast enough to match the steam production rate at the new demand level. When the pressure in the drum drops, it has a dramatic effect on the natural convection within the boiler. The drop in pressure causes a small fraction of the saturated water in the boiler to immediately vaporize, producing a large amount of boil-up from most of the tubes in the boiler. During the transient, most of the tubes temporarily become risers. The result is that the level in the steam drum above the combustion chamber rises.

However, this rise in level is actually an inverse response to the load change. Since, the net steam draw rate has gone up, the net flow of water to the boiler needs to increase, because the total mass of water in the boiler is falling. However, the level controller senses a rise in the level of the steam drum and calls for a reduction in the flow of feedwater to the boiler.

This inverse response to a sudden load increase is dynamic swell. Dynamic shrink is also observed when a sudden load decrease occurs. However, the dynamic shrink phenomenon does not disrupt the natural convection circulation of the boiler as completely as the dynamic swell effect. Consequently, the reduction in level produced by a sudden decrease in load is typically much smaller and of shorter duration than the effect produced by dynamic swell.

Control Strategy for Shrink/Swell

When a sudden load (steam demand) increase occurs, the feed forward portion of the strategy will produce an increase in the set point for the feedwater flow controller. This increase in feedwater flow controller set point will be countered to varying degrees by the level controller response to the temporary rise in level produced by the dynamic swell.

The standard tool used to minimize the impact of the swell phenomenon on the level in a three-element level control system is the lead-lag relay in the feed forward signal from the flow difference relay. This is the traditional means of dealing with mismatched disturbance and manipulated variable dynamics in feed forward systems, and is certainly applicable in this control strategy. When used in the three-element level control strategy, the lead-lag relay is commonly termed the "shrink/swell relay."

There are two significant limitations to the use of the lead-lag relay for shrink/swell compensation. To begin with, the response of most boilers to a load increase (swell event) is much more dramatic than the response to a load decrease (shrink event). In other words, the system response is very asymmetric. The lead-lag relay is perfectly symmetrical in responding to load changes in each direction and cannot be well matched to both directions.

Furthermore, the standard method of establishing the magnitudes of the lead time constant and lag time constant involves open loop tests of the process response to the disturbance (steam load) and to the manipulated variable (feedwater flow). A step test of the manipulated variable is generally not too difficult to conduct. However, changing the firing rate upward fast enough to actually produce significant swell is difficult without seriously upsetting the steam system, an event that is to be avoided in most operating plants. Therefore, the practitioner's only choice is to gather accurate data continuously and wait for a disturbance event that will exercise the shrink/swell relay's function.

When a lead-lag relay is to be added to an existing three-element boiler control scheme, operator knowledge of the boiler behavior in sudden load increase situations can guide the initial settings. For example, if the operators indicate that they must manually lead the feedwater valve by opening it faster than the control system will open it automatically, it is clear that a lead time constant larger than the lag time is required. Similarly, if the operator must retard the valve response to prevent excessively high level, the lead time constant must be less than the lag time. The lag time constant will typically fall in the range of one minute to three minutes. The ratio of the lead time constant to the lag time constant determines the magnitude of the initial response to the disturbance. If the ratio is one to one, the system behaves the same as a system with no lead-lag relay.

Ultimately, the system must be adjusted by watching the response to actual steam system upsets that require sudden firing increases. If the initial observed response of level to an upset is a rising level, the ratio of lead time to lag time should be decreased. The inverse is similarly true. If the recovery from an initial rise in level is followed by significant overshoot below the level target, the lag time should be reduced. If the recovery from an initial level drop is followed by a large overshoot above the level target, the lag time should be increased.

Chapter 14

DISTILLATION COLUMN CONTROL

DISTILLATION : INTRODUCTION TO CONTROL

Background

Approximately 40,000 distillation columns are operated in the U.S. chemical process industries and they comprise 95% of the separation processes for these industries. Because distillation operation directly affects product quality, process production rates and utility usage, the economic importance of distillation control is clear. Distillation control is a challenging problem because of the following factors :

- Process nonlinearity
- Multivariable coupling
- Severe disturbances
- Nonstationary behavior.

Distillation columns exhibit static nonlinearity because impurity levels asymptotically approach zero. The impurity level in the overhead product is the concentration of the heavy key, and the impurity level in the bottoms product is the concentration of the light key. Nonlinear dynamics, *i.e.*, variations in time constants with the size and direction of an input change, and static nonlinearity are much more pronounced for columns that produce high-purity products, *e.g.*, columns that have impurity levels less than 1%.

Coupling is significant when the composition of both overhead and bottoms products are being controlled. Columns are affected by a variety of disturbances, particularly feed composition and flow upsets. Nonstationary behavior stems from changes in tray efficiencies caused by entertainment or fouling.

Improved distillation control is characterized by a reduction in the variability of the impurities in the products. Meeting the specification requirements on the variability of final products can make the difference between the product being a high value-added product with large market demand and being a low-valued product with a small market demand.

For customers who purchase the products produced by distillation columns as feedstock for their processes, the variability of the feedstock can directly affect the quality of the products they produce, *e.g.*, the variability in the monomer feed to a polymerization process can directly affect the mechanical properties of the resulting polymer produced.

In addition, control performance can affect plant processing rates and utility usage. After the variability of a product has been reduced, the set point for the impurity in the product can be increased, moving the set point closer to the specification limit. If this column is the bottleneck for the process, then increasing the average impurity level, *i.e.*, moving the impurity set point closer to the specification limit, allows greater plant processing rates.

Even if the column in question is not a bottleneck, moving the impurity set point closer to the specification limit reduces the utility usage for the column. While each of these factors can be economically important for large-scale processes, the order of economic importance is usually product quality first, followed by process throughput and finally utility reductions.

Column Schematic

A schematic of a binary distillation column with one feed and two products is shown in Figure :

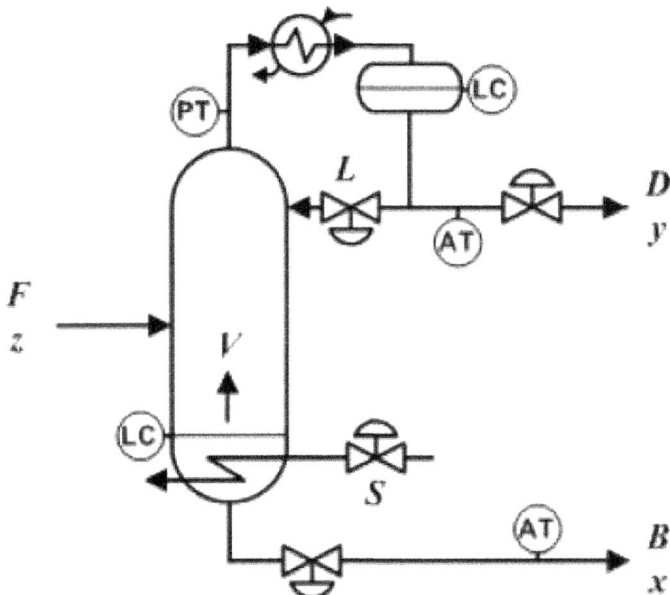

Figure : Schematic of a two product distillation column. Note that the valves on this figure represent flow control loops.

Material Balance and Energy Balance Effects

Combining an overall steady-state material balance with the light component material balance for a binary separation yields :

$$\frac{D}{F} - \frac{z-x}{y-x}$$

Rearranging results in :

$$y = x + \frac{z-x}{D/F}$$

This equation indicates that as the flow rate of the distillate product, D, decreases while keeping F, z and x constant, the purity of the overhead product, y, increases. Likewise, as D increases, its purity decreases.

Because the sum of the product flows must equal the feed rate at steady state, when one product becomes more pure, the other product must get less pure. This is shown graphically in Figure. This is an example of the material balance effect in which the product impurity level varies directly with the flow rate of the corresponding product.

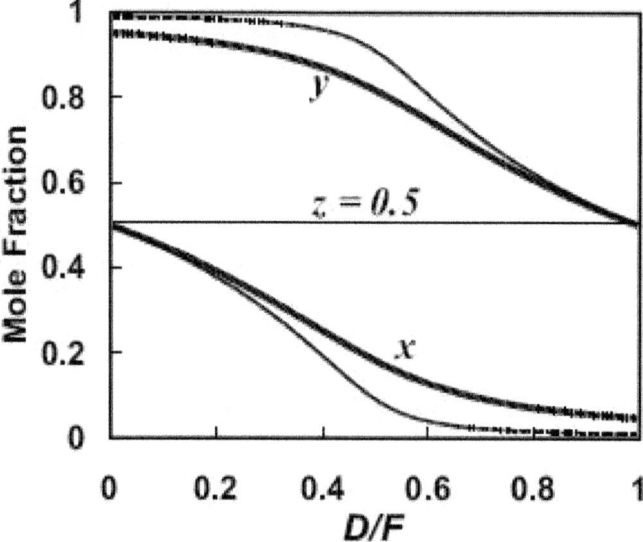

Figure : Effect of D/F and energy input on product purities. The thin line corresponds to a case in which the energy input is increased.

Another key factor that affects product purities is the energy input to the column, which determines the vapor rate, V, up the column. As the energy input to the column increases, the separation of the light and heavy components usually increases (Figure). One measure of the separation is the separation factor, S, which is given by —

$$S = \frac{y}{1-y} \frac{1-x}{x}$$

As the impurity levels in the products decrease (*i.e.*, y → 1 and/or x → 0), S increases.

Another way to understand the effect of an increase in energy input to the column is to consider the vapor/liquid traffic inside the column. If V increases while D and B are kept constant, the reflux, L, increases by the same amount as V. As a result, the reflux ratio, L/D, increases. This increase in vapor/liquid traffic inside the column causes a decrease in the impurities in the products for the same D/F ratio (Figure). When evaluating how a column responds in a control situation, it is helpful to remember that *the energy input to the column generally determines the degree of separation that the column can achieve while the material balance (i.e., D/F) determines how the separation is allocated between the two products.*

Vapor and Liquid Dynamics

The difference between vapor and liquid dynamics in a distillation column can contribute to interesting composition dynamic behavior. For all but very-high-pressure columns, *i.e.*, those operating near the critical pressure of the light key, a change in V in the reboiler can be observed in the overhead in just a few seconds while a change in the reflux flow rate requires several minutes to reach the reboiler.

The hydraulic response of a tray depends on the accumulation or depletion of liquid on it. The hydraulic time constant for flow from a tray typically ranges between 3 and 10 seconds. As a result, for a column with 50 or more trays, the overall hydraulic response time is on the order of several minutes.

As an example of the effect of the difference between liquid and vapor dynamics, consider the overhead product purity for an increase in V for a column in which the accumulator level sets the reflux flow rate and the distillate rate is held constant. Initially, the increase in vapor flow moves up the column rapidly while the liquid flow down the column remains relatively constant because the reflux rate is set by the level controller on the accumulator.

In the rectifying section, the L/V ratio determines the separating power of that section. As a result of the increase in V, the concentration of the impurity in the overhead increases initially. The increase in V begins to increase the level in the accumulator, which, after some time, leads to an increase in the reflux flow. As the increased reflux flow slowly makes its way down the rectifying section, L/V increases, causing a decrease in the impurity level in the overhead product. Therefore, for this scenario, an increase in V results in an inverse response in the concentration of the overhead product due to the difference in vapor and liquid dynamic in the column.

Entrainment

For columns operating at pressures less than about 165 psia, as V increases above 80% of flood conditions, droplets of liquid from the active area of the tray are blown in the vapor to the tray above, thus reducing the separation efficiency of the column. For certain vacuum columns, the tray efficiency can drop as much as 30% as the boilup rate increases above 80% of flood. 100% of flood corresponds to the condition for which an increase in vapor rate results in a decrease in separation for the column.

As the tray efficiency decreases because of increased entrainment, the process gain decreases, requiring larger changes in the manipulated variables to obtain the same change in the product impurity levels.

Structure Packed Columns

Columns that use sections of structured packing offer significant efficiency advantages over trayed columns for low-pressure applications because there is less pressure drop across the structured packing than across a corresponding set of trays. Because of the low liquid holdup on structured packing, these columns have faster composition dynamics than trayed columns.

The liquid holdup on the structured packing is low enough that the composition profile through the packing reaches its steady-state profile much more quickly than the reboiler and accumulator. For a column with structured packing, the dynamic lag of the accumulator and the reboiler primarily determine the dynamic response of the product compositions.

DISTILLATION : INFERENTIAL TEMPERATURE CONTROL & SINGLE-ENDED CONTROL

Product Composition Measurements

Product impurity levels (measured either on-line or in the laboratory) are used by feedback controllers to adjust column operation to meet product specifications. In addition, tray temperatures are used to infer product compositions, where feasible.

On-line Analyzers

There is a range of on-line analyzers commonly used in the chemical process industries (*e.g.*, gas chromatographs). A key issue with analyzers is their associated analyzer delay.

For columns that have slow-responding composition dynamics, analyzer delay is usually less of an issue. For a propylene/propane splitter (*i.e.*, a high reflux ratio column), the composition dynamics for the primary product have a time constant of about 2 hours. As the cycle time for the analyzer increases from 5 minutes to 10 minutes, it does not significantly affect the feedback control performance.

For an ethane/propane splitter (*i.e.*, a low reflux ratio column), even a five-minute analyzer delay seriously undermines composition control performance because the time constant for this process is less than five minutes. Fortunately, most fast-acting columns have a significant temperature drop across them so that product composition can be effectively inferred from tray temperature measurements.

For columns that have a low relative volatility (*i.e.*, less than 1.4), inferential temperature control is not feasible and feedback based on an on-line analyzer is required, but these columns generally have slow composition dynamics compared to the analyzer delay. For columns with a high relative volatility (*i.e.*, greater than 2.0), inferential temperature control is usually effective.

Inferential Temperature Control

Inferential temperature control is an effective means of maintaining composition control for columns from a control and economic standpoint. That is, it greatly reduces the analyzer dead time for feedback control and is much less expensive to install and maintain than an on-line composition analyzer. Because of their superior repeatability, RTDs (resistance temperature detectors) and thermistors are usually used for inferential temperature applications.

For multicomponent separations, tray temperatures do not uniquely determine the product composition. As a result, for these cases it is essential that an on-line analyzer or, at least, periodic laboratory analysis be used to adjust the tray temperature set point to the proper level.

If feedback based on laboratory analysis is not used, offset between the desired product composition and the actual product composition can result. Column pressure also significantly affects tray temperature measurements. For most systems, a simple linear correction compensates for variations in column pressure :

$$T_{pc} = T_{meas} - K_{pr}(P - P_0)$$

where T_{pc} is the pressure-compensated temperature that should be used for feedback control, T_{meas} is the measured tray temperature, K_{pr} is pressure correction factor, P is the operating pressure and P_0 is the reference pressure.

K_{pr} can be estimated by applying a steady-state column simulator for two different pressures within the normal operating range and using the following equation :

$$K_{pr} = \frac{T_i(P_1) - T_i(P_2)}{P_1 - P_2}$$

where T_i is the value of the tray temperature for tray i predicted by the column simulator.

A steady-state column model can also be used to determine the best tray locations for inferential control by finding the trays whose temperatures show the strongest correlation with product composition. The following procedure is recommended :

1. Apply the steady-state column model at the base-case conditions (*i.e.*, X^{BC} and y^{BC}) and record the temperature of each tray as T_i^{BC}.

2. Run the column model for an increase in the impurity level in the bottoms product (*i.e.*, $x^{BC} + \Delta x$, y^{BC}) and record the tray temperatures as $T_i^{\Delta x}$. Δx should be approximately 25–50% of the impurity level at the base-case conditions.

3. Run the column model for an increase in the impurity level in the overhead product (*i.e.*, x^{BC}, $y^{BC} + \Delta y$) and record the tray temperatures as $T_i^{\Delta y}$ should be approximately 25–50% of the impurity level at the base-case conditions.

4. To locate the best tray for inferential control for the stripping section, determine tray number i that maximizes

$$\Delta T_i^{net} = \left| T_i^{\Delta x} - T_i^{BC} \right| - \left| T_i^{\Delta y} - T_i^{BC} \right|$$

Because the proper tray should not be particularly sensitive to variations in the composition of the overhead product, the effect of Δy is subtracted from the temperature difference caused by changes in the bottom product composition.

5. To locate the best tray for inferential control for the rectifying section, determine the tray number i that maximizes

$$\Delta T_i^{net} = \left| T_i^{\Delta y} - T_i^{BC} \right| - \left| T_i^{\Delta y} - T_i^{BC} \right|$$

6. Finally, repeat this procedure for a representative range of feed compositions. Then, select the tray that provides the best overall results. Remember that it may not be possible to locate the tray temperature measurement precisely at the optimum tray location.

Table shows an example of this approach used to locate the best tray temperature for inferential temperature control in the stripping section of a depropanizer.

Table : Example for Choosing the Proper Tray Location for Inferential Temperature Control

	BC	Δx	Δy	
x	0.04	0.02	0.04	
y	0.03	0.03	0.15	
	T_i^{BC}	$T_i^{\Delta x}$	$T_i^{\Delta y}$	ΔT_i^{NET}
T_1	112.4	114.1	111.8	1.1
T_2	105.1	107.5	104.6	1.9
T_3	102.8	105.6	102.2	2.3
T_4	100.8	103.9	100.2	2.5
T_5	99.0	102.4	98.5	2.9
T_6	97.3	101.1	96.8	3.3
T_7	95.6	99.7	95.0	3.5

(Contd...)

	BC	Δx	Δy	
T_8	94.0	98.4	93.6	4.0
T_9	92.4	97.1	91.7	4.2
T_{10}	90.9	95.8	90.2	4.2
T_{11}	89.4	94.4	88.6	4.2
T_{12}	87.4	93.0	85.6	4.2
T_{13}	86.5	91.6	87.0	4.2
T_{14}	85.1	90.0	85.6	4.0
T_{15}	83.6	88.4	84.6	3.8
T_{16}	82.2	86.7	81.2	3.5
T_{17}	80.7	84.9	79.7	3.2
T_{18}	79.2	83.0	78.1	2.7

Note that a temperature measurement anywhere between tray 7 and tray 16 should work well for this application for the assumed feed composition.

For certain columns, the bulk of the temperature change occurs in a few trays, resulting in a very steep temperature profile. If a single tray temperature is used to infer the product composition in such a case, feed composition changes can move the location of the steep temperature change away from the tray selected for control, leading to a situation in which the chosen tray temperature is insensitive to changes in product impurity levels.

This problem can be handled by controlling the average of several tray temperatures that bracket the area where the steep temperature changes occur for feed composition changes. In this manner, when feed changes cause the temperature profile in the column to move, at least one of the tray temperatures used in the average of the temperatures is located on the steep temperature front. The average of the tray temperatures should still be sensitive to product impurity changes over the full range of feed composition changes.

Single Composition Control

Here the composition of one product is controlled while the composition of the other product is allowed to float. In the chemical industry over 90% of the columns are operated under single composition control compared to dual composition control, which controls both the overhead and the bottoms product compositions.

Figure shows single composition control using the reflux to control the purity of the overhead product while maintaining a fixed reboiler duty. The bottom composition is not controlled directly and can vary significantly as the feed composition changes.

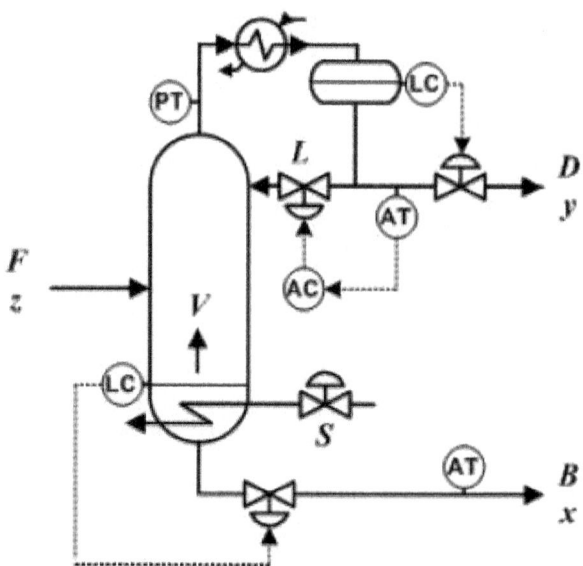

Figure : Single composition control using L to control the purity of the overhead product. The symbol for a control valve represents a flow control loop.

The control performance of the overhead product is generally best when reflux, L, rather than either the distillate flow rate, D, or the reflux ratio, L/D is the MV. L is the fastest-acting MV for the overhead and the least sensitive to feed composition changes. Because the reboiler duty is fixed, coupling is not an issue.

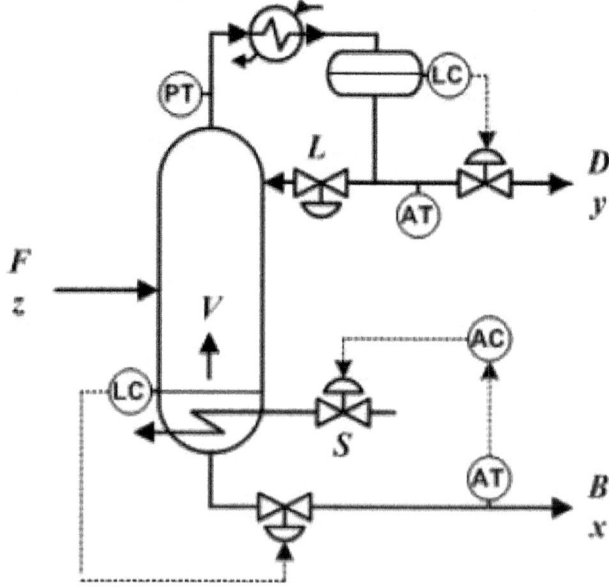

Figure : Single composition control using V to control the purity of the bottoms product. The symbol for a control valve represents a flow control loop.

When the bottom product is controlled by single composition control, the control configuration shown in Figure is recommended. Because the boilup rate, V, is faster acting and less sensitive to disturbances than either the bottoms product rate, B, or the boilup ratio, V/B, V is used to control the bottom product composition with the reflux rate fixed, which allows the overhead composition to float.

Single composition control is much easier to implement, tune and maintain than dual composition control. The choice between single and dual composition control is based on the trade off between the additional cost associated with dual composition control (analyzer costs, increased controller maintenance, etc.) and the economic benefit of dual composition control (increased product recovery and reduced utility costs).

While single composition control is in widespread use in the chemical industry, dual composition control is generally preferred for refinery columns and columns that produce high-volume chemical intermediates because the energy usage for these columns is much larger.

DISTILLATION : DUAL COMPOSITION CONTROL & CONSTRAINT CONTROL

Dual Composition Control

The choice of the proper configuration for dual composition control is a more challenging problem than for single composition control because there are more viable approaches and the analysis of performance is more complex.

There is a variety of choices for the manipulated variables, or MVs, including L, D, L/D, V, B, V/B, B/L, and D/V, that can be paired to the four control objectives (x, y, reboiler level and accumulator level). As a result, there are a large number of possible configuration choices although most of them are not practical.

It is assumed that the choice for the control configuration for the column pressure (*Figures*) is made separately from the selection of the composition control configuration.

If we limit our choices to L, D and L/D for controlling y and V, B and V/B for controlling x, there are nine possible configurations to consider : (L,V), (D,V), (L/D,V), (L,B), (D,B), (L/D,B), (L,V/B), (D,V/B) and (L/D,V/B). In each configuration, the first term is the MV used to control y and the second term is used to control x.

Figure shows the (L,V) configuration. The set point for the reflux flow controller is set by the overhead composition controller and the set point for the flow controller on the reboiler duty is set by the bottom composition controller. This leaves D to control the accumulator level and B to control the reboiler level.

Figure shows the (D,B) configuration where D is adjusted to control y and B is changed to control x, which leaves L for the accumulator level and V for the reboiler level.

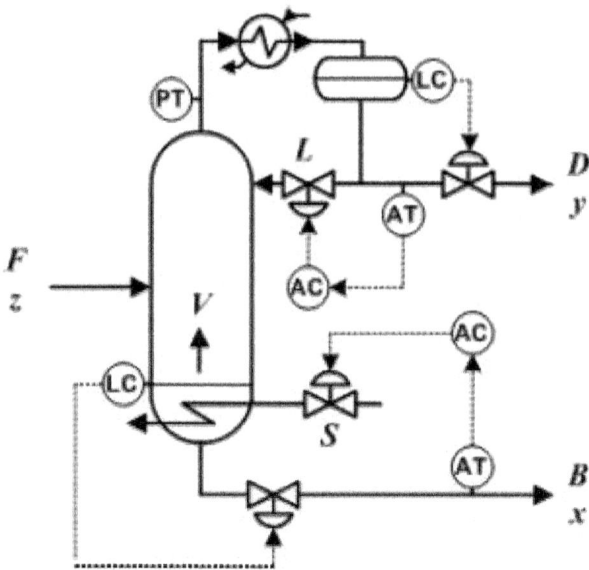

Figure : Control diagram for the (L, V) configuration for dual composition control. The symbol for a control valve represents a flow control loop.

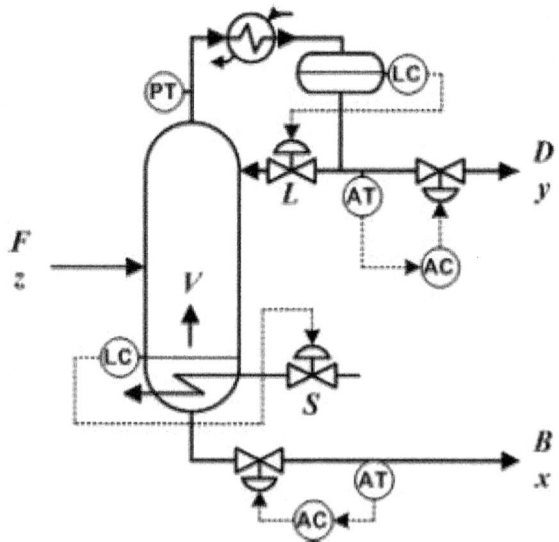

Figure : Control diagram for the (D, B) configuration for dual composition control. The symbol for a control valve represents a flow control loop.

Consider the classifications of these nine control configurations. The five configurations that use either D or B as a MV for composition control are referred to as "material balance configurations" because they use the overall material balance for the column to adjust product compositions. In fact, the (D,B) configuration has been referred to as the super material balance configuration.

The four configurations that do not use D or B as MVs are known as "energy balance configurations" because they directly adjust the vapor/liquid traffic in the column for composition control. The (L/D,V/B) configuration is known as the "double ratio configuration."

The major factors affecting the composition control performance of a particular configuration are coupling, sensitivity to disturbances and the response time for changes in the MV. The most commonly used configuration is the (L,V) configuration because it provides good dynamic response, is the least sensitive to feed composition disturbances and is the easiest to implement, even though it is highly susceptible to coupling.

On the other hand, the (L/D,V/B) configuration is, in general, the least affected by coupling and has good dynamic response, but is quite sensitive to feed composition disturbances and is more difficult to implement. The (D,B) configuration has advantages for certain high-purity columns if the levels are tuned tightly, but is open-loop unstable, *i.e.*, non-self-regulating.

As a result, there is no clear choice for the best configuration for dual composition control of distillation columns. In fact, there are specific cases for which each of the nine potential configurations listed earlier provides the best control performance.

While it is not possible to a *priori* choose the optimum configuration, there are some guidelines that can reduce the possibility of choosing a poor configuration for a particular column. In general, for high reflux ratio cases (L/D > 8), configurations that use material balance MVs (D and B) or ratios (L/D and V/B) are preferred while for low reflux ratio cases (L/D < 5), configurations that use energy balance MVs (L and V) or ratios are preferred.

In many cases, the control of one of the two products is more important than control of the other. For these cases, when the overhead product is more important, L is usually the best MV. When the bottoms product is more important, V is the proper MV. If the column is a high reflux ratio column, the MV for the less important product should be a material balance knob (D or B) or a ratio (L/D or V/B).

Because a C3 splitter is a high reflux ratio column and control of the overhead product is more important, the (L,B) or (L,V/B) configuration is preferred, which is consistent with simulation studies that have been performed. If the column is a low reflux ratio column, the less important product should be controlled by an energy balance knob (V or L) or a ratio (L/D or V/B). For example, for a low reflux column for which the bottom product is more important, the (L,V) or (L/D,V) configuration is preferred.

Table summarizes the recommended control configurations for columns in which one product is more important than the other.

Table : Recommended control configurations when one product is more important than the other

	x more important	y more important
Low reflux ratio columns (L/D <5)	$(L/D,V)$; (L,V)	$(L,V/B)$; (L,V)
High reflux ratio columns (L/D >8)	$(L/D, V)$; (D, V)	$(L, V/B)$; (L, B)

Composition Controller Tuning

For most distillation columns, the effective dead time-to-time constant ratio is relatively small. Therefore, derivative action is not necessary and PI composition controllers are commonly used. When inferential temperature control is used, fast-acting temperature loops with significant sensor lag may require derivative action because of their effective dead time-to-time constant ratios. Because most composition and temperature control loops are relatively slow responding, ATV identification with on-line tuning is recommended.

When one product is more important than the other, it is best to first tune the less important loop loosely (*i.e.*, less aggressively tuned, *e.g.*, critically damped) and then tune the important loop to control to set point tightly (*e.g.*, a 1/6 decay ratio). This dynamically decouples the multivariable control problem by providing relatively fast closed-loop dynamics for the important loop and considerably slower closed-loop dynamics for the less important loop.

In this manner, the coupling effects of the less important loop are slow enough that the important loop can easily absorb them. As a result, the variability in the important loop can be maintained consistently at a relatively low level. In effect, this approach approximates the performance of single composition control without allowing the less important product composition to suffer large offsets from set point.

When the importance of the control of both products is approximately equal, both loops need to be tuned together. In this case, both loops must be detuned equally to the point where the effects of coupling are at an acceptable level.

Constraint Control

Some of the most common column constraints include :

Maximum reboiler duty : This constraint can result from (1) an increase in the column pressure that reduces the temperature difference for heat transfer in the reboiler; (2) fouled or plugged heat-exchanger tubes in the reboiler that reduce the maximum heat transfer-rate; (3) an improperly sized steam trap that causes condensate to back up into the reboiler tubes; (4) an improperly sized control valve on the steam to the reboiler that limits the maximum steam flow to the reboiler; or (5) an increase in the column feed rate such that the required reboiler duty exceeds the maximum duty of the reboiler.

Maximum condenser duty : This constraint can be due to (1) an increase in the ambient air temperature that decreases the temperature difference for heat transfer; (2) fouled or plugged tubes in the condenser that reduce its maximum heat duty; (3) an improperly sized coolant flow control valve; (4) an increase in coolant temperature; or (5) an increase in column feed rate such that the required condenser duty exceeds the maximum duty of the condenser.

Flooding : Kister discusses in detail the three types of flooding, showing that each type results from excessive levels of vapor/liquid traffic in the column.

Weeping : Weeping results when the vapor flow rate is too low to keep the liquid from draining through a tray onto the tray below.

Maximum reboiler temperature : For certain systems, elevated temperatures in the reboiler can promote polymerization reactions to the level that excessive fouling of the reboiler results. A reboiler duty constraint can be identified when the steam flow control valve remains fully open and the requested steam flow is consistently above the measured flow rate.

Condenser duty constraints are usually identified when the column pressure reaches a certain level or when the reflux temperature rises to a certain value. The onset of flooding or weeping is generally correlated to the pressure drop across the column or across a portion of the column. It should be emphasized that it is the reboiler duty that is adjusted to honor each of these constraints, *i.e.*, prevent the process from violating the constraints.

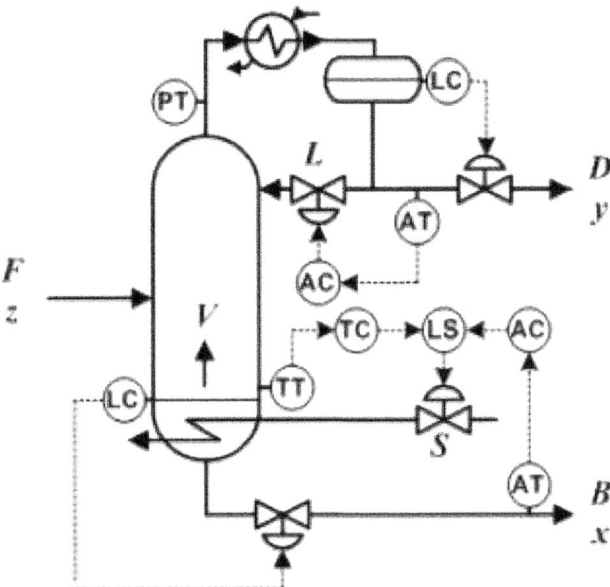

Figure : Schematic of the (L, V) configuration with LS constraint controls designed to maintain control of the overhead purity when a maximum reboiler temperature constraints is encountered.

Three approaches can be used for constraint control of a distillation column :

1. Convert from dual composition control to single composition control.
2. Reduce the column feed rate to maintain the purity of the products.
3. Reduce the product impurity setpoints for both products.

Figure shows how a low select can be used to switch between constraint control and unconstrained control for a maximum reboiler temperature constraint when the overhead is the more important product. When the bottom product is more important, the combined constrained and unconstrained configuration is more complicated, with more potential configuration choices.

Figure shows a control configuration for observing the maximum reboiler temperature constraint when the bottoms product is more important. Note that in this case, the reflux flow rate is used to maintain the bottom product composition while the reboiler duty is used to maintain the reboiler temperature. In this configuration, the reboiler temperature control loop acts relatively quickly while the bottoms composition loop is slower acting because reflux flow is used as the MV.

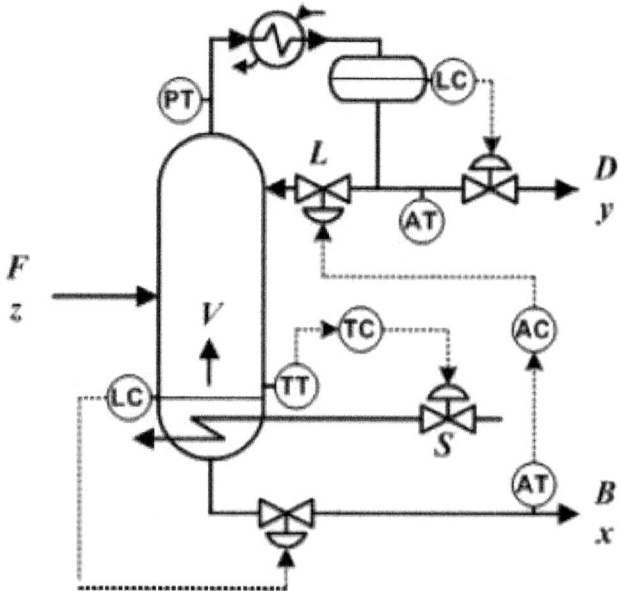

Figure : Schematic for the configuration of a constraint controller for a maximum reboiler constraint for which the bottom product is most important.

Multivariable Control

The advantage of model predictive control (MPC) applied to distillation columns is greatest when MPC is applied to a series of columns, *e.g.*, an entire separation train. This results because the MPC controller can efficiently operate the system of columns against the operative constraints to maximize the throughput for the system.

For a typical industrial separation train, there is a large complex set of constraints that limit the overall process throughput. As a result, there is a large number of combinations of operative constraints that must be controlled if the controller is to maximize the throughput for the full range of operation. Advanced PID constraint controllers applied to this problem require a separate control configuration for each combination of operative constraints, resulting in an excessively large number of separate control configurations.

On the other hand, an MPC controller can directly handle the full range of combinations of operative constraints with a single MPC controller. In addition, the MPC controller is much easier to maintain than a custom-built advanced PID controller for such a complex system. MPC can provide significant control improvements over PID control for a single column in most cases, but these improvements pale in comparison to the economic advantages offered by applying MPC to large-scale processes.

Keys to Effective Distillation Control

For effective distillation control, it is imperative to take care of the basics first.

1. Ensure that the regulatory controls are functioning properly.
2. Evaluate the analyzer dead time, reliability and accuracy.
3. Check that RTDs or thermistors are being used to measure tray temperatures for composition inference and that they are correctly located. Also, ensure that pressure-corrected tray temperatures are used.
4. Use internal reflux controls for changes in reflux temperature.
5. When L, D, V and B are used as MVs for composition control, ratio them to the measured feed rate when column feed rate changes are a common disturbance.

For configuration selection, use the (L,V) configuration for single composition control. For dual composition control, use an energy balance configuration for low reflux ratio cases and use material balance or ratio configurations for high reflux ratio columns. For many dual composition control cases, the control of one product is much more important than the other. For such cases, you should use L as the MV when the important product is produced in the overhead and V when it is produced in the bottoms.

Additionally, the less important product should be controlled using an energy balance knob (L or V) or a ratio knob (L/D or V/B) for low reflux cases or D, L/D, B or V/B for high reflux columns. For these cases, it is important to tune the important loop tightly and tune the less important loop much less aggressively.

Finally, override and select control should be applied to ensure that all column constraints are satisfied when they become operative.

Chapter 15

A DISCRETE TIME LINEAR MODEL OF THE HEAT EXCHANGER

Building the Discrete Time Model

Learning how to model a first order plus dead time (FOPDT) system is the easiest place to start. This model simply uses the last process variable (PV) and the new CO (control output) to estimate what the new value of PV will be in the next time period.

The form of the discrete time model is :

$$PV(n+1) = A \cdot PV(n) + B \cdot Kp \cdot CO(n-\theta p) + C$$

The FOPDT model is basically a simple single pole filter similar to an RC circuit in electronics, except here :

1. The system has gain, Kp
2. The input at time n affects the output at time n+1
3. The system has an initial steady state, C, that is not 0
4. The system has dead time, θp

The first thing that must be done is to build the basic model by specifying the A and B constants. Note that A and B are constants for a first order system and are arrays for higher order systems. So in this study :

$$A = \exp(-T/Tp)$$
$$B = (1-A)$$

T is the sample period and Tp is the plant time constant. A and B are coefficients in the range from 0 to 1, and coefficient A tends to be close to 1.

Calculating coefficient C in the above model is the tricky part. The heat exchanger steady state is not 0°C. If coefficient C is left out the above model equation or it is set to zero, than PV will approach 0. This does not match the observed behavior of the heat exchanger. Therefore, a value for C must be calculated that will provide the correct value of PV when CO is set to 0.

This is done as follows. From the *heat exchanger data plot*, we can see that the heat exchanger temperature (PV) is 140°C degrees when the controller output (CO) is 39%. If these values are plugged into the formula y=mx+b, we can solve for the steady state temperature, PVss :

140 = 39Kp + PVss

And this lets us calculate coefficient C :

C = (1–A) PVss

The data below is from the *Validating Our Heat Exchanger FOPDT Model* tutorial. The heat exchanger is modeled as a reverse acting FOPDT system with Laplace domain transfer function :

$$Gp = \frac{Kpe^{-\theta ps}}{Tp\,s+1}$$

The model parameter values for Kp, Tp and θp where provided in the tutorial as :

- Plant gain, Kp =–0.533 °C/%
- Time constant, Tp = 1.3 min
- Dead time, θp = 0.8 min

    Also,

- Sample time, T = 1/60 min
- Dead time in samples = nθp = θp/T = 48

    Steady state PV can now be computed assuming the system is linear :

    PVss = 140–39Kp = 160.8 °C

Next generate the discrete time transition matrix. In this simple case it is a single element :

A = exp(–T/Tp) = 0.987

Also generate the discrete time input coupling matrix, which again is a single element. Notice that the plant gain is negative :

B Kp = (1–A) Kp

= (1–0.987)(–0.533)

= –0.00679

Using the above information, coefficient C becomes :

C = (1–A) PVss = 2.048

Substitute our values for A, B and C into the discrete time model form above to calculate temperature (PV) as a function of controller output (CO) delayed by dead time, or :

- PV(n+1) = 0.987 PV(n)–0.00679 CO(n–48) + 2.048

Open Loop Dynamics

The easiest way to test this discrete time model is to execute an open loop step test that is similar to the example given in a previous tutorial.

Initialize :

temperature as : $PV(0) = 140$

controller output as : $CO = 39$ for $n \leq 48$

Implement controller output step at time $t = 25.5$ min, or $n = 1530$:

$CO = 39$ for $n < 1530$

$CO = 42$ for $n \geq 1531$

Simulate for 60 min, or : $n = 0,1,\ldots,3600$. Display results in the plot below after 20 min :

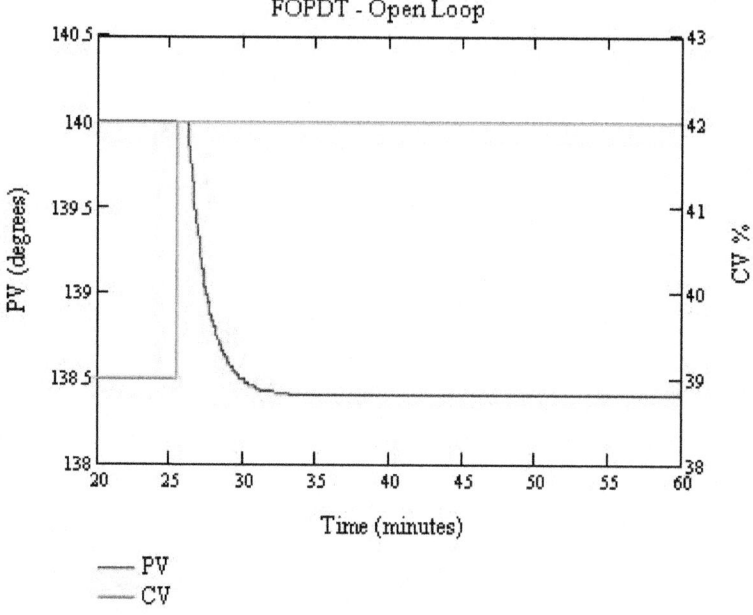

Controller Tuning

The formulas for calculating the PI tuning values can now be tested using the discrete time process model that was generated above.

The tuning values are for the PI controller that is expressed in the continuous time or analog form as :

$$CO = CO_{bias} + Kc \cdot e(t) + \frac{Kc}{Ti} \int e(t)dt$$

The PI controller tuning values are calculated using the equations provided in previous tutorials. Following that procedure, the desired closed loop time constant, Tc, must be chosen to determine how quickly or how aggressive the desired response should be.

For an aggressive response, Tc should be the larger of 0.1Tp or 0.8θp. As listed above, Tp = 1.3 min and θp = 0.8 min for the heat exchanger. Using these model parameters, then :

Tc = max(0.1Tp, 0.8θp) = 0.64

A moderate Tc of the larger of 1.0Tp or 8.0θp will produce a response with no overshoot. We will use this rule in the closed loop simulation below :

Tc = max(1.0Tp, 8.0θp) = 6.4

Using this moderate Tc, we can now compute the controller gain, Kc, and integrator term, Ti :

- $Kc = \dfrac{Tp}{Kp(Tc+\theta p)} = \dfrac{1.3}{-0.533(0.8+6.4)} = -0.339\dfrac{\%}{°C}$

- Ti = Tp = 1.3 min

Closed Loop PI Control

The function logic that follows implements the PI controller and the discrete time FOPDT model's response. The controller uses the PV to calculate a CO that is used later by the plant model. How much later is determined by the dead time. The model then responds by changing the PV. This new PV is then used in the next iteration.

- compute controller error :
  ERR = SP–PV
- compute COp, the proportional contribution to CO :
  COp = Kc·ERR
- compute COi, the integral contribution to CO :
  COi = COi + (Kc·T·ERR)/Ti
- compute the total controller output :
  CO = COp + COi
- compute the process response at the next step n using the time-delayed CO :
  PV(n+1) = A·PV(n) + B·Kp·CO(n–θp) + C

Aside : The integral contribution to CO can include the integrator limiting technique talked about in the *Integral (Reset) Windup and Jacketing Logic* with the expression :

$$COi = \max\left(\min\left(COi + \frac{Kc \cdot T}{Ti}ERR, 100 - COp \right), 0 - COp \right)$$

We initialize the variables at time t = 0 by assuming that there is no error and the system is in a quiescent state. Thus, all of the control output, CO, is due only to the integrator term, COi :

PV(0) = 140; CO(0) = 39; COi(0) = CO(0)

The plot below shows the response of the closed loop system over a simulation time of 60 minutes. The set point is stepped from 140 to 138.4 at time t = 25.5 minutes.

The function logic above calculates the PV and CO for each time period in the 60 minute simulation. Note that there are three CO type variables. CO is the current control value, COi is the current integrator contribution to the CO, and CO(n-θp) is the delayed CO that simulates the dead time.

When programming the functions, additional logic is required keep from using indexes less than 0.

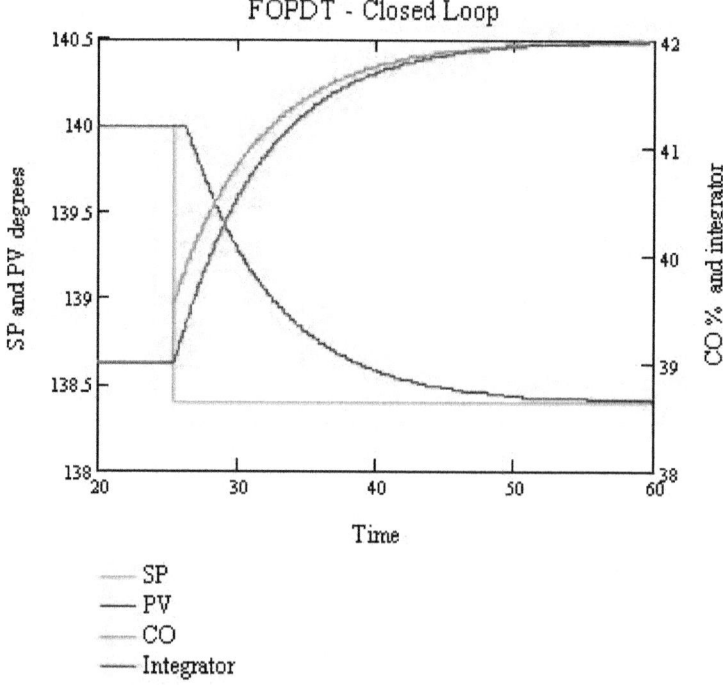

Notice that to prevent overshoot, the rate at which the controller's integrator output, COi, responds or winds up must be limited so the total control output does not exceed the final steady state control output, and final steady state control output is not reached before the system starts to respond. The way I think about it is that the function is calculating the value for Kc so this goal can be achieved.

Also notice that with moderate tuning, the closed loop response is much slower than the open loop response.

Chapter 16

ENVELOPE OPTIMIZATION AND CONTROL USING FUZZY LOGIC

Background

The basic building tool for this optimization methodology is the fuzzy logic Envelope Controller (EC) function block. Such EC blocks are often chained (or sequenced) one after the other. As shown below, each EC in the chain sequence focuses on a single process variable, with the complete chain forming an efficient and sophisticated optimization strategy.

Envelope Controllers in a Chain Sequence

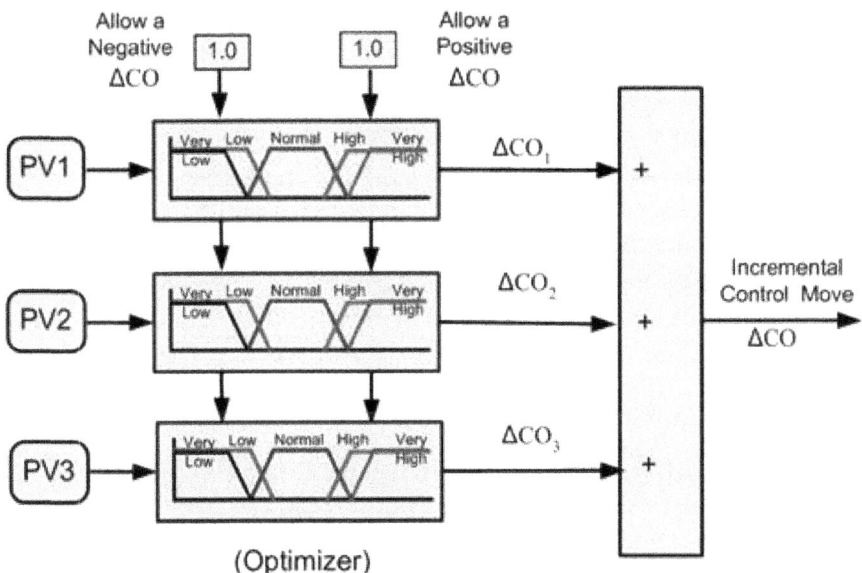

(Optimizer)

With such a chain sequence construction, it is possible to incorporate many constraints. It is not difficult to add or remove a constraint or change its priority, even in real time.

To better understand the operation of the chain sequence shown in the figure above, use the following analogy. Three operators are sitting side-by-side in front of their own operator's console. Each operator can only view a single PV. There are operating constraints for the PV [Very Low, Low, Normal, High, Very High]. Each operator can make incremental control moves to the single output variable. They can make a move once a minute.

Operator #1 monitors PV1 : If PV1 is Very High, a fixed size negative incremental move will be made. If PV1 is Very Low, a positive move will be made. Otherwise no control move is made. Control moves are made to insure the constraints are not exceeded.

Operator #2 monitors PV2 : If PV2 is Very High, a fixed size negative incremental move needs to be made. If PV2 is Very Low, a positive move needs to be made. Otherwise no control move needs to be made. But first, permission from Operator #1 is required. If PV1 is Very High, a positive control move is not allowed. If PV1 is Very Low, a negative move is not allowed. Otherwise permission is granted for any move.

Permission is a fuzzy variable which varies from 0.0 to 1.0. Permission of 0.50 allows Operator # 2 to make half of the fixed control move.

Operator #3 monitors PV3 : This operator performs the optimization function. In this example, it will be to maximize production rate. Therefore PV3 is the production rate. If PV3 is not High, a positive control move will be made. This operator only makes positive control moves. But first permission from both Operator #1 and Operator #2 is required. It will insure that the constraints are not violated. This requires that neither PV1 or PV2 are High. Operator #3 will stop making control moves when PV3 becomes High.

It is up to Operator #1 and Operator #2 to reduce the production rate if a constraint is exceeded.

The first deals with minimizing the amount of combustion air for an industrial power boiler. The second deals with minimizing the outlet pressure of a forced draft fan of an industrial power boiler to minimize energy losses.

This controller is best suited to applications where there are multiple measured input variables that must stay within prescribed limits and there is a single output variable.

The Envelope Controller Block

Before discussing how the fuzzy logic Envelope Controller (EC) function block can be chained in a sequence to form an optimization strategy, we first consider the function of a single EC block.

A schematic of the basic EC block is shown in the figure below :

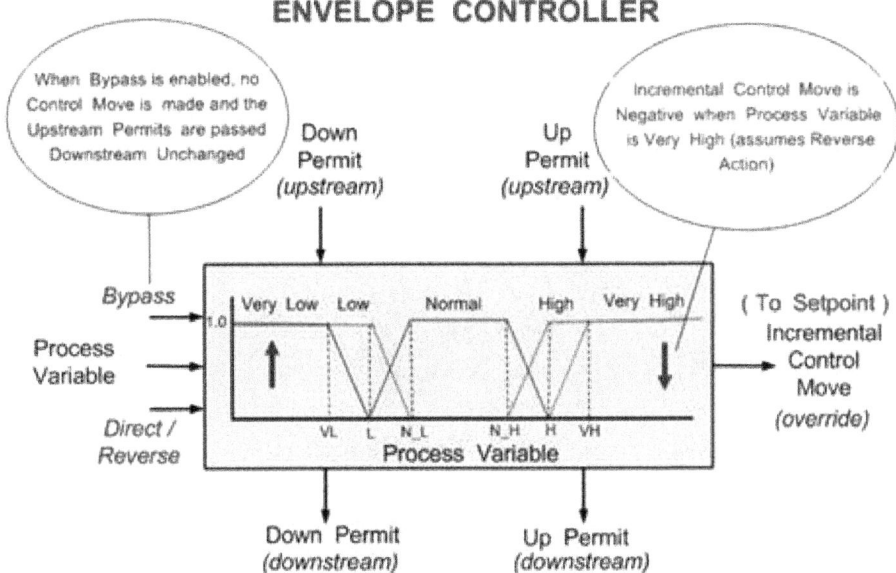

The terms "upstream" and "downstream" refer to the flow of information in the chain sequence of EC's in the optimization strategy and not the flow of material through processing units.

The basic construction is to use the chain of ECs to oversee constraints in order of priority. The last EC in the chain is used as an optimizer.

Envelope Controller Inputs, Outputs and Parameters

Inputs :

1. Process Variable (continuous PV)
2. Down-Permit (0–1.0)
3. Up-Permit (0–1.0)

Outputs :

1. Incremental Control Output (continuous CO)
2. Down-Permit (0–1.0)
3. Up-Permit (0–1.0)

Parameters to Specify :

1. NEG-(size of negative move)
2. POS-(size of positive move)
3. Six breakpoints (VL, L, N_L, N_H, H, VH) on the PV axis which uniquely identifies all five membership functions
4. G-(Overall Gain)

Switches to Set :

1. Bypass (on/off)
2. Control Action (Reverse or Direct)

Membership Functions

Five membership functions are shown superimposed in the Envelope Controller (EC) figure above. Each membership function is a fuzzifier. A continuous process variable (PV) is mapped to 5 sets of fuzzy variables whose values are between 0.0 and 1.0. They are :

1. $PV\text{-}VL = \mu_1(PV)$ VL : Very Low
2. $PV\text{-}L = \mu_2(PV)$ L : Low
3. $PV\text{-}N = \mu_3(PV)$ N : Normal
4. $PV\text{-}H = \mu_4(PV)$ H : High
5. $PV\text{-}VH = \mu_5(PV)$ VH : Very High

Where μ means degree of membership. When PV is midway between VL and L in the above EC diagram, for example, then $\mu_1 = 0.5$ and $\mu_2 = 1.0$. All the rest are 0.0.

The fuzzy variables, such as PV-N, are called linguistic variables. Here, PV-N means "PV is Normal." Similarly, PV-VH means "PV is very high."

Control Action Switch

This switch has two positions as shown in the figure above : "Reverse" and "Direct." In most processes, when the process variable (PV) exceeds the high limit, the controller output (CO) must be decreased to bring it back into the safe operating envelope. The control action switch must be in the "Reverse" position for this to occur.

However, there are processes where the CO must be increased to bring it back into the safe operating envelope if the PV exceeds the high limit. The position of the control action switch must be "Direct" for these processes.

Thus, the control action switch of the EC is analogous to the direct and reverse action of PID controllers.

Bypass Selector Switch

As shown in the figure above, an EC has a bypass switch. When the EC is in bypass mode, the EC is switched out of the chain sequence and the Up Permit-In and Down Permit-In signals pass through the EC unchanged as if it does not exist.

Controller Functionality

The membership functions of each EC define the safe operating region for the incoming PV. When the PV is in the "PV Normal Region," the incremental control move (override) produced by the EC is 0.0. The Up Permit-In and Down Permit-In values are passed unmodified to the next Envelope Controller in the chain sequence.

To simplify the discussion, we assume the EC uses reverse action in the following logic. A direct action EC uses analogous, though opposite, logic.

When the PV entering an EC is in the "High" or "Low" regions, the Up Permit-Out or the Down Permit-Out variable is modified by the fuzzy variable Correction Factors (CF) :

1. Up Permit-CF = NOT (H)
2. Down Permit-CF = NOT (L) where NOT $(x) = 1-x$

The product operator $(\mu_A[PV]^*\mu_B[PV])$ is used in the EC computation. This is one version of the fuzzy "AND." The Up Permit-In and Down Permit-In are multiplied by the appropriate correction factors to become :

1. Up Permit-Out = (Up Permit-In)* (Up Permit-CF)
2. Down Permit-Out = (Down Permit-In)*(Down Permit-CF)

When the PV enters the "Very High" or "Very Low" regions, an incremental control move (override) is made to nudge the PV back into the safe region :

1. ΔCO = (Down Permit-In)*(PV-VH)*NEG*G
2. ΔCO = (Up Permit-In)*(PV-VL)*POS*G

This is a simple two-rule controller. For a EC with reverse action, these rules can be summarized :

[R1] if there is permission to make a negative control move **and** the PV is very high (VH), then make a negative incremental control move.

[R2] if there is permission to make a positive control move **and** the PV is very low (VL), then make a positive incremental control move.

NEG*G and POS*G are singletons. ΔCO is the defuzzified value. It is simply the product of the singleton and the truth of the rule. There are many defuzzification techniques. A simple but highly effective method uses singletons and a simple center of gravity calculation. This is the one used in these applications.

Chaining in Sequence

ECs are chained in sequence to form an efficient Envelope Optimization controller. With a proper number of constraints in proper priority order, they form a safe operating envelope for the process. ECs can be implemented in many modern process control system to create Envelope Optimization logic.

Example : Minimum Excess Air Optimization

This application of Envelope Optimization is for an industrial power boiler. As shown below, it uses five ECs chained in sequence. Like many optimization problems, the solution for minimum excess air optimization is found on the constraint boundary.

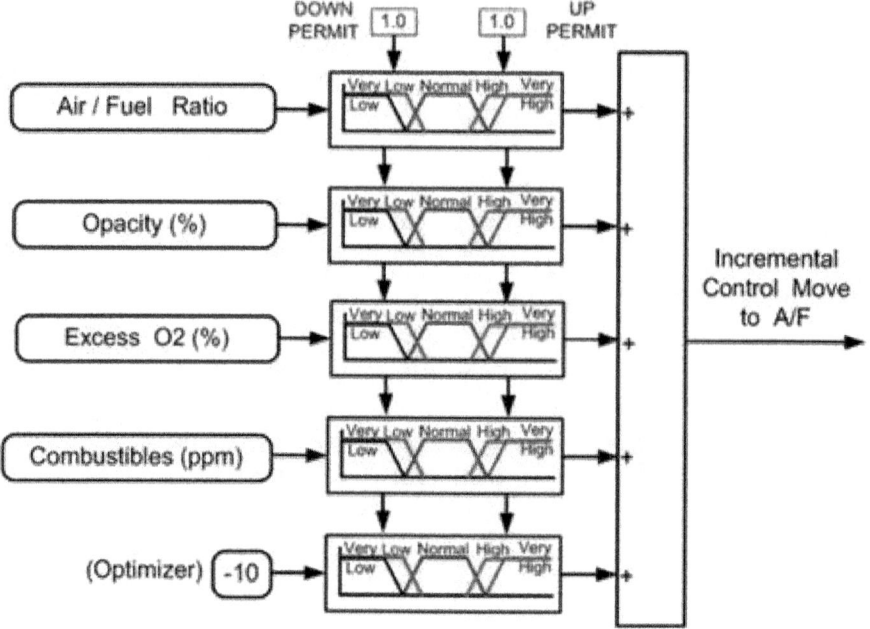

Control Objectives

1. Maintain combustion air at minimum flow required for safe operation and maximum boiler efficiency.
2. Stay within the safe operating envelope as defined by the limits of the first four ECs.
3. Take immediate corrective action when any of the prioritized constraints are violated.
4. Optimization must proceed slowly to avoid upsetting the process.

Optimization Functionality

The air-to-fuel (A/F) ratio is slowly reduced by the negative incremental moves made by the Optimizer EC (the last in the chain sequence) until a limit is reached on one of the first four ECs. This limit condition would then reduce the Down Permit entering the Optimizer EC to 0.0.

Notice that the Up and Down Permits are set to 1.0 at the top of the chain. If the PVs of the first four ECs are all in the "Normal" regions. The Up Permit and Down Permit will pass down the chain unmodified as 1.0.

The first EC sets the high and low limits for the A/F ratio. Note that this EC has the highest priority. If the A/F reaches its low limit, the Down Permit goes to 0.0. The second EC sets the high limit for Opacity (no low limit). The third EC sets the high and low limits for O_2. The fourth EC sets the high limit for combustibles (unburned hydrocarbons). There is no low limit. Normally this control operates at the low limit for A/F ratio with low O_2 and very low combustibles.

Aside : The optimization function is automatically disabled when the combustibles meter goes into self-calibration and resumes normal operation when the calibration is finished. The operator can also turn the optimization "off" at any time. The Minimum Excess Air Optimization can function with just O_2 if the combustibles meter is out of service. The Combustibles EC would need to be placed on "bypass" for this to occur.

While the optimizing function is slow, that is not the case for the constraint override action. For example, if the O_2 fell below the low limit, the A/F ratio would quickly increase to bring O_2 back within the "Normal" region.

Tuning this control is not simple. It requires selecting numerous gains and membership breakpoints. The slope of the membership functions are part of the loop gain when override action is taking place.

While tuning the control is difficult, the good news is that once it is tuned it does not seem to require any tweaking afterwards. This particular application has used the same tuning parameters since it was commissioned several years ago with no apparent degradation in performance.

Example : Forced Draft Fan Outlet Pressure Optimization

This application of Envelope Optimization is for an industrial power boiler. It uses five ECs chained together as shown in the graphic below :

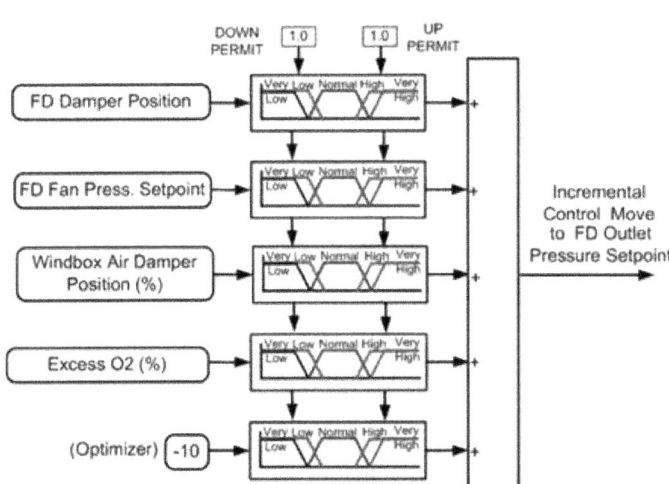

FD Fan Outlet Pressure Optimization

On this power boiler, the Forced Draft (FD) Fan Outlet Pressure is controlled with a PID controller. The PID controller accomplishes this by adjusting the inlet vanes. This provides a constant pressure source for the windbox air dampers. At high steaming rates the pressure needs to be fairly high (about 10.0 inches). At lower steaming rates it doesn't need to be this high and results in a large pressure drop across the windbox dampers. This results in more energy being used than is necessary.

Control Objectives

1. Maintain the FD fan outlet pressure at the minimum pressure required to provide the amount of air needed for combustion.
2. Maintain the windbox air dampers in a mid-range position for maximum controllability.
3. Stay within the "Safe Operating Envelope" as defined by the limits of the first four Envelope Controllers.

Optimization Functionality

The FD fan outlet pressure set point is slowly reduced by negative incremental moves made by the last Envelope Controller (Optimizer) until a limit is reached on one of the first four ECs. Normally it is the FD Fan Pressure set point EC that reaches its low limit and reduces the Down Permit to 0.0 entering the Optimizer EC.

The Down Permit does not suddenly go to 0.0 as the low limit is reached. It gradually decreases because of the slope of the PV-L membership function. When the truth of PV-L is 0.5, the Down Permit becomes 0.5 and only half of the Optimizer's regular incremental move is made. Likewise, when PV-L is 0.1 only one tenth of the Optimizer's regular move is made.

The first EC sets the high and low limits for the FD damper position. The actuator always has the highest priority. The second EC sets the high and low limits for the FD outlet pressure set point. The third EC sets the high limit for the windbox air dampers (no low limit). The fourth EC sets the low limit for O_2 (no high limit).

At low boiler steaming rates the FD fan outlet pressure is held at minimum by the second EC. As the boiler steaming rate increases, the windbox dampers open further to provide more combustion air. The damper position reaches its very high limit causing a positive incremental override control move from the third EC. This causes the pressure set point to increase. As the FD fan outlet pressure rises, the damper moves back to a lower value.

Note that the total incremental control move is made up of the sum of the moves from all of the ECs. In this situation, the new equilibrium occurs when the sum of the third and fifth ECs equal 0.0. The third EC is making a positive move and the Optimizer EC is making a negative move. If the gains and slopes of the membership functions of the third EC are not carefully selected, cycling (otherwise known as instability) will occur. In control jargon this feature could be called mid-ranging. It is desired to always have the windbox damper operating in its normal range.

If for some reason the O_2 goes below its very low limit, positive incremental override control moves will be made by the fourth EC raising the FD fan outlet pressure. This increases air flow to the windbox. This is a safety feature.

Tuning this control is not simple. This is particularly true for the interaction between the third EC and the Optimizer. In this application none of the PVs have significant time constants or delays. Therefore predicted PVs were not used. While tuning the control is difficult, the good news is that once it is tuned it does not seem to require any tweaking afterwards. This particular application has used the same tuning parameters since it was commissioned several years ago with no apparent degradation in performance.

Final Thoughts

Constraints can be prioritized with those affecting safety and environmental concerns at the top of the list. The Envelope Controller is at the heart of the approach. Linking these controllers together in a chain allows one to form a "Safe Operating Envelope" which is the constraint boundary. Optimization proceeds as the process is kept within this envelope. Two examples were shown that use this methodology.

The reader should study some of the fundamental concepts of fuzzy logic if they are interested in an in-depth understanding of this approach. Understanding how to tune these controls is the most difficult part of the application. This can only be done during the startup phase or with a dynamic simulation.

The method described and the examples presented are valid for situations where there is only one variable to manipulate. This approach can be extended to problems where there are multiple variables to manipulate. The author has worked on problems where two variables were manipulated. It usually requires a sophisticated predictive model to provide inputs to some of the Envelope Controllers. It actually becomes a type of Model Predictive Control (MPC).

Chapter 17

PROCESS CONTROL PRELIMINARIES

Processes with streams comprised of gases, liquids, powders, slurries and melts tend to exhibit variations in behavior as operating level changes. This, in fact, is the very nature of a nonlinear process. For this reason, our recipe for controller design and tuning begins by specifying our design level of operation.

Controller Design and Tuning Recipe :

1. Establish the design level of operation (DLO), which is the normal or expected values for set point and major disturbances.
2. Bump the process and collect controller output (CO) to process variable (PV) dynamic process data around this design level.
3. Approximate the process data behavior with a first order plus dead time (FOPDT) dynamic model.
4. Use the model parameters from step 3 in rules and correlations to complete the controller design and tuning.

NONLINEAR BEHAVIOR OF THE GRAVITY DRAINED TANKS

The dynamic behavior of the *gravity drained tanks* process is reasonably intuitive. Increase or decrease the inlet flow rate into the upper tank and the liquid level in the lower tank rises or falls in response.

One challenge this process presents is that its *dynamic behavior* is nonlinear. That is, the process gain, K_p; time constant, T_p; and/or dead time, θ_p; changes as operating level changes. This is evident in the open loop response plot below.

As shown above, the CO is stepped in equal increments, yet the response behavior of the PV changes as the level in the tank rises. The consequence of nonlinear behavior is that a controller designed to give desirable performance at one operating level may not give desirable performance at another level.

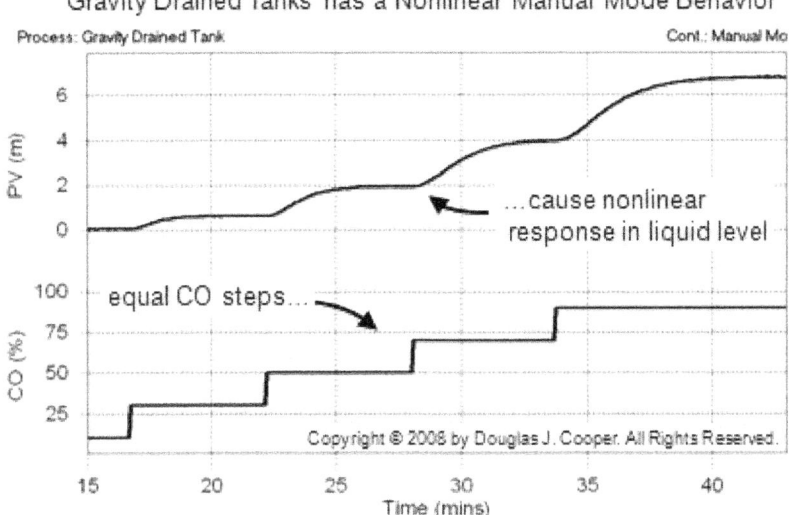

Gravity Drained Tanks has a Nonlinear Manual Mode Behavior

NONLINEAR BEHAVIOR OF THE HEAT EXCHANGER

When tuned for a moderate response as shown in the first set point step from 140°C to 155°C in the plot below, the process variable (PV) responds in a manner consistent with our design goals. That is, the PV moves to the new set point (SP) reasonably quickly but does not overshoot the set point.

Heat Exchanger Under PI Control Shows Nonlinear Behavior

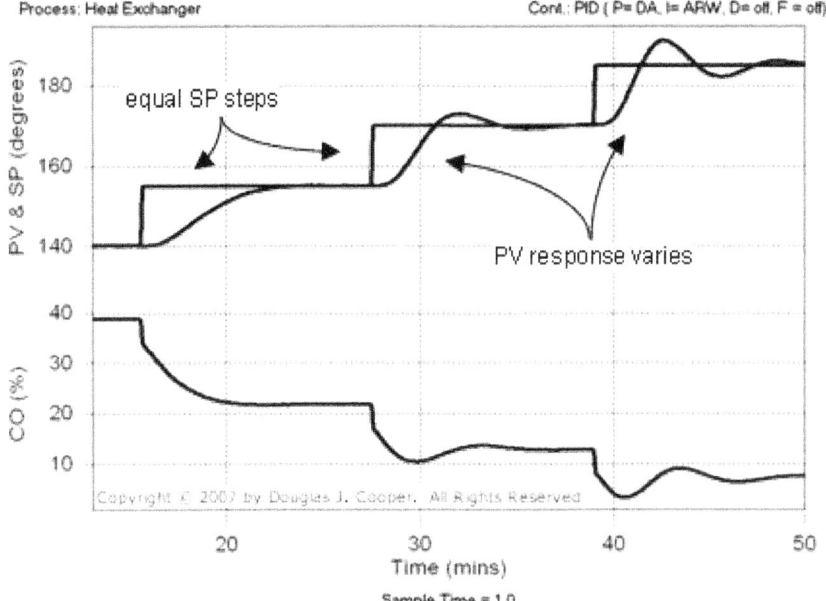

The consequence of a nonlinear process character is apparent as the set point steps continue to higher temperatures. In the third set point step from 170°C to 185°C, the same controller that had given a desired moderate performance now produces a PV response with a clear overshoot and some oscillation.

Such a change in performance with operating level may be tolerable in some applications and unacceptable in others."Best" performance is something we judge for ourselves based on the goals of production, capabilities of the process, impact on down stream units and the desires of management.

Nonlinear behavior should not catch us by surprise. It is something we can know about our process in advance. And this is why we should choose a design level of operation as a first step in our controller design and tuning procedure.

STEP 1 : ESTABLISH THE DESIGN LEVEL OF OPERATION (DLO)

Because, as shown in the examples above, processes have *process gain, Kp*; *time constant, Tp*; and/or *dead time, θp* values that change as operating level changes, and these *FOPDT model parameter* values are used to complete the controller design and tuning procedure, it is important that dynamic process test data be collected at a pre-determined level of operation.

Defining this design level of operation (DLO) includes specifying where we expect the set point (SP) and measured process variable (PV) to be during normal operation, and the range of values the SP and PV might typically assume. This way we know where to explore the dynamic process behavior during controller design and tuning.

The DLO also considers our major disturbances (D). We should know the normal or typical values for our major disturbances. And we should be reasonably confident that the *disturbances are quiet* so we may proceed with a bump test to generate and record dynamic process data.

STEP 2 : COLLECT DYNAMIC PROCESS DATA AROUND THE DLO

The next step in our recipe is to collect dynamic process data as near as practical to our design level of operation. We do this with a bump test, where we step or pulse the CO and collect data as the PV responds.

It is important to wait until the CO, PV and D have settled out and are as near to constant values as is possible for our particular operation before we start a bump test. The point of bumping a process is to learn about the cause and effect relationship between the CO and PV.

With the process at steady state, we are starting with a clean slate. As the PV responds to the CO bumps, the dynamic cause and effect behavior is isolated and evident in the data. On a practical note, be sure the data capture routine is enabled before the initial bump is implemented so all relevant data is collected.

Two popular open loop (manual mode) methods are the step test and the doublet test.

For either method, the CO must be moved far enough and fast enough to force a response in the PV that dominates the *measurement noise*.

Also, our bump should move the PV both above and below the DLO during testing. With data from each side of the DLO, the model (step 3) will be able to average out the nonlinear effects as discussed above.

- *Step Test* : To collect data that will "average out" to our design level of operation, we start the test at steady state with the PV on one side of (either above or below) the DLO. Then, as shown in the plot below, we step the CO so that the measured PV moves across to settle on the other side of the DLO.

Manual Mode Step Test from Gravity Drained Tanks Process

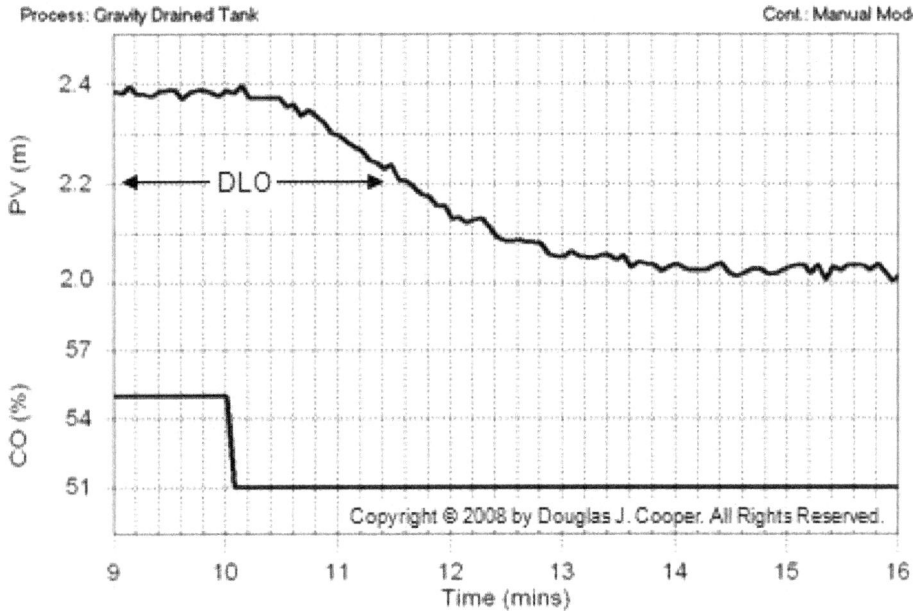

We can either start high and step the CO down (as shown above), or start low and step the CO up. Both methods produce dynamic data of equal value for our design and tuning recipe.

- *Doublet Test* : A doublet test, as shown below, is two CO pulses performed in rapid succession and in opposite direction. The second pulse is implemented as soon as the process has shown a clear response to the first pulse that dominates the noise in the PV. It is not necessary to wait for the process to respond to steady state for either pulse.

The doublet test offers attractive benefits, including that it starts from and quickly returns to the DLO, it produces data both above and below the design level to "average out" the nonlinear effects, and the PV always stays close to the DLO, thus minimizing off-spec production. Such data does require *commercial software* for model fitting, however.

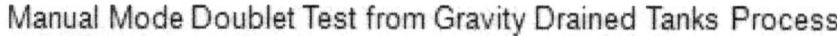

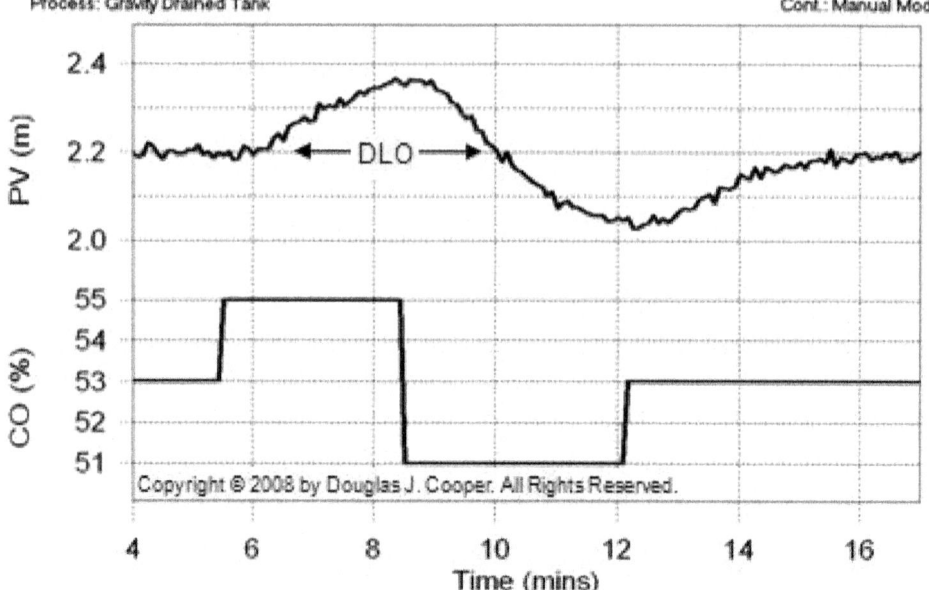

STEP 3 : FIT A FOPDT DYNAMIC MODEL TO PROCESS DATA

In fitting a first order plus dead time (FOPDT) model, we approximate those essential features of the dynamic process behavior that are fundamental to control.

$$Tp \frac{dPV(t)}{dt} + PV(t) - Kp \cdot CO(t - \theta p)$$

Where :

PV(t) = measured process variable as a function of time

CO(t−θp) = controller output signal as a function of time and shifted by θp

θp = process dead time

t = time

When the FOPDT dynamic model is fit to process data, the results describe how PV will respond to a change in CO via the model parameters. In particular :

- *Process gain, Kp,* describes the direction and how far PV will travel,
- *Time constant, Tp,* states how fast PV moves after it begins its response,
- *Dead time, θp,* is the delay from when CO changes until when PV begins to respond.

An example study that compares dynamic process data from the heat exchanger with a FOPDT model prediction can be *found here.* Comparisons between data and model for the gravity drained tanks can be found *here* and *here.*

STEP 4 : USE THE MODEL PARAMETERS TO COMPLETE THE DESIGN AND TUNING

In step 4, the three FOPDT model parameters are used in correlations to compute controller tuning values. For example, the chart below lists internal model control (IMC) tuning correlations for the *PI controller* and *dependent ideal PID controller*, and *dependent ideal PID with CO filter* forms :

	Controller Gain Kc	Reset Time Ti	Deriv Time Td	Filter Const α
PI	$\dfrac{1}{Kp}\dfrac{Tp}{(\theta p + Tc)}$	Tp		
Ideal PID	$\dfrac{1}{Kp}\left(\dfrac{Tp+0.5\theta p}{Tc+0.5\theta p}\right)$	$Tp + 0.5\,\theta p$	$\dfrac{Tp\,\theta p}{2Tp+\theta p}$	
PID W/ CO Flter	$\dfrac{1}{Kp}\left(\dfrac{Tp+0.5\,\theta p}{Tc+\theta p}\right)$	$Tp + 0.5\,\theta p$	$\dfrac{Tp\,\theta p}{2Tp+\theta p}$	$\dfrac{Tc(Tp+0.5\,\theta p)}{Tp(Tc+\theta p)}$

The closed loop time constant, Tc, in the IMC correlations is used to specify the desired speed or quickness of our controller in responding to a set point change or rejecting a disturbance. The closed loop time constant is computed :

- *Aggressive performance :* Tc is the larger of $0.1\cdot Tp$ or $0.8\,\theta p$
- *Moderate performance :* Tc is the larger of $1\cdot Tp$ or $8\,\theta p$
- *Conservative performance :* Tc is the larger of $10\cdot Tp$ or $80\,\theta p$.

USE THE RECIPE-IT IS BEST PRACTICE

The FOPDT dynamic model of step 3 also provides us the information we need to decide other controller design issues, including :

- *Controller Action :* Before implementing our controller, we must input the proper direction our controller should move to correct for growing errors. Some vendors use the term "reverse acting" and "direct acting." Others use terms like "up-up" and "up-down" (as CO goes up, then PV goes up or down). This specification is determined solely by the sign of the process gain, Kp.
- *Loop Sample Time, T :* Process time constant, Tp, is the clock of a process. The size of Tp indicates the maximum desirable loop sample time. *Best practice* is to set loop sample time, T, at 10 times per time constant or faster ($T \leq 0.1Tp$). Faster may provide modestly improved performance. Slower than five times per time constant leads to significantly degraded performance.
- *Dead Time Problems :* As dead time grows greater than the process time constant (when $\theta p > Tp$), controller performance can benefit from a model based dead time compensator such as the Smith predictor.

- *Model Based Control* : If we choose to employ a Smith predictor, or perhaps a dynamic feed forward element, a multivariable decoupler, or any other model based controller, we need a dynamic model of the process to enter into the control computer. The FOPDT model from step 3 of the recipe is usually appropriate for this task.

A Controller's "Process" Goes from Wire Out to Wire in

A controller seeks to maintain the measured process variable (PV) at set point (SP) in spite of unplanned and unmeasured disturbances. Since $e(t) = \text{SP} - \text{PV}$, this is equivalent to saying that a controller seeks to maintain controller error, $e(t)$, equal to zero.

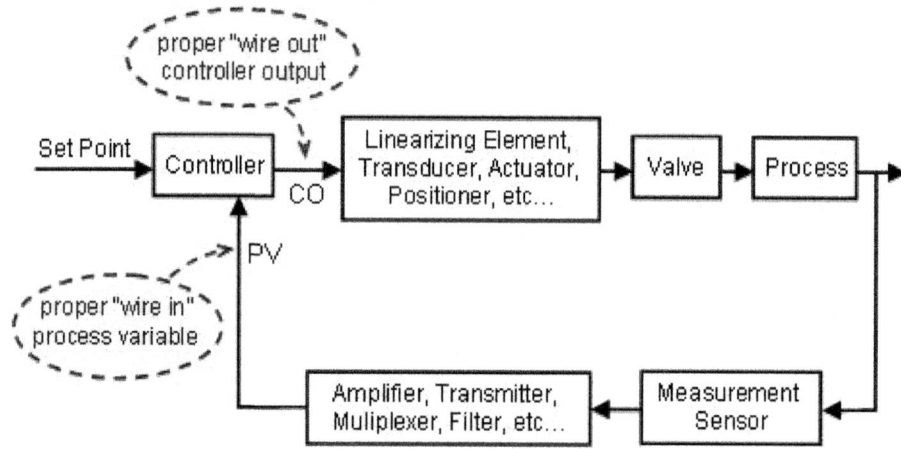

Proper CO to PV Bump Test Data is Recorded at Controller

A controller repeats a measurement-computation-action procedure at every loop sample time, T. Starting at the far right of the control loop block diagram above :

- A sensor measures a temperature, pressure, concentration or other property of interest from our process.

- The sensor signal is transmitted to the controller. The pathway from sensor to controller might include : a *transducer*, an *amplifier*, a *scaling* element, *quantization*, a *signal filter*, a *multiplexer*, and other operations that can add delay and change the size, sign, and/or units of the measurement.

- After all electronic and digital operations, the result terminates at our controller as the **"wire in"** measured process variable (PV) signal.

- This "wire in" process variable is subtracted from set point in the controller to compute error, $e(t) = \text{SP} - \text{PV}$, which is then used in an algorithm (examples *here* and *here*) to compute a controller output (CO) signal.

- The computed CO signal is transmitted on the **"wire out"** from the controller on a path to the final control element (FCE).

- Similar to the measurement path, the signal from the controller to FCE might include filtering, scaling, linearization, amplification, multiplexing, transducing and other operations that can add delay and change the size, sign, and/ or units of our original CO signal.

- After any electronic and digital operations, the signal reaches the valve, pump, compressor or other FCE, causing a change in the manipulated variable (a liquid or gas stream flow rate, for example).

- The change in the manipulated variable causes a change in our temperature, pressure, concentration or other process property of interest, all with the goal of making $e(t) = 0$.

DESIGN BASED ON CO TO PV DYNAMICS

The steps of the controller *design and tuning recipe* include : bumping the CO signal to generate CO to PV dynamic process data, approximating this test data with a first order plus dead time (FOPDT) model, and then using the model parameters in rules and correlations to complete the controller design and tuning.

The recipe provides a *proven basis* for controller design and tuning that avoids wasteful and expensive trial-and-error experiments. But for success, controller design and tuning must be based on process data as the controller sees it.

The controller only knows about the state of the process from the PV signal arriving on the "wire in" *after* all operations in the signal path from the sensor. It can only impact the state of the process with the CO signal it sends on the "wire out" *before* any such operations are made in the path to the final control element.

As indicated in the diagram, the proper signals that describe our complete "process" from the controller's view is the "wire out" CO and the "wire in" PV.

COMPLETE THE CIRCUIT

Sometimes we find ourselves unable to proceed with an orderly controller design and tuning. Perhaps our controller interface does not make it convenient to directly record process data. Maybe we find a vendor's documentation to be so poorly written as to be all but worthless. There are a host of complications that can hinder progress.

Being resourceful, we may be tempted to move the project forward by using portable instrumentation. It seems reasonable to collect, say, temperature in a vessel during a bump test by inserting a spare thermocouple into the liquid. Or maybe we feel we can be more precise by standing right at the valve and using a portable signal generator to bump the process rather than doing so from a remote control panel.

As shown below, such an approach cuts out or short circuits the complete control loop pathway. External or portable instrumentation will not be recording the actual CO or PV as the controller sees it, and the data will not be appropriate for controller design or tuning.

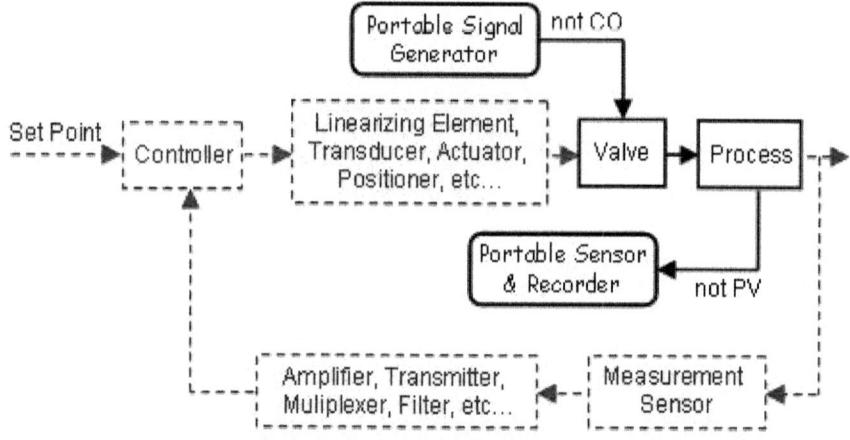

Influence of Dashed Items Not Refelected in Data

EVERY ITEM COUNTS

The illustration above is extreme in that it shows many items that are not included in the control loop. But please recognize that it can be problematic to leave out even a single step in the complete signal pathway.

A simple scaling element that multiplies the signal by a constant value, for example, may seem reasonably unimportant to the overall loop dynamics. But this alone can change the size and even the sign of Kp, thus having dramatic impact on best tuning and final controller performance.

From a controller's view, the complete loop goes from "wire out" to "wire in" as shown below.

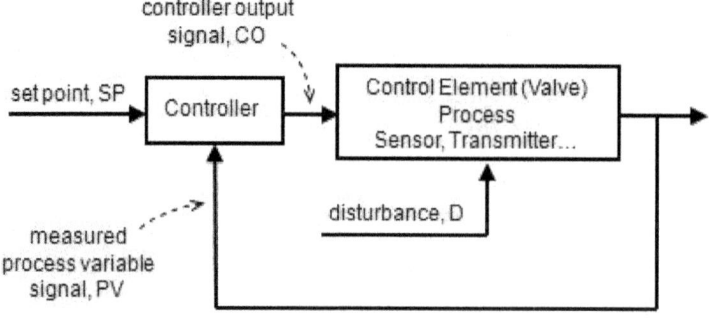

Controller's View: Wire Out (CO) to Wire In (PV) Loop

Every item in the loop counts. Always use the complete CO to PV data for process control analysis, design and tuning.

PAY ATTENTION TO UNITS

Signals can appear in a control loop in electronic units (*e.g.*, volts, mA), in engineering units (*e.g.* °C, Lb/hr), as percent of scale (*e.g.*, 0% to 100%), or as discrete or digital counts (*e.g.* 0 to 4095 counts).

It is critical that we remain aware of the units of a signal when working with a particular instrument or device. All values entered and computations performed must be consistent with the form of the data at that point in the loop.

Beyond the theory and methods discussed in this e-book, such "accounting confusion" can be one of the biggest challenges for the process control practitioner.

THE NORMAL OR STANDARD PID ALGORITHM

The question arises quite often, "What is the normal or standard form of the PID (proportional-integral-derivative) algorithm?"

The answer is both simple and complex. Before we explore the answer, consider the screen displays shown below :

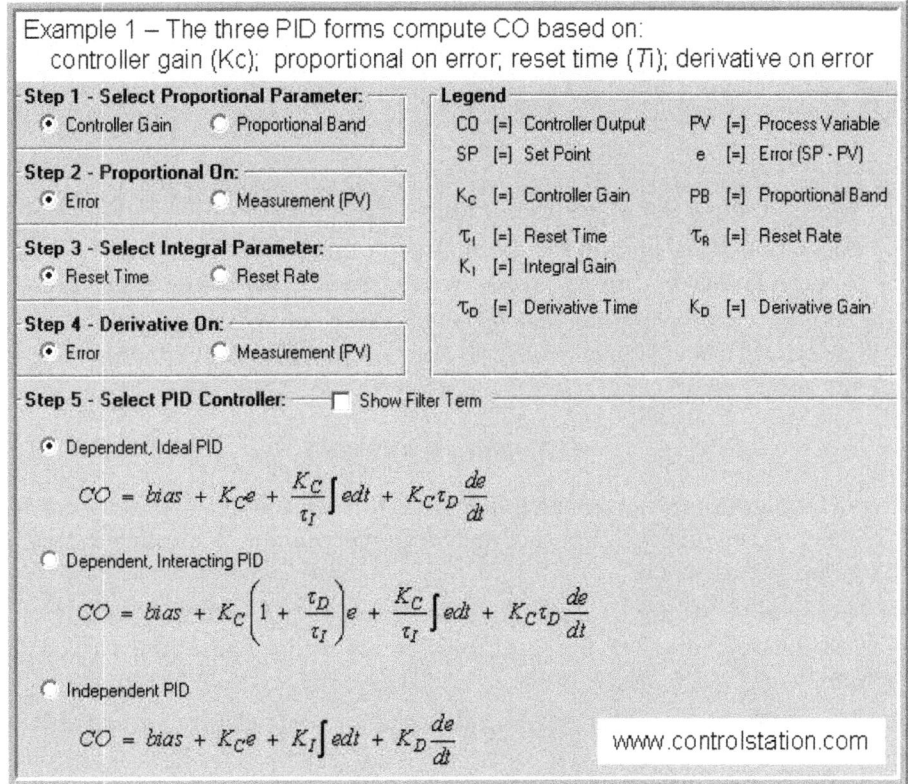

Example 1 – The three PID forms compute CO based on:
controller gain (Kc); proportional on error; reset time (Ti); derivative on error

Step 1 - Select Proportional Parameter:
- ⦿ Controller Gain   ○ Proportional Band

Step 2 - Proportional On:
- ⦿ Error   ○ Measurement (PV)

Step 3 - Select Integral Parameter:
- ⦿ Reset Time   ○ Reset Rate

Step 4 - Derivative On:
- ⦿ Error   ○ Measurement (PV)

Legend

CO [=] Controller Output	PV [=] Process Variable	
SP [=] Set Point	e [=] Error (SP - PV)	
K_C [=] Controller Gain	PB [=] Proportional Band	
τ_I [=] Reset Time	τ_R [=] Reset Rate	
K_I [=] Integral Gain		
τ_D [=] Derivative Time	K_D [=] Derivative Gain	

Step 5 - Select PID Controller: ☐ Show Filter Term

⦿ Dependent, Ideal PID

$$CO = bias + K_C e + \frac{K_C}{\tau_I}\int edt + K_C\tau_D\frac{de}{dt}$$

○ Dependent, Interacting PID

$$CO = bias + K_C\left(1 + \frac{\tau_D}{\tau_I}\right)e + \frac{K_C}{\tau_I}\int edt + K_C\tau_D\frac{de}{dt}$$

○ Independent PID

$$CO = bias + K_C e + K_I\int edt + K_D\frac{de}{dt}$$

www.controlstation.com

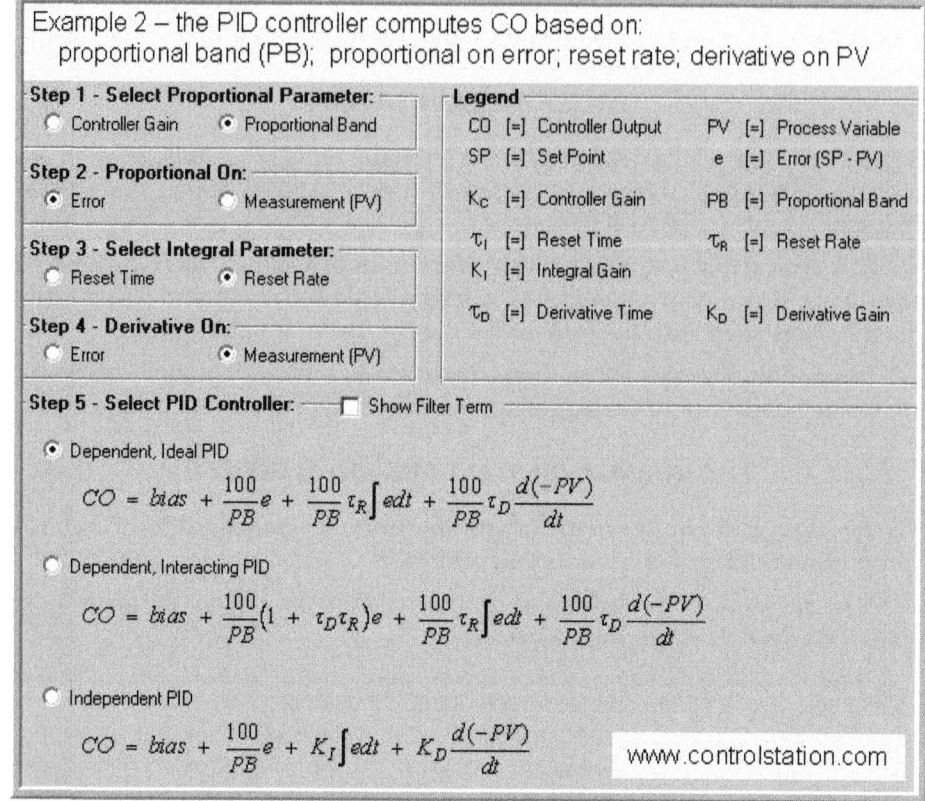

As shown in the screen displays :

- There are three popular PID algorithm forms.
- Each of the three algorithms has tuning parameters and algorithm variables that can be cast in different ways (see steps 1–4 in the large image views).

So your vendor might be using one of dozens of possible algorithm forms. And if you add a *filter term* to your controller, the number of possibilities increases substantially.

THE SIMPLE ANSWER

Any of the algorithms can deliver the same performance as any of the others. There is no control benefit from choosing one form over another. They are all standard or normal in that sense.

If you are considering a purchase, select the vendor that serves your needs the best and don't dwell on the specifics of the algorithm. Some things to consider include :

- Compatibility with existing controllers and associated hardware and software
- Cost

- Ease of installation and maintenance
- Reliability
- Your operating environment (is it clean? cool? dry?)

A MORE COMPLETE ANSWER

Most of the different controller algorithm forms can be found in one vendor's product or another. Some vendors even use different forms within their own product lines.

And while the various forms are equally capable, each must be tuned (values for the adjustable parameters must be specified) using tuning correlations specifically designed for that particular control algorithm.

Commercial software makes it straightforward to get desired performance from any of them. But it is essential that you know your vendor and controller model number to ensure a correct match between controller form and computed tuning values.

The alternative to an orderly design methodology is a "guess and test" approach. While used by some practitioners, such trial and error tuning squanders valuable production time, consumes more feedstock and utilities than is necessary, generates additional waste and off-spec product, and can even present safety concerns.

We use some variation of the dependent, ideal PID controller form :

$$CO = CO_{bias} + Kc \cdot e(t) + \frac{Kc}{Ti}\int e(t)dt + Kc \cdot Td\frac{de(t)}{dt}$$

Where :

CO = controller output signal

CO_{bias} = controller bias

$e(t)$ = current controller error, defined as SP–PV

SP = set point

PV = measured process variable

Kc = controller gain, a tuning parameter

Ti = reset time, a tuning parameter

Td = derivative time, a tuning parameter.

To reinforce that the controllers all are equally capable, we occasionally use variations of the dependent, interacting form :

$$CO = CO_{bias} + Kc\left(1+\frac{Td}{Ti}\right)e(t) + \frac{Kc}{Ti}\int e(t)dt + Kc \cdot Td\frac{de(t)}{dt}$$

or variations of the independent PID form :

$$CO = CO_{bias} + Kc \cdot e(t) + Ki\int e(t)dt + kd\frac{de(t)}{dt}$$

FINAL THOUGHTS

The discussion above glosses over some of the subtle differences in algorithm form that we can exploit to improve control performance. We will learn about these details as we progress in our learning.

For example, derivative on error behaves different from *derivative on measured PV*. This is true for all of the algorithms. Derivative on error can "kick" after set point steps and this is rarely considered desirable behavior. Thus, derivative on PV is recommended for industrial applications.

And if you are considering programming the controller yourself, it is not the algorithm form that is the challenge. The big hurdle is properly accounting for the *anti-reset windup and jacketing logic* to allow bumpless transition between operating modes.

Chapter 18

EXAMPLES OF CONTROL SYSTEMS

AUTOMATION

Automation or *automatic control*, is the use of various *control systems* for operating equipment such as machinery, processes in factories, boilers and heat treating ovens, switching in telephone networks, steering and stabilization of ships, aircraft and other applications with minimal or reduced human intervention. Some processes have been completely automated.

The biggest benefit of automation is that it saves labor, however, it is also used to save energy and materials and to improve quality, accuracy and precision.

The term *automation*, inspired by the earlier word *automatic* (coming from *automaton*), was not widely used before 1947, when General Motors established the automation department. It was during this time that industry was rapidly adopting *feedback controllers*, which were introduced in the 1930s.

Automation has been achieved by various means including mechanical, hydraulic, pneumatic, electrical, electronic and computers, usually in combination. Complicated systems, such as modern factories, airplanes and ships typically use all these combined techniques.

TYPES OF AUTOMATION

Two common types of automation are feedback control, which is usually continuous and involves taking measurements using a sensor and making calculated adjustments to keep the measured variable within a set range, and sequence control, in which a programmed sequence of discrete operations is performed, often based on system logic. Cruise control is an example of the former while an elevator or an automated teller machine (ATM) is an example of the latter.

The theoretical basis of feedback control is *control theory*, which also covers *servomechanisms*, which are often part of an automated system. Feedback control is called "closed loop" while non-feedback control is called "open loop."

FEEDBACK CONTROL

Feedback control is accomplished with a *controller*. To function properly, a controller must provide correction in a manner that maintains stability. Maintaining stability is a principal objective of control theory.

As an example of feedback control, consider a steam coil air heater in which a temperature sensor measures the temperature of the heated air, which is the measured variable. This signal is constantly "fed back" to the controller, which compares it to the desired setting (set point). The controller calculates the difference (error), then calculates a correction and sends the correction signal to adjust the air pressure to a diaphragm that moves a positioner on the steam valve, opening or closing it by the calculated amount. All the elements constituting the measurement and control of a single variable are called a *control loop*.

The complexities of this are that the quantities involved are all of different physical types; the temperature sensor signal may be electrical or pressure from an enclosed fluid, the controller may employ pneumatic, hydraulic, mechanical or electronic techniques to sense the error and send a signal to adjust the air pressure that moves the valve.

The first controllers used *analog* methods to perform their calculations. Analog methods were also used in solving differential equations of control theory. The electronic *analog computer* was developed to solve control type problems and electronic analog controllers were also developed. Analog computers were displaced by digital computers when they became widely available.

Common applications of feedback control are control of temperature, pressure, flow, and speed.

SEQUENTIAL CONTROL AND LOGICAL SEQUENCE CONTROL

Sequential control may be either to a fixed sequence or to a logical one that will perform different actions depending on various system states. An example of an adjustable but otherwise fixed sequence is a timer on a lawn sprinkler. An elevator is an example that uses logic based on the system states to perform certain actions in response to operator input.

A development of sequential control was *relay logic*, by which electrical relays engage electrical contacts which either start or interrupt power to a device. Relays were first used in telegraph networks before being developed for controlling other devices, such as when starting and stopping industrial-sized electric motors or opening and closing solenoid valves. Using relays for control purposes allowed event-driven control, where actions could be triggered out of sequence, in response to external events. These were more flexible in their response than the rigid single-sequence cam timers. More complicated examples involved maintaining safe sequences for devices such as swing bridge controls, where a lock bolt needed to be disengaged before the bridge could be moved, and the lock bolt could not be released until the safety gates had already been closed.

The total number of relays, cam timers and drum sequencers can number into the hundreds or even thousands in some factories. Early *programming* techniques and languages were needed to make such systems manageable, one of the first being *ladder logic*, where diagrams of the interconnected relays resembled the rungs of a ladder. Special computers called *programmable logic controllers* were later designed to replace these collections of hardware with a single, more easily re-programmed unit.

In a typical hard wired motor start and stop circuit (called a *control circuit*) a motor is started by pushing a "Start" or "Run" button that activates a pair of electrical relays. The "lock-in" relay locks in contacts that keep the control circuit energized when the push button is released. (The start button is a normally open contact and the stop button is normally closed contact.) Another relay energizes a switch that powers the device that throws the motor starter switch (three sets of contacts for three phase industrial power) in the main power circuit. (Note : Large motors use high voltage and experience high in-rush current, making speed important in making and breaking contact. This can be dangerous for personnel and property with manual switches.) All contacts are held engaged by their respective electromagnets until a "stop" or "off" button is pressed, which de-energizes the lock in relay.

Commonly *interlocks* are added to a control circuit. Suppose that the motor in the example is powering machinery that has a critical need for lubrication. In this case an interlock could be added to insure that the oil pump is running before the motor starts. Timers, limit switches and electric eyes are other common elements in control circuits.

Solenoid valves are widely used on compressed air or hydraulic fluid for powering *actuators* on mechanical components. While motors are used to supply continuous rotary motion, actuators are typically a better choice for intermittently creating a limited range of movement for a mechanical component, such as moving various mechanical arms, opening or closing valves, raising heavy press rolls, applying pressure to presses.

COMPUTER CONTROL

Computers can perform both sequential control and feedback control, and typically a single computer will do both in an industrial application. *Programmable logic controllers* (PLCs) are a type of special purpose microprocessor that replaced many hardware components such as timers and drum sequencers used in *relay logic*. General purpose process control computers have increasingly replaced stand alone controllers, with a single computer able to perform the operations of hundreds of controllers. Process control computers can process data from a network of PLCs, instruments and controllers in order to implement typical (such as *PID*) control of many individual variables or, in some cases, to implement complex control *algorithms* using multiple inputs and mathematical manipulations. They can also analyze data and create real time graphical displays for operators and run reports for engineers and management.

Control of an *automated teller machine* (ATM) is an example of an interactive process in which a computer will perform a logic derived response to a user selection based on information retrieved from a networked database. The ATM process has a lot of similarities to other online transaction processes. The different logical responses are called *scenarios*. Such processes are typically designed with the aid of *use cases* and *flowcharts*, which guide the writing of the software code.

HISTORY

The earliest feedback control mechanisms was used to tent the sails of windmills. It was patented by Edmund Lee in 1745.

The *centrifugal governor*, which dates to the last quarter of the 18th century, was used to adjust the gap between *millstones*. The centrifugal governor was also used in the automatic flour mill developed by *Oliver Evans* in 1785, making it the first completely automated industrial process. The governor was adopted by James Watt for use on a steam engine in 1788 after Watt's partner Boulton saw one at a flour mill *Boulton & Watt* were building.

The governor could not actually hold a set speed; the engine would assume a new constant speed in response to load changes. The governor was able to handle smaller variations such as those caused by fluctuating heat load to the boiler. Also, there was a tendency for oscillation whenever there was a speed change. As a consequence, engines equipped with this governor were not suitable for operations requiring constant speed, such as cotton spinning.

Several improvements to the governor, plus improvements to valve cut-off timing on the steam engine, made the engine suitable for most industrial uses before the end of the 19th century. Advances in the steam engine stayed well ahead of science, both thermodynamics and control theory.

The governor received relatively little scientific attention until *James Clerk Maxwell* published a paper that established the beginning of a theoretical basis for understanding control theory. Development of the electronic amplifier during the 1920s, which was important for long distance telephony, required a higher signal to noise ratio, which was solved by negative feedback noise cancellation. This and other telephony applications contributed to control theory. Military applications during the Second World War that contributed to and benefited from control theory were *fire-control systems* and aircraft controls. The so-called classical theoretical treatment of control theory dates to the 1940s and 1950s.

Relay logic was introduced with factory *electrification*, which underwent rapid adaption from 1900 though the 1920s. Central electric power stations were also undergoing rapid growth and operation of new high pressure boilers, steam turbines and electrical substations created a large demand for instruments and controls.

Central control rooms became common in the 1920s, but as late as the early 1930s, most process control was on-off. Operators typically monitored charts drawn by recorders that plotted data from instruments. To make corrections, operators manually opened or closed valves or turned switches on or off. Control rooms also used color coded lights to send signals to workers in the plant to manually make certain changes.

Controllers, which were able to make calculated changes in response to deviations from a set point rather than on-off control, began being introduced the 1930s. Controllers allowed manufacturing to continue showing productivity gains to offset the declining influence of factory electrification.

In 1959 Texaco's Port Arthur refinery became the first chemical plant to use *digital control*. Conversion of factories to digital control began to spread rapidly in the 1970s as the price of computer hardware fell.

SIGNIFICANT APPLICATIONS

The automatic telephone switchboard was introduced in 1892 along with dial telephones. By 1929, 31.9% of the Bell system was automatic. Automatic telephone switching originally used electro-mechanical switches, which consumed a large amount of electricity. Call volume eventually grew so fast that it was feared the telephone system would consume all electricity production, prompting *Bell Labs* to begin research on the *transistor*.

The logic performed by telephone switching relays was the inspiration for the digital computer.

The first commercially successful glass bottle blowing machine was an automatic model introduced in 1905. The machine, operated by a two man crew working 12 hour shifts, could produce 17,280 bottles in 24 hours, compared to 2,880 bottles made by a crew of six men and boys working in a shop for a day. The cost of making bottles by machine was 10 to 12 cents per gross compared to $1.80 per gross by the manual glassblowers and helpers.

Sectional electric drives were developed using control theory. Sectional electric drives are used on different sections of a machine where a precise differential must be maintained between the sections. In steel rolling, the metal elongates as it passes through pairs of rollers, which must run at successively faster speeds. In paper making the paper sheet shrinks as it passes around steam heated drying arranged in groups, which must run at successively slower speeds. The first application of a sectional electric drive was on a paper machine in 1919. One of the most important developments in the steel industry during the 20th century was continuous wide strip rolling, developed by Armco in 1928.

Before automation many chemicals were made in batches. In 1930, with the widespread use of instruments and the emerging use of controllers, the founder of Dow Chemical Co. was advocating *continuous production*.

Self-acting machine tools that displaced hand dexterity so they could be operated by boys and unskilled laborers were developed by *James Nasmyth* in the 1840s. *Machine tools* were automated with *Numerical control* (NC) using punched paper tape in the 1950s. This soon evolved into computerized numerical control (CNC).

Today extensive automation is practiced in practically type of manufacturing and assembly process. Some of the larger processes include electrical power generation, oil refining, chemicals, steel mills, plastics, cement plants, fertilizer plants, pulp and paper mills, automobile and truck assembly, aircraft production, glass manufacturing, natural gas separation plants, food and beverage processing, canning and bottling and manufacture of various kinds of parts. Robots are especially useful in hazardous applications like automobile spray painting. Robots are also used to assemble electronic circuit boards. Automotive welding is done with robots and automatic welders are used in applications like pipelines.

ADVANTAGES AND DISADVANTAGES

The main advantages of automation are :

- Increased through output or productivity.
- Improved quality or increased predictability of quality.
- Improved robustness (consistency), of processes or product.
- Increased consistency of output.
- Reduced direct human labor costs and expenses.

The following methods are often employed to improve productivity, quality, or robustness :

- Install automation in operations to reduce cycle time.
- Install automation where a high degree of accuracy is required.
- Replacing human operators in tasks that involve hard physical or monotonous work.
- Replacing humans in tasks done in dangerous environments (*i.e.* fire, space, volcanoes, nuclear facilities, underwater, etc.)
- Performing tasks that are beyond human capabilities of size, weight, speed, endurance, etc.
- *Economic improvement* : Automation may improve in economy of enterprises, society or most of humanity. For example, when an enterprise invests in automation, technology recovers its investment; or when a state or country increases its income due to automation like *Germany* or *Japan* in the 20th Century.
- Reduces operation time and work handling time significantly.
- Frees up workers to take on other roles.
- Provides higher level jobs in the development, deployment, maintenance and running of the automated processes.

The main disadvantages of automation are :

- Causing *unemployment* and *poverty* by replacing human labor.
- *Security Threats/Vulnerability* : An automated system may have a limited level of intelligence, and is therefore more susceptible to committing errors outside of its immediate scope of knowledge (*e.g.*, it is typically unable to apply the rules of simple logic to general propositions).

- Unpredictable/excessive development costs : The *research and development* cost of automating a process may exceed the cost saved by the automation itself.
- *High initial cost* : The automation of a new *product* or *plant* typically requires a very large initial investment in comparison with the unit cost of the product, although the cost of automation may be spread among many products and over time.

In manufacturing, the purpose of automation has shifted to issues broader than productivity, cost, and time.

LIGHTS OUT MANUFACTURING

Lights out manufacturing is when a production system is 100% or near to 100% automated (not hiring any workers). In order to eliminate the need for labor costs all together.

HEALTH AND ENVIRONMENT

The costs of automation to the environment are different depending on the technology, product or engine automated. There are automated engines that consume more energy resources from the Earth in comparison with previous engines and those that do the opposite too. Hazardous operations, such as *oil refining*, the manufacturing of *industrial chemicals*, and all forms of *metal working*, were always early contenders for automation.

CONVERTIBILITY AND TURNAROUND TIME

Another major shift in automation is the increased demand for flexibility and convertibility in manufacturing processes. Manufacturers are increasingly demanding the ability to easily switch from manufacturing Product A to manufacturing Product B without having to completely rebuild the *production lines*. Flexibility and distributed processes have led to the introduction of *Automated Guided Vehicles* with Natural Features Navigation.

Digital electronics helped too. Former analogue-based instrumentation was replaced by digital equivalents which can be more accurate and flexible, and offer greater scope for more sophisticated configuration, parametrization and operation. This was accompanied by the *fieldbus* revolution which provided a networked (*i.e.* a single cable) means of communicating between control systems and field level instrumentation, eliminating hard-wiring.

Discrete manufacturing plants adopted these technologies fast. The more conservative process industries with their longer plant life cycles have been slower to adopt and analogue-based measurement and control still dominates. The growing use of *Industrial Ethernet* on the factory floor is pushing these trends still further, enabling manufacturing plants to be integrated more tightly within the enterprise, via the internet if necessary. Global competition has also increased demand for *Reconfigurable Manufacturing Systems*.

AUTOMATION TOOLS

Engineers can now have *numerical control* over automated devices. The result has been a rapidly expanding range of applications and human activities. *Computer-aided technologies* (or CAx) now serve the basis for mathematical and organizational tools used to create complex systems. Notable examples of CAx include *Computer-aided design* (CAD software) and *Computer-aided manufacturing* (CAM software). The improved design, analysis, and manufacture of products enabled by CAx has been beneficial for industry.

Information technology, together with *industrial machinery* and *processes*, can assist in the design, implementation, and monitoring of control systems. One example of an *industrial control system* is a *programmable logic controller* (PLC). PLCs are specialized hardened computers which are frequently used to synchronize the flow of inputs from (physical) *sensors* and events with the flow of outputs to actuators and events.

An *automated online assistant* on a website, with an *avatar* for enhanced *human–computer interaction*.

Human-machine interfaces (HMI) or *computer human interfaces* (CHI), formerly known as *man-machine interfaces*, are usually employed to communicate with

PLCs and other computers. Service personnel who monitor and control through HMIs can be called by different names. In industrial process and manufacturing environments, they are called operators or something similar. In boiler houses and central utilities departments they are called stationary engineers.

Different types of automation tools exist :

* ANN-*Artificial neural network*
* DCS-*Distributed Control System*
* HMI-*Human Machine Interface*
* SCADA-*Supervisory Control and Data Acquisition*
* PLC-*Programmable Logic Controller*
* *Instrumentation*
* *Motion control*
* *Robotics*

LIMITATIONS TO AUTOMATION

* Current technology is unable to automate all the desired tasks.
* Many operations using automation have large amounts of invested capital and produce high volumes of product, making malfunctions extremely costly and potentially hazardous. Therefore, some personnel are needed to insure that the entire system functions properly and that safety and product quality are maintained.
* As a process becomes increasingly automated, there is less and less labor to be saved or quality improvement to be gained. This is an example of both *diminishing returns* and the *logistic function*.
* As more and more processes become automated, there are fewer remaining non-automated processes. This is an example of exhaustion of opportunities. New technological paradigms may however set new limits that surpass the previous limits.

CURRENT LIMITATIONS

Many roles for humans in industrial processes presently lie beyond the scope of automation. Human-level *pattern recognition*, *language comprehension*, and language production ability are well beyond the capabilities of modern mechanical and computer systems. Tasks requiring subjective assessment or synthesis of complex sensory data, such as scents and sounds, as well as high-level tasks such as strategic planning, currently require human expertise. In many cases, the use of humans is more cost-effective than mechanical approaches even where automation of industrial tasks is possible. Overcoming these obstacles is a theorized path to *post-scarcity* economics.

RECENT AND EMERGING APPLICATIONS

Automated Retail

Food and Drink

The food retail industry has started to apply automation to the ordering process; *McDonald's* has introduced touch screen ordering and payment systems in many of its restaurants, reducing the need for as many cashier employees. University of Texas has introduced *fully automated machine cafe* retail locations. Some Cafes and restaurants have utilized mobile and tablet "*apps*" to make the ordering process more efficient by customers ordering and paying on their device. Some restaurants have automated food delivery to customers tables using a *Conveyor belt system*. The use of *robots* is sometimes employed to replace *waiting staff*.

Stores

Many *Supermarkets* and even smaller stores are rapidly introducing *Self checkout* systems reducing the need for employing checkout workers.

    Online shopping could be considered a form of automated retail as the payment and checkout are through an automated *Online transaction processing* system. Other forms of automation can also be an integral part of online shopping, for example the deployment of automated warehouse robotics such as that applied by *Amazon* using *Kiva Systems*.

Automated Mining

Involves the removal of human labor from the *mining* process. The *mining industry* is currently in the transition towards Automation. Currently it can still require a large amount of *human capital*, particularly in the *third world* where labor costs are low so there is less incentive for increasing efficiency through automation.

Automated Video Surveillance

The Defense Advanced Research Projects Agency (*DARPA*) started the research and development of automated visual surveillance and monitoring (VSAM) program, between 1997 and 1999, and airborne video surveillance (AVS) programs, from 1998 to 2002. Currently, there is a major effort underway in the vision community to develop a fully automated *tracking surveillance* system. Automated video surveillance monitors people and vehicles in real time within a busy environment. Existing automated surveillance systems are based on the environment they are primarily designed to observe, *i.e.*, indoor, outdoor or airborne, the amount of sensors that the automated system can handle and the mobility of sensor, *i.e.*, stationary camera vs. mobile camera. The purpose of a surveillance system is to record properties and trajectories of objects in a given area, generate warnings or notify designated authority in case of occurrence of particular events.

Automated Highway Systems

As demands for safety and mobility have grown and technological possibilities have multiplied, interest in automation has grown. Seeking to accelerate the development and introduction of fully automated vehicles and highways, the *United States Congress* authorized more than $650 million over six years for *intelligent transport systems* (ITS) and demonstration projects in the 1991 *Intermodal Surface Transportation Efficiency Act* (ISTEA). Congress legislated in ISTEA that "the *Secretary of Transportation* shall develop an automated highway and vehicle prototype from which future fully automated intelligent vehicle-highway systems can be developed. Such development shall include research in human factors to ensure the success of the man-machine relationship. The goal of this program is to have the first fully automated highway roadway or an automated test track in operation by 1997. This system shall accommodate installation of equipment in new and existing motor vehicles." [ISTEA 1991, part B, Section 6054(b)].

Full automation commonly defined as requiring no control or very limited control by the driver; such automation would be accomplished through a combination of sensor, computer, and communications systems in vehicles and along the roadway. Fully automated driving would, in theory, allow closer vehicle spacing and higher speeds, which could enhance traffic capacity in places where additional road building is physically impossible, politically unacceptable, or prohibitively expensive. Automated controls also might enhance road safety by reducing the opportunity for driver error, which causes a large share of motor vehicle crashes. Other potential benefits include improved air quality (as a result of more-efficient traffic flows), increased fuel economy, and spin-off technologies generated during research and development related to automated highway systems.

Automated Waste Management

Automated waste collection trucks prevent the need for as many workers as well as easing the level of Labor required to provide the service.

Home Automation

Home automation (also called *domotics*) designates an emerging practice of increased automation of household appliances and features in residential dwellings, particularly through electronic means that allow for things impracticable, overly expensive or simply not possible in recent past decades.

Industrial Automation

Industrial automation deals primarily with the automation of manufacturing, quality control and material handling processes. General purpose controllers for industrial processes include *Programmable logic controllers* and computers. One trend is increased use of *Machine vision* to provide automatic inspection and robot guidance functions, another is a continuing increase in the use of *robots*.

Energy efficiency in industrial processes has become a higher priority. Semi-conductor companies like *Infineon Technologies* are offering 8-bit micro-controller applications for example found in *motor controls*, general purpose pumps, fans, and e-bikes to reduce energy consumption and thus increase efficiency.

Agriculture

Now that we're moving towards automated orange-sorting and autonomous tractors, the next step in automated agriculture is robotic strawberry pickers.

Agent-assisted Automation

Agent-assisted Automation refers to automation used by call center agents to handle customer inquiries. There are two basic types : desktop automation and automated voice solutions. Desktop automation refers to software programming that makes it easier for the call center agent to work across multiple desktop tools. The automation would take the information entered into one tool and populate it across the others so it did not have to be entered more than once, for example. Automated voice solutions allow the agents to remain on the line while disclosures and other important information is provided to customers in the form of pre-recorded audio files. Specialized applications of these automated voice solutions enable the agents to process credit cards without ever seeing or hearing the credit card numbers or CVV codes

The key benefit of agent-assisted automation is compliance and error-proofing. Agents are sometimes not fully trained or they forget or ignore key steps in the process. The use of automation ensures that what is supposed to happen on the call actually does, every time.

Relationship to Unemployment

Based on a formula by *Gilles Saint-Paul*, an economist at Toulouse 1 University, the demand for unskilled human capital declines at a slower rate than the demand for skilled human capital increases. In the long run and for society as a whole it has led to cheaper products, *lower average work hours*, and new industries forming (*i.e*, robotics industries, computer industries, design industries). These new industries provide many high salary skill based jobs to the economy.

RESEARCH PROBLEM

Dead beat controllers are often used in *process control* due to their good dynamic properties. They are a classical *feedback controller* where the control gains are set using a table based on the plant system order and normalized natural frequency.

The deadbeat response has the following characteristics :

1. Zero steady-state error
2. Minimum rise time
3. Minimum settling time

4.   Less than 2% overshoot/undershoot
5.   Very high control signal output

DISTRIBUTED PARAMETER SYSTEM

A **distributed parameter system** (as opposed to a *lumped parameter system*) is a *system* whose *state space* is infinite-*dimensional*. Such systems are therefore also known as infinite-dimensional systems. Typical examples are systems described by *partial differential equations* or by *delay differential equations*.

Linear Time-Invariant Distributed Parameter Systems

Abstract Evolution Equations

Discrete-Time

With U, X and Y *Hilbert spaces* and $A \in L(X)$, $B \in L(U, X)$, $C \in L(X, Y)$ and $D \in L(U, Y)$ the following equations determine a discrete-time linear time-invariant system :

$$x(k + 1) = Ax(k) + Bu(k)$$
$$y(k) = Cx(k) + Du(k)$$

with x (the state) a sequence with values in X, u (the input or control) a sequence with values in U and Y (the output) a sequence with values in Y.

Continuous-Time

The continuous-time case is similar to the discrete-time case but now one considers differential equations instead of difference equations :

$$\dot{x}(t) = Ax(t) + Bu(t),$$
$$y(t) = Cx(t) + Du(t).$$

An added complication now however is that to include interesting physical examples such as partial differential equations and delay differential equations into this abstract framework, one is forced to consider *unbounded operators*. Usually A is assumed to generate a *strongly continuous semigroup* on the state space X. Assuming B, C and D to be bounded operators then already allows for the inclusion of many interesting physical examples, but the inclusion of many other interesting physical examples forces unboundedness of B and C as well.

Example : a partial differential equation

The partial differential equation with $t > 0$ and $\xi \in [0, 1]$ given by

$$\frac{\partial}{\partial t} w(t, \xi) = -\frac{\partial}{\partial \xi} w(t, \xi) + u(t),$$

$$w(0, \xi) = w\delta(\xi),$$
$$w(t, 0) = 0,$$

$$y(t) = \int_0^1 w(t, \xi) d\xi,$$

fits into the abstract evolution equation framework described above as follows. The input space U and the output space Y are both chosen to be the set of complex numbers. The state space X is chosen to be $L^2(0, 1)$. The operator A is defined as

$$Ax = -x', \quad D(A) = \{x \in X : x \text{ absolutely continuous}, x' \in L^2(0, 1) \text{ and } x(0) = 0\}.$$

It can be shown that A generates a strongly continuous semi-group on X. The bounded operators B, C and D are defined as

$$Bu = u, \quad Cx = \int_0^1 x(\xi) d\xi, \quad D = 0.$$

Example : a delay differential equation

The delay differential equation

$$\dot{w}(t) = w(t) + w(t - t) + u(t),$$
$$y(t) = w(t),$$

fits into the abstract evolution equation framework described above as follows. The input space U and the output space Y are both chosen to be the set of complex numbers. The state space X is chosen to be the product of the complex numbers with $L^2(-\tau, 0)$. The operator A is defined as :

$$A \begin{pmatrix} r \\ f \end{pmatrix} = \begin{pmatrix} r + f(-\tau) \\ f' \end{pmatrix}, D(A)$$

$$= \left\{ \begin{pmatrix} r \\ f \end{pmatrix} \in X : f \text{ absolutely continuous}, f' \in L^2([-\tau, 0]) \text{ and } r = f(0) \right\}.$$

It can be shown that A generates a strongly continuous semi-group on X. The bounded operators B, C and D are defined as :

$$Bu = \begin{pmatrix} u \\ 0 \end{pmatrix}, C \begin{pmatrix} r \\ f \end{pmatrix} = r, \quad D = 0.$$

Transfer Functions

As in the finite-dimensional case the *transfer function* is defined through the *Laplace transform* (continuous-time) or *Z-transform* (discrete-time). Whereas in the finite-dimensional case the transfer function is a proper rational function, the infinite-dimensionality of the state space leads to irrational functions (which are however still *holomorphic*).

Discrete-time

In discrete-time the transfer function is given in terms of the state space parameters by $D + \sum_{k=0}^{\infty} CA^k Bz^k$ and it is holomorphic in a disc centered at the origin.

In case $1/z$ belongs to the resolvent set of A (which is the case on a possibly smaller disc centered at the origin) the transfer function equals $D + Cz(I - zA)^{-1} B$. An interesting fact is that any function that is holomorphic in zero is the transfer function of some discrete-time system.

Continuous-Time

If A generates a strongly continuous semi-group and B, C and D are bounded operators, then the transfer function is given in terms of the state space parameters by $D + C(sI - A)^{-1}B$ for s with real part larger than the exponential growth bound of the semi-group generated by A. In more general situations this formula as it stands may not even make sense, but an appropriate generalization of this formula still holds. To obtain an easy expression for the transfer function it is often better to take the Laplace transform in the given differential equation than to use the state space formulas as illustrated below on the examples given above.

Transfer Function for the Partial Differential Equation Example

Setting the initial condition w_0 equal to zero and denoting Laplace transforms with respect to t by capital letters we obtain from the partial differential equation given above :

$$sW(s, \xi) = -\frac{d}{d\xi}W(s,\xi)+U(s),$$

$$W(s, 0) = 0,$$

$$Y(s) = \int_0^1 W(s,\xi)d\xi.$$

This is an inhomogeneous linear differential equation with ξ as the variable, s as a parameter and initial condition zero. The solution is $W(s, \xi) = U(s) (1 - e^{-s\xi})/s$. Substituting this in the equation for Y and integrating gives $Y(s) = U(s)(e^{-s} + s - 1)/s^2$ so that the transfer function is $(e^{-s} + s - 1)/s^2$.

Transfer Function for the Delay Differential Equation Example

Proceeding similarly as for the partial differential equation example, the transfer function for the delay equation example is $1/(s - 1 - e^{-s})$.

Controllability

In the infinite-dimensional case there are several non-equivalent definitions of *controllability* which for the finite-dimensional case collapse to the one usual notion of controllability. The three most important controllability concepts are :

- Exact controllability,
- Approximate controllability,
- Null controllability.

Controllability in Discrete-Time

An important role is played by the maps Φ_n which map the set of all U valued sequences into X and are given by $\Phi_n u = \sum_{k=0}^{n} A^k B u_k$. The interpretation is that $\Phi_n u$ is the state that is reached by applying the input sequence u when the initial condition is zero. The system is called :

- Exactly controllable in time n if the range of Φ_n equals X,
- Approximately controllable in time n if the range of Φ_n is dense in X,
- Null controllable in time n if the range of Φ_n includes the range of A^n.

Controllability in Continuous-Time

In controllability of continuous-time systems the map Φ_t given by $\int_0^t e^{As} Bu(s)ds$ plays the role that Φ_n plays in discrete-time. However, the space of control functions on which this operator acts now influences the definition. The usual choice is $L^2(0, \infty; U)$, the space of (equivalence classes of) U-valued square integrable functions on the interval $(0, \infty)$, but other choices such as $L^1(0, \infty; U)$ are possible. The different controllability notions can be defined once the domain of Φ_t is chosen. The system is called :

- Exactly controllable in time t if the range of Φ_t equals X,
- Approximately controllable in time t if the range of Φ_t is dense in X,
- Null controllable in time t if the range of Φ_t includes the range of e^{At}.

Observability

As in the finite-dimensional case, *observability* is the dual notion of controllability. In the infinite-dimensional case there are several different notions of observability which in the finite-dimensional case coincide. The three most important ones are :

- Exact observability (also known as continuous observability),
- Approximate observability,
- Final state observability.

Observability in Discrete-Time

An important role is played by the maps Ψ_n which map X into the space of all Y valued sequences and are given by $(\Psi_n x)_k = CA^k x$ if $k \leq n$ and zero if $k > n$. The interpretation is that $\Psi_n x$ is the truncated output with initial condition x and control zero. The system is called :

- Exactly observable in time n if there exists a $k_n > 0$ such that $||\Psi_n x|| \geq k_n ||x||$ for all $x \in X$,
- Approximately observable in time n if Ψ_n is *injective*,
- Final state observable in time n if there exists a $k_n > 0$ such that $||\Psi_n x|| \geq k_n ||A^n x||$ for all $x \in X$.

Observability in Continuous-Time

In observability of continuous-time systems the map Ψ_t given by $(\Psi_t)(s) = Ce^{As}x$ for $s \in [0,t]$ and zero for $s > t$ plays the role that Ψ_n plays in discrete-time. However, the space of functions to which this operator maps now influences the definition. The usual choice is $L^2(0, \infty, Y)$, the space of (equivalence classes of) Y-valued square integrable functions on the interval $(0,\infty)$, but other choices such as $L^1(0, \infty, Y)$ are possible. The different observability notions can be defined once the co-domain of Ψ_t is chosen. The system is called

- Exactly observable in time t if there exists a $k_t > 0$ such that $||\Psi_t x|| \geq k_t ||x||$ for all $x \in X$,
- Approximately observable in time t if Ψ_t is *injective*,
- Final state observable in time t if there exists a $k_t > 0$ such that $||\Psi_t x|| \geq k_t ||e^{At}x||$ for all $x \in X$.

Duality between Controllability and Observability

As in the finite-dimensional case, controllability and observability are dual concepts (at least when for the domain of Φ and the co-domain of Ψ the usual L^2 choice is made). The correspondence under duality of the different concepts is :

- Exact controllability $\leftrightarrow$ Exact observability,
- Approximate controllability $\leftrightarrow$ Approximate observability,
- Null controllability $\leftrightarrow$ Final state observability.

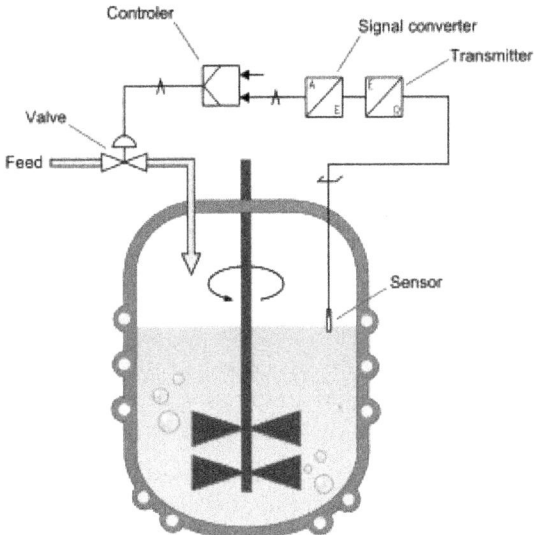

Figure : Example of control system of a *continuous stirred-tank reactor*.

PROCESS CONTROL

Process control is an *engineering* discipline that deals with *architectures, mechanisms* and *algorithms* for maintaining the output of a specific *process* within a desired range. For instance, the temperature of a chemical reactor may be controlled to maintain a consistent product output.

Process control is extensively used in industry and enables mass production of consistent products from continuously operated processes such as oil refining, paper manufacturing, chemicals, power plants and many others. Process control enables *automation,* by which a small staff of operating personnel can operate a complex process from a central control room.

Background

Process control may either use *feedback* or it may be open loop. Control may also be continuous (automobile cruise control) or cause a sequence of discrete events, such as a timer on a lawn sprinkler (on/off) or controls on an elevator (logical sequence).

A thermostat on a heater is an example of control that is on or off. A temperature sensor turns the heat source on if the temperature falls below the set point and turns the heat source off when the set point is reached. There is no measurement of the difference between the set point and the measured temperature (*e.g.* no error measurement) and no adjustment to the rate at which heat is added other than all or none.

A familiar example of feedback control is cruise control on an automobile. Here speed is the **measured variable**. The operator (driver) adjusts the desired speed **set point** (*e.g.* 100 km/hr) and the controller monitors the speed sensor and compares the measured speed to the set point. Any deviations, such as changes in grade, drag, wind speed or even using a different grade of fuel (for example an ethanol blend) are corrected by the controller making a compensating adjustment to the fuel valve open position, which is the **manipulated variable**. The controller makes adjustments having information only about the error (magnitude, rate of change or cumulative error) although settings known as *tuning* are used to achieve stable control. The operation of such controllers is the subject of *control theory*.

A commonly used control device called a *programmable logic controller,* or a PLC, is used to read a set of digital and analog inputs, apply a set of logic statements, and generate a set of analog and digital outputs.

For example, if an adjustable valve were used to hold level in a tank the logical statements would compare the equivalent pressure at depth setpoint to the pressure reading of a sensor below the normal low liquid level and determine whether more or less valve opening was necessary to keep the level constant. A PLC output would then calculate an incremental amount of change in the valve position. Larger more complex systems can be controlled by a *Distributed Control System* (DCS) or *SCADA* system.

Types of Processes Using Process Control

In practice, processes can be characterized as one or more of the following forms:

- *Discrete*: Found in many manufacturing, motion and packaging applications. Robotic assembly, such as that found in automotive production, can be characterized as discrete process control. Most discrete manufacturing involves the production of discrete pieces of product, such as metal stamping.

- *Batch*: Some applications require that specific quantities of raw materials be combined in specific ways for particular durations to produce an intermediate or end result. One example is the production of adhesives and glues, which normally require the mixing of raw materials in a heated vessel for a period of time to form a quantity of end product. Other important examples are the production of food, beverages and medicine. Batch processes are generally used to produce a relatively low to intermediate quantity of product per year (a few pounds to millions of pounds).

- *Continuous*: Often, a physical system is represented through variables that are smooth and uninterrupted in time. The control of the water temperature in a heating jacket, for example, is an example of continuous process control. Some important continuous processes are the production of fuels, chemicals and plastics. Continuous processes in manufacturing are used to produce very large quantities of product per year (millions to billions of pounds).

Applications having elements of discrete, batch and continuous process control are often called *hybrid* applications.

Examples

- An *anti-lock braking system* (ABS) is a complex example, consisting of multiple inputs, conditions and outputs.
- Aircraft stability control is a highly complex example using multiple inputs and outputs.

ROBUST CONTROL

Robust control is a branch of *control theory* that explicitly deals with uncertainty in its approach to controller design. Robust control methods are designed to function properly so long as uncertain parameters or disturbances are within some (typically *compact*) set. Robust methods aim to achieve robust performance and/ or *stability* in the presence of bounded modeling errors.

The early methods of *Bode* and others were fairly robust; the state-space methods invented in the 1960s and 1970s were sometimes found to lack robustness, prompting research to improve them. This was the start of the theory of Robust Control, which took shape in the 1980s and 1990s and is still active today.

In contrast with an *adaptive control* policy, a robust control policy is static; rather than adapting to measurements of variations, the controller is designed to work assuming that certain variables will be unknown but, for example, bounded.

When is a Control Method Said to be Robust?

Informally, a controller designed for a particular set of parameters is said to be robust if it would also work well under a different set of assumptions. High-gain feedback is a simple example of a robust control method; with sufficiently high gain, the effect of any parameter variations will be negligible. High-gain feedback is the principle that allows simplified models of *operational amplifiers* and emitter-degenerated *bipolar transistors* to be used in a variety of different settings. This idea was already well understood by *Bode* and *Black* in 1927.

The Modern Theory of Robust Control

The theory of robust control began in the late 1970s and early 1980s and soon developed a number of techniques for dealing with bounded system uncertainty.

Probably the most important example of a robust control technique is *H-infinity loop-shaping*, which was developed by *Duncan McFarlane* and *Keith Glover* of *Cambridge University*; this method minimizes the *sensitivity* of a system over its frequency spectrum, and this guarantees that the system will not greatly deviate from expected trajectories when disturbances enter the system.

An emerging area of robust control from application point of view is Sliding Mode Control (SMC) which is a variation of variable structure control (VSS). Robustness property of SMC towards matched uncertainty as well as the simplicity in design attracted a variety of application.

Another example is loop transfer recovery (LQG/LTR), which was developed to overcome the robustness problems of *LQG* control.

Other robust techniques includes Quantitative Feedback Theory (QFT), Gain scheduling etc.

VECTOR CONTROL (MOTOR)

Vector control, also called **field-oriented control** (FOC), is a *variable frequency drive* (VFD) control method which controls *three-phase AC electric motor* output by means of two controllable VFD inverter output variables :

- Voltage magnitude
- *Frequency* : (Voltage angle, or phase, is only indirectly controlled)

FOC is a control technique that is used in *AC synchronous* and *induction motor* applications that was originally developed for high-performance motor applications which can operate smoothly over the full *speed* range, can generate full *torque* at zero speed, and is capable of fast *acceleration* and *deceleration* but that is becoming increasingly attractive for lower performance applications as well due to FOC's motor size, cost and *power consumption* reduction superiority.

Not only is FOC very common in induction motor control applications due to its traditional superiority in high-performance applications, but the expectation is that it will eventually nearly universally displace single-variable *scalar volts-per-Hertz* (V/f) control.

Development History

Technical University Darmstadt's K. Hasse and Siemens' F. Blaschke pioneered *vector* control of AC motors starting in 1968 and in the early 1970s, Hasse in terms of proposing indirect vector control, Blaschke in terms of proposing direct vector control. Technical University Braunschweig's Werner Leonhard further developed FOC techniques and was instrumental in opening up opportunities for *AC drives* to be a competitive alternative to *DC drives*.

Yet it was not until after the commercialization of *microprocessors*, that is in the early 1980s, that general purpose AC drives became available. Barriers to use of FOC for AC drive applications included higher cost and complexity and lower maintainability compared to DC drives, FOC having until then required many electronic components in terms of sensors, amplifiers and so on.

The *Park transformation* has long been widely used in the analysis and study of *synchronous* and induction machines. The transformation is by far the single most important concept needed for an understanding of how FOC works, the concept having been first conceptualized in a 1929 paper authored by *Robert H. Park*. Park's paper was ranked second most important in terms of impact from among all power engineering related papers ever published in the twentieth century. The novelty of Park's work involves his ability to transform any related machine's linear *differential equation* set from one with time varying coefficients to another with time *invariant* coefficients.

Technical Overview

Overview of key competing VFD control platforms :

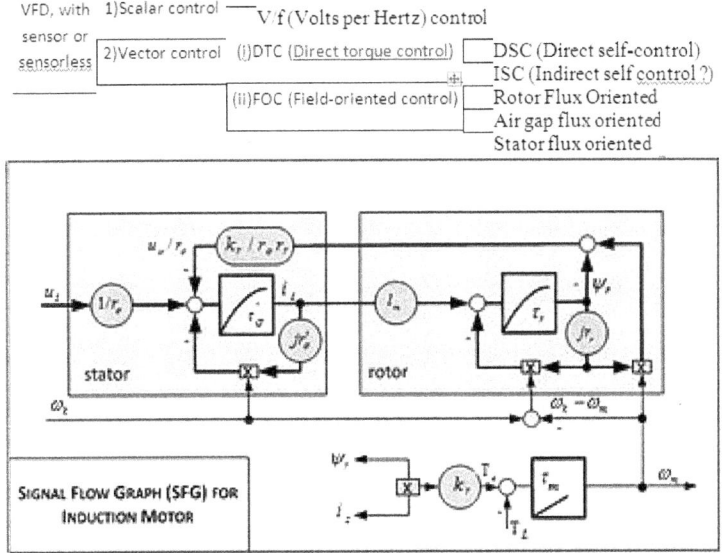

Figure : Signal Flow Graph (SFG) for Induction Motor

Induction motor model equations

$$t'_e \frac{di_s}{d\tau} + i_s = -j\omega_k t'_e i_s + \frac{k_r}{t_r r_e}(1 - jr_r\omega_m)\psi_r + \frac{1}{r_e}u_s \quad (1)$$

$$t_r \frac{d\psi_r}{d\tau} + \psi_r = -j(\omega_k - \omega_m)t_r\psi_r + l_m i_s \quad (2)$$

where

$$\sigma'_r = \sigma l_s/r_e \quad r_e = r_s + k_r^2 r_r \quad k_r = l_m/l_r \quad \tau = \omega_{sR}$$

$\sigma = 1 - l_m^2/l_r l_s = $ *total leakage coefficient*

$\omega_{sR} = $ nominal *stator frequency*

Basic parameter symbols	Subscripts and superscripts
i - current	*e - electromechanical*
k - coupling factor of respective winding	*i - induced voltage*
l - inductance	*k - referred to k-coordinates*
r - resistance	*L - load*
t - time	*m - mutual (inductance)*
T - torque	*m - mechanical (T.C., angular velocity)*
u - voltage	*r - rotor*
ψ - flux linkage	*R - rated value*
τ - normalized time	*s - stator*
t - time constant (T.C.) with subscript	*' - denotes transient time constant*
ω - angular velocity	
σl_s - total leakage inductance	

SFG Equations

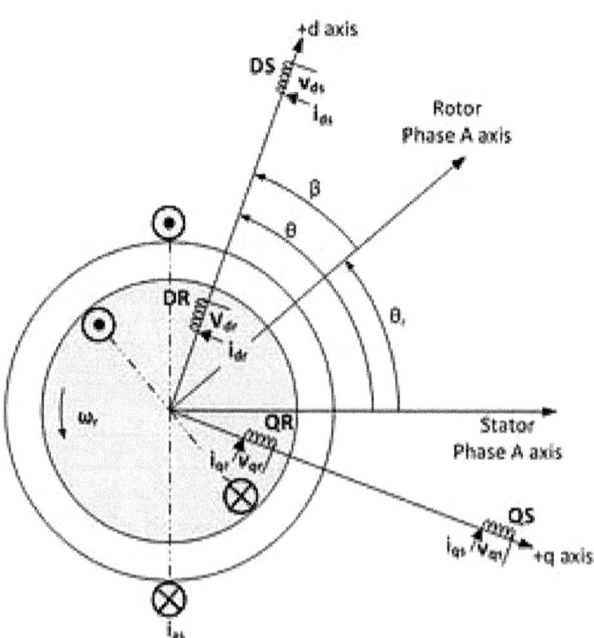

Figure : (d,q) Coordinate System Superimposed on Three-Phase Induction Motor

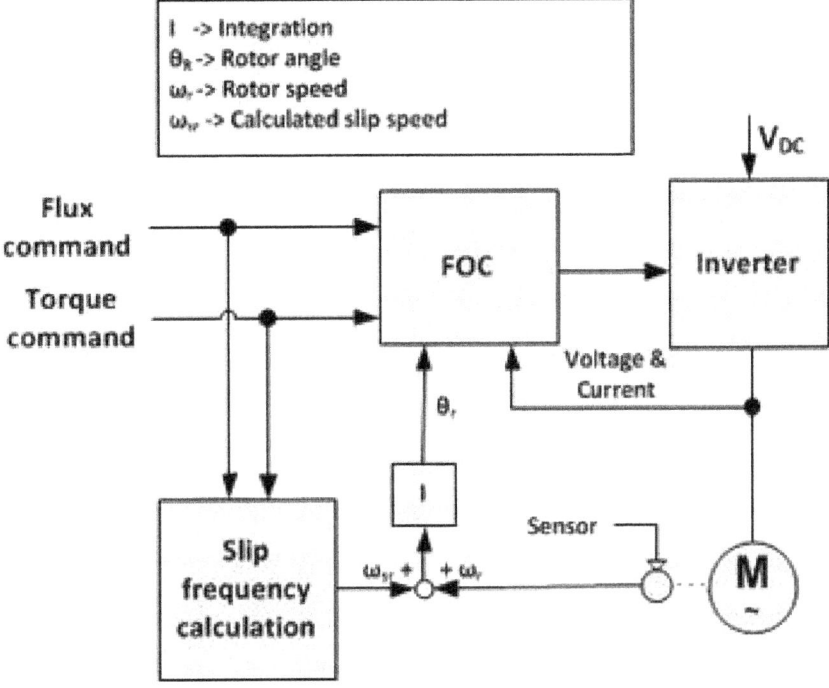

Figure : Simplified Indirect FOC Block Diagram.

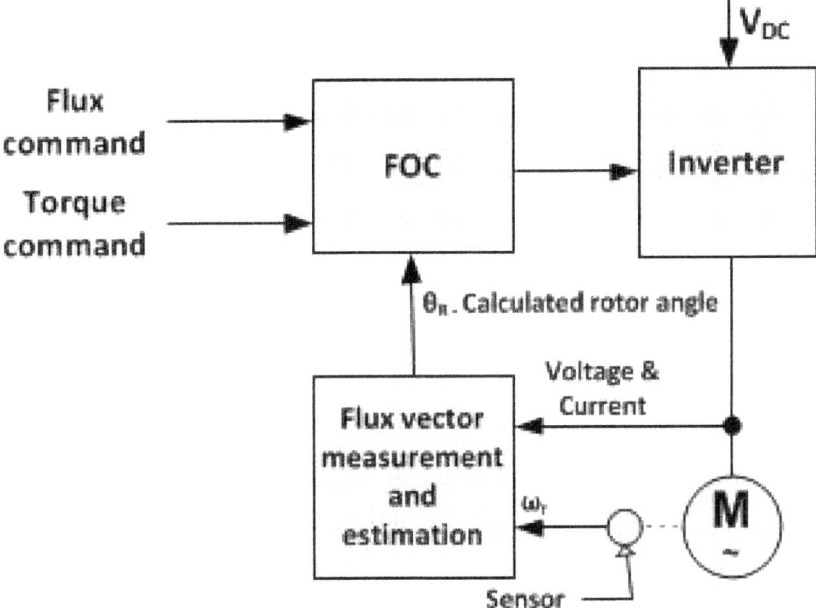

Figure : Simplified Direct FOC Block Diagram.

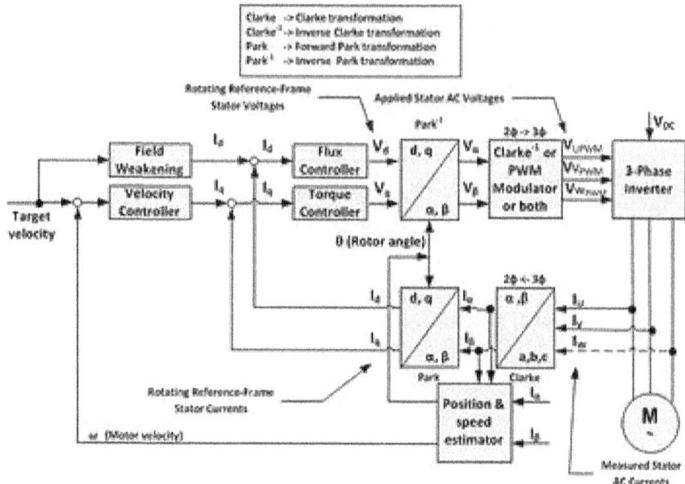

Figure : Sensorless FOC Block Diagram.

While the analysis of AC drive controls can be technically quite involved, such analysis invariably starts with modeling of the drive-motor circuit involved along the lines of accompanying *signal flow graph* and equations.

In vector control, an AC induction or synchronous motor is controlled under all operating conditions like a separately *excited* DC motor. That is, the AC motor behaves like a DC motor in which the *field flux linkage* and *armature* flux linkage created by the respective field and armature (or torque component) currents are *orthogonally* aligned such that, when torque is controlled, the field flux linkage is not affected, hence enabling dynamic torque response.

Vector control accordingly generates a three-phase *PWM* motor voltage output derived from a *complex* voltage vector to control a complex current vector derived from motor's three-phase motor stator current input through *projections* or *rotations* back and forth between the three-phase speed and time dependent system and these vectors' rotating reference-frame two-*coordinate* time invariant system.

Such complex *stator* motor current space vector can be defined in a (d,q) co-ordinate system with orthogonal components along d (direct) and q (quadrature) axes such that field flux linkage component of current is aligned along the d axis and torque component of current is aligned along the q axis. The induction motor's (d,q) coordinate system can be superimposed to the motor's instantaneous (a,b,c) three-phase *sinusoidal* system as shown in accompanying image (phases a & b not shown for clarity). Components of the (d,q) system current vector, allow conventional control such as proportional and integral, or *PI, control*, as with a DC motor.

Projections associated with the (d,q) coordinate system typically involve :

* Forward projection from instantaneous currents to (a,b,c) complex *stator* current space vector representation of the three-phase *sinusoidal* system.

- Forward three-to-two phase, (a,b,c)-to-(α, β) projection using the *Clarke* transformation. Vector control implementations usually assume ungrounded motor with balanced three-phase currents such that only two motor current phases need to be sensed. Also, backward two-to-three phase, (α, β)-to-(a,b,c) projection uses space vector PWM modulator or inverse Clarke transformation and one of the other PWM modulators.

- Forward and backward two-to-two phase,(α, β)-to-(d,q) and (d,q)-to-(α, β) projections using the Park and inverse Park transformations, respectively.

However, it is not uncommon for sources to use three-to-two, (a,b,c)-to-(d,q) and inverse projections.

While (d,q) coordinate system rotation can arbitrarily be set to any speed, there are three preferred speeds or reference frames :

- Stationary reference frame where (d,q) coordinate system does not rotate;
- Synchronously rotating reference frame where (d,q) coordinate system rotates at synchronous speed;
- Rotor reference frame where (d,q) coordinate system rotates at rotor speed.

Decoupled torque and field currents can thus be derived from raw stator current inputs for control algorithm development.

Whereas magnetic field and torque components in DC motors can be operated relatively simply by separately controlling the respective field and armature currents, economical control of AC motors in variable speed application has required development of microprocessor-based controls with all AC drives now using powerful DSP (*digital signal processing*) technology.

Inverters can be implemented as either *open-loop* sensorless or closed-loop FOC, the key limitation of open-loop operation being mimimum speed possible at 100% torque, namely, about 0.8 Hz compared to standstill for closed-loop operation.

There are two vector control methods, direct or *feedback* vector control (DFOC) and indirect or *feedforward* vector control (IFOC), IFOC being more commonly used because in closed-loop mode such drives more easily operate throughout the speed range from zero speed to high-speed field-weakening. In DFOC, flux magnitude and angle feedback signals are directly calculated using so-called voltage or current models. In IFOC, flux space angle feedforward and flux magnitude signals first measure stator currents and *rotor* speed for then deriving flux space angle proper by summing the rotor angle corresponding to the rotor speed and the calculated reference value of *slip* angle corresponding to the slip frequency.

Sensorless control AC drives is attractive for cost and reliability considerations. Sensorless control requires derivation of rotor speed information from measured stator voltage and currents in combination with open-loop estimators or closed-loop observers.

Application Recap

1. Stator phase currents are measured, converted to complex space vector in (a,b,c) coordinate system.

2. Current vector is converted to (α, β) coordinate system. *Transformed to a coordinate system* rotating in *rotor* reference frame, rotor position being derived by *integrating* the speed by means of *speed measurement* sensor.

3. Rotor *flux linkage* vector is estimated by multiplying the stator current vector with magnetizing inductance L_m and *low-pass filtering* the result with the rotor no-load *time constant* L_r/R_r, namely, the rotor inductance to rotor resistance ratio.

4. Current vector is converted to (d,q) coordinate system.

5. d-axis component of the stator current vector is used to control the rotor flux linkage and the imaginary q-axis component is used to control the motor torque. While PI controllers can be used to control these currents, *bang-bang* type current control provides better dynamic performance.

6. PI controllers provide (d,q) coordinate voltage components. A decoupling term is sometimes added to the controller output to improve control performance to mitigate cross coupling or big and rapid changes in speed, current and flux linkage. PI-controller also sometimes need *low-pass filtering* at the input or output to prevent the current ripple due to transistor switching from being amplified excessively and destabilizing the control. However, such filtering also limits the dynamic control system performance. High switching frequency (typically more than 10 kHz) is typically required to minimize filtering requirements for high-performance drives such as servo drives.

7. Voltage components are transformed from (d,q) coordinate system to (α, β) coordinate system.

8. Voltage components are transformed from (α, β) coordinate system to (a,b,c) coordinate system or fed in *Pulse Width Modulation (PWM)* modulator, or both, for signaling to the power inverter section.

Significant aspects of vector control application :

*   Speed or position measurement or some sort of estimation is needed.

*   Torque and flux can be changed reasonably fast, in less than 5–10 milliseconds, by changing the references.

*   The *step response* has some *overshoot* if PI control is used.

*   The switching frequency of the transistors is usually constant and set by the modulator.

*   The accuracy of the torque depends on the accuracy of the motor parameters used in the control. Thus large errors due to for example rotor temperature changes often are encountered.

*   Reasonable processor performance is required; typically the control algorithm has to be calculated at least every millisecond.

Although the vector control algorithm is more complicated than the *Direct Torque Control* (DTC), the algorithm is not needed to be calculated as frequently as the DTC algorithm. Also the current sensors need not be the best in the market. Thus the cost of the processor and other control hardware is lower making it suitable for applications where the ultimate performance of DTC is not required.

CUT-INSERTION THEOREM

The **Cut-insertion theorem**, also known as **Pellegrini's theorem**, is a linear network theorem that allows transformation of a generic network N into another network N' that makes analysis simpler and for which the main properties are more apparent.

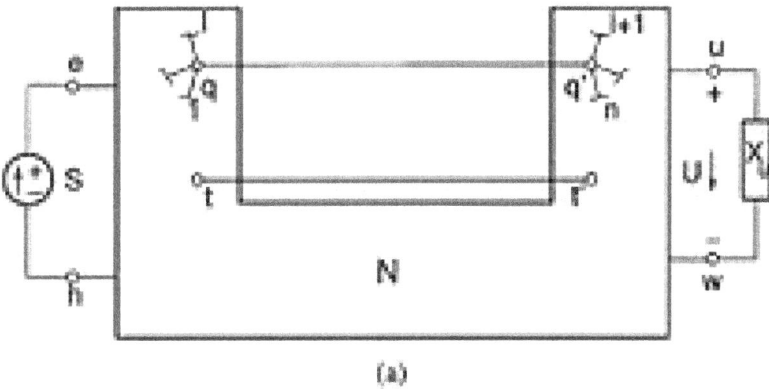

Figure : Generic linear network N.

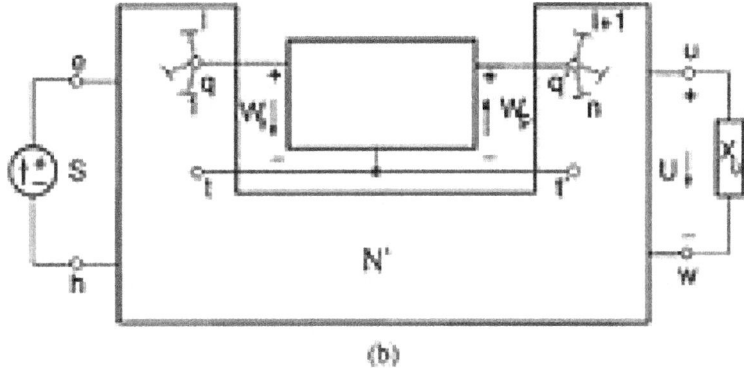

Figure : Equivalent linear network N'.

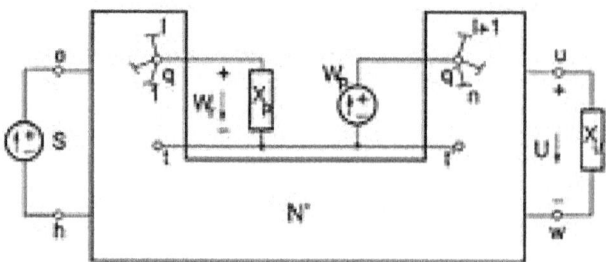

Statement

Implementation of the three-terminal circuit by means of an independent source W_r and an immittance X_p.

Let $e, h, u, w, q=q'$, and $t=t'$ be six arbitrary nodes of the network N and S be an independent voltage or current source connected between e and h, while U is the output quantity, either a voltage or current, relative to the branch with *immittance* X_u, connected between u and w. Let us now cut the qq' connection and insert a three-terminal circuit ("TTC") between the two nodes q and q' and the node $t=t'$, as in figure b (W_r and W_p are homogeneous quantities, voltages or currents, relative to the ports qt and $q'q't'$ of the TTC).

In order for the two networks N and N' to be equivalent for any S, the two constraints $W_r = W_p$ and $W_r = W_{p'}$, where the overline indicates the dual quantity, are to be satisfied.

The above mentioned three-terminal circuit can be implemented, for example, connecting an ideal independent voltage or current source W_p between q' and t', and an immittance X_p between q and t.

Network Functions

With reference to the network N', the following *network functions* can be defined :

$$A \equiv \frac{U}{W_p}\Big|_{S=0} ; \beta \equiv \frac{W_r}{U}\Big|_{s=0} ; X_i \equiv \frac{W_p}{W_p}\Big|_{s=0}$$

$$\gamma \equiv \frac{U}{S}\Big|_{W_p=0} ; \alpha \equiv \frac{W_r}{S}\Big|_{W_p=0} ; \rho \equiv \frac{W_p}{S}\Big|_{W_p=0}$$

from which, exploiting the *Superposition theorem*, we obtain :

$$W_r = \alpha S + \beta A W_p$$

$$W_p = \rho S + \frac{W_p}{X_i}$$

Therefore the first constraint for the equivalence of the networks is satisfied

if $W_p = \dfrac{\alpha}{1-\beta A} S$.

Furthermore,

$$W_r = \frac{W_r}{X_p}$$

$$W_p = \left(\frac{1}{X_i} + \frac{\rho}{\alpha}(1 - \beta A) \right) W_r$$

therefore the second constraint for the equivalence of the networks holds if

$$\frac{1}{X_p} = \left(\frac{1}{X_1} + \frac{\rho}{\alpha}(1 - \beta A) \right).$$

Transfer Function

If we consider the expressions for the network functions γ and A, the first constraint for the equivalence of the networks, and we also consider that, as a result of the superposition principle, $U = \gamma S + A W_p$, the transfer function $A_f \equiv \dfrac{U}{S}$ is given by

$$A_f = \frac{\alpha A}{1 - \beta A} + \gamma.$$

For the particular case of a *feedback amplifier*, the network functions α, γ and ρ take into account the nonidealities of such amplifier. In particular :

- α takes into account the nonideality of the comparison network at the input
- γ takes into account the non unidirectionality of the feedback chain
- ρ takes into account the non unidirectionality of the amplification chain.

If the amplifier can be considered ideal, *i.e.* if $\alpha = 1$, $\rho = 0$ and $\gamma = 0$, the transfer function reduces to the known expression deriving from classical feedback theory :

$$A_f = \frac{A}{1 - \beta A}$$

Evaluation of the Impedance and of the Admittance between Two Nodes

The evaluation of the *impedance* (or of the *admittance*) between two nodes is made somewhat simpler by the cut-insertion theorem.

Impedance

Let us insert a generic source S between the nodes $j=e=q$ and $k=h$ between which we want to evaluate the impedance Z. By performing a cut as shown in the figure, we notice that the immittance X_p is in series with S and the current through it is thus the same as that provided by S. If we choose an input voltage source $V_s = S$ and, as a consequence, a current $I_s = S$, and an impedance $Z_p = X_p$, we can write the following relationships :

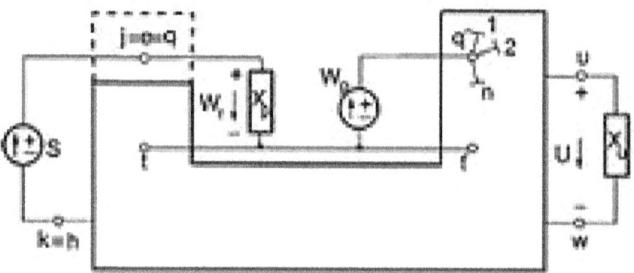

Figure : Cut for the evaluation of the impedance between the nodes $k=h$ and $j=e=q$.

$$Z = \frac{V_s}{I_s} = \frac{V_s}{I_s} = Z_p \frac{V_s}{V_r} = Z_p \frac{V_s}{V_p} = Z_p \frac{1-\beta A}{\alpha}.$$

Considering that $\alpha = \frac{V_r}{V_s}\Big|_{V_{p=0}} = \frac{Z_p}{Z_p + Z_b}$, where Z_b is the impedance seen be-

tween the nodes $k=h$ and t if remove Z_b and short-circuit the voltage sources, we obtain the impedance Z between the nodes j and k in the form :

$$Z = (Z_p + Z_b)\,(1 - \beta A)$$

Admittance

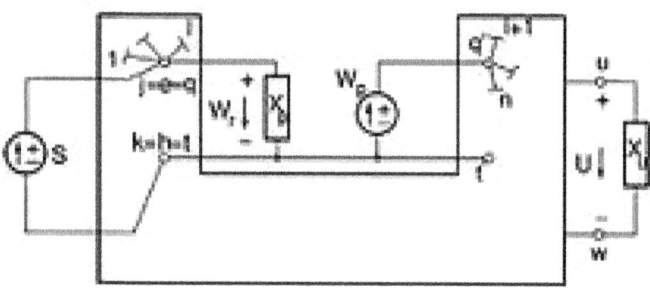

Figure : Cut for the evaluation of the impedance between the nodes $k=h=t$ and $j=e=q$.

We proceed in a way analogous to the impedance case, but this time the cut will be as shown in the figure to the right, noticing that S is now in parallel to X_p. If we consider an input current source $I_s = S$ (as a result we have a voltage $V_s = S$) and an admittance $Y_p = X_p$, the admittance Y between the nodes j and k can be computed as follows :

$$Y = \frac{I_s}{V_s} = \frac{I_s}{V_r} = Y_p \frac{I_s}{I_r} = Y_p \frac{I_s}{I_p} = Y_p \frac{1-\beta A}{\alpha}$$

Considering that $\alpha = \dfrac{I_r}{I_s}\big|_{I_p=0} = \dfrac{Y_p}{Y_p+Y_b}$, where Y_b is the admittance seen between

the nodes $k=h$ and t if we remove Y_p and open the current sources, we obtain the admittance Y in the form :

$$Y = (Y_p + Y_b)(1 - \beta A)$$

Comments

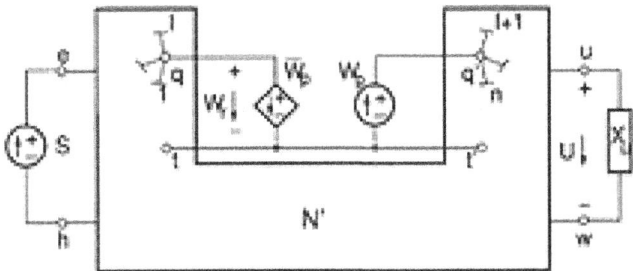

Implementation of the three-terminal circuit by means of an independent source W_p and a dependent source $W_{p'}$.

The implementation of the TTC with an independent source W_p and an immittance X_p is useful and intuitive for the calculation of the impedance between two nodes, but involves, as in the case of the other network functions, the difficulty of the calculation of X_p from the equivalence equation. Such difficulty can be avoided using a dependent source W_p in place of X_p and using the Blackman formula for the evaluation of X. Such an implementation of the TTC allows finding a feedback topology even in a network consisting of a voltage source and two impedances in series.

COEFFICIENT DIAGRAM METHOD

Coefficient diagram method (CDM), developed and introduced by *Prof. Shunji Manabe* in 1991. CDM is an *algebraic* approach applied to a *polynomial* loop in the parameter space, where a special diagram called a *"coefficient diagram"* is used as the vehicle to carry the necessary information, and as the criteria of good design. The performance of the closed loop system is monitored by the coefficient diagram.

The most important properties of the method are : the adaptation of the polynomial representation for both the plant and the controller, the use of the two-degree of freedom (2DOF) control system structure, the non-existence (or very small) of the overshoot in the step response of the closed loop system, the determination of the settling time at the start and to continue the design accordingly, the good robustness for the control system with respect to the plant parameter changes, the sufficient gain and phase margins for the controller. The most considerable advantages of CDM can be listed as follows :

1. The design procedure is easily understandable, systematic and useful. Therefore, the coefficients of the CDM controller polynomials can be determined more easily than those of the PID or other types of controller. This creates the possibility of an easy realisation for a new designer to control any kind of system.

2. There are explicit relations between the performance parameters specified before the design and the coefficients of the controller polynomials as described in. For this reason, the designer can easily realize many control systems having different performance propirties for a given control problem in a wide range of freedom.

3. The development of different tuning methods is required for time delay processes of different properties in PID control. But it is sufficient to use the single design procedure in the CDM technique. This is an outstanding advantage.

4. It is particularly hard to design robust controllers realizing the desired performance propefties for unstable, integrating and oscillatory processes having poles near the imaginary axis. It has been reported that successful designs can be achieved even in these cases by using CDM.

5. It is theoretically proven that CDM design is equivalent to LQ design with proper state augmentation. Thus, CDM can be considered an "improved LQG", because the order of the controller is smaller and weight selection rules are also given.

It is usually required that the controller for a given plant should be designed under some practical limitations. The controller is desired to be of minimum degree, minimum phase (if possible) and stable. It must have enough bandwidth and power rating limitations. If the controller is designed without considering these limitations, the robustness property will be very poor, even though the stability and time response requirements are met. CDM controllers designed while considering all these problems is of the lowest degree, has a convenient bandwidth and results with a unit step time response without an overshoot. These properties guarantee the robustness, the sufficient damping of the disturbance effects and the low economic property.

Although the main principles of CDM have been known since the 1950s, the first systematic method was proposed by Shunji Manabe. He developed a new method that easily builds a target characteristic polynomial to meet the desired time response. CDM is an algebraic approach combining classical and modern control theories and uses polynomial representation in the mathematical expression. The advantages of the classical and modern control techniques are integrated with the basic principles of this method, which is derived by making use of the previous experience and knowledge of the controller design. Thus, an efficient and fertile control method has appeared as a tool with which control systems can be designed without needing much experience and without confronting many problems.

Many control systems have been designed successfully using CDM. It is very easy to design a controller under the conditions of stability, time domain performance and robustness. The close relations between these conditions and coefficients of the characteristic polynomial can be simply determined. This means that CDM is effective not only for control system design but also for controller parameters tuning.

CONTROL RECONFIGURATION

Control reconfiguration is an active approach in *control theory* to achieve *fault-tolerant control* for *dynamic systems*. It is used when severe *faults*, such as actuator or sensor outages, cause a break-up of the *control loop*, which must be restructured to prevent *failure* at the system level. In addition to loop restructuring, the *controller* parameters must be adjusted to accommodate changed plant dynamics. Control reconfiguration is a building block toward increasing the *dependability* of systems under *feedback* control.

Re-configuration problem

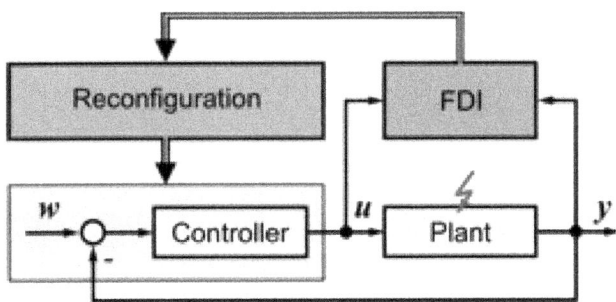

Schematic diagram of a typical active fault-tolerant control system. In the nominal, *i.e.* fault-free situation, the lower control loop operates to meet the control goals. The fault-detection (FDI) module monitors the closed-loop system to detect and isolate faults. The fault estimate is passed to the reconfiguration block, which modifies the control loop to reach the control goals in spite of the fault.

Fault Modelling

The figure to the right shows a plant controlled by a controller in a standard control loop.

The nominal linear model of the plant is —

$$\begin{cases} \dot{x} = Ax + Bu \\ \quad y = Cx \end{cases}$$

The plant subject to a fault (indicated by a red arrow in the figure) is modelled in general by —

$$\begin{cases} \dot{x}_f = A_f x_f + B_f u \\ \quad y_f = C_f x_f \end{cases}$$

where the subscript f indicates that the system is faulty. This approach models multiplicative faults by modified system matrices. Specifically, actuator faults are represented by the new input matrix B_f, sensor faults are represented by the output map C_f, and internal plant faults are represented by the system matrix A_f.

The upper part of the figure shows a supervisory loop consisting of *fault detection and isolation* (FDI) and *reconfiguration* which changes the loop by —

1.   Choosing new input and output signals from $\{u, y\}$ to reach the control goal,
2.   Changing the controller internals (including dynamic structure and parameters),
3.   Adjusting the reference input w.

To this end, the vectors of inputs and outputs contain *all available signals*, not just those used by the controller in fault-free operation.

Alternative scenarios can model faults as an additive external signal f influencing the state derivatives and outputs as follows :

$$\begin{cases} \dot{x}_f = Ax_f + Bu + Ef \\ \quad y_f = C_f x_f + Ff \end{cases}$$

Reconfiguration Goals

The goal of reconfiguration is to keep the reconfigured control-loop performance sufficient for preventing plant shutdown. The following goals are distinguished :

1.   Stabilization
2.   Equilibrium recovery
3.   Output trajectory recovery
4.   State trajectory recovery
5.   Transient time response recovery.

Internal stability of the reconfigured closed loop is usually the minimum requirement. The equilibrium recovery goal (also referred to as weak goal) refers to the steady-state output equilibrium which the reconfigured loop reaches after a given constant input. This equilibrium must equal the nominal equilibrium under the same input (as time tends to infinity). This goal ensures steady-state reference tracking after reconfiguration. The output trajectory recovery goal (also referred to as strong goal) is even stricter. It requires that the dynamic response to an input must equal the nominal response at all times. Further restrictions are imposed by the state trajectory recovery goal, which requires that the state trajectory be restored to the nominal case by the reconfiguration under any input.

Usually a combination of goals is pursued in practice, such as the equilibrium-recovery goal with stability.

The question whether or not these or similar goals can be reached for specific faults is addressed by *reconfigurability* analysis.

Reconfiguration Approaches

Fault Hiding

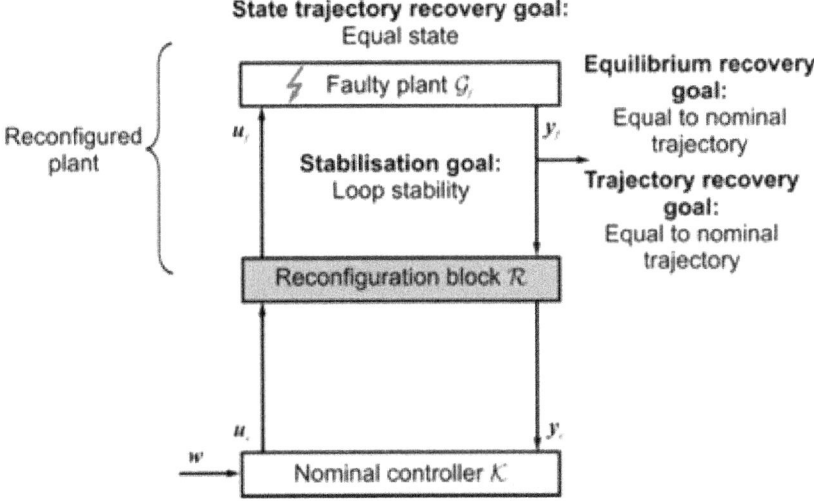

Fault hiding principle. A reconfiguration block is placed between faulty plant and nominal controller. The reconfuigured plant behaviour must match the nominal behaviour. Furthermore, the reconfiguration goals are pointed out.

This paradigm aims at keeping the nominal controller in the loop. To this end, a reconfiguration block can be placed between the faulty plant and the nominal controller. Together with the faulty plant, it forms the reconfigured plant. The reconfiguration block has to fulfill the requirement that the behaviour of the reconfigured plant matches the behaviour of the nominal, that is fault-free plant.

Linear Model Following

In linear model following, a formal feature of the nominal closed loop is attempted to be recovered. In the classical pseudo-inverse method, the closed loop system matrix $\bar{A} = A - BK$ of a state-feedback control structure is used. The new controller K_f is found to approximate $\bar{A}$ in the sense of an induced matrix norm.

In perfect model following, a dynamic compensator is introduced to allow for the exact recovery of the complete loop behaviour under certain conditions.

In eigenstructure assignment, the nominal closed loop eigenvalues and eigenvectors (the eigenstructure) is recovered to the nominal case after a fault.

Optimisation-based Control Schemes

Optimisation control schemes include : Linear-quadratic regulator design (LQR), model predictive control (MPC) and eigenstructure assignment methods.

Probabilistic Approaches

Some probabilistic approaches have been developed.

Learning Control

There are learning automata, neural networks, etc.

Mathematical Tools and Frameworks

The methods by which reconfiguration is achieved differ considerably. The following list gives an overview of mathematical approaches that are commonly used.

- *Adaptive control* (AC)
- Disturbance decoupling (DD)
- Eigenstructure assignment (EA)
- *Gain scheduling* (GS)/linear parameter varying (LPV)
- Generalised internal model control (GIMC)
- Intelligent control (IC)
- *Linear matrix inequality* (LMI)
- *Linear-quadratic regulator* (LQR)
- Model following (MF)
- *Model predictive control* (MPC)
- *Pseudo-inverse* method (PIM)
- *Robust control* techniques.

Chapter 19

Study of a QCM Dimethyl Methylphosphonate Sensor Based on a ZnO-Modified Nanowire-Structured Manganese Dioxide Film

Zhifu Pei[1], Xingfa Ma[2,3], Pengfei Ding[1], Wuming Zhang[1,*], Zhiyuan Luo[4] and Guang Li[1]

[1] National Key Laboratory of Industrial Control Technology, Department of Control Science and Engineering, Zhejiang University, Hangzhou 310027, China;
E-Mails: florapzf@yahoo.com.cn(Z.P); dpfzjuv@126.com(P.D.); guangli@zju.edu.cn (G.L.)

[2] State Key Laboratory of Silicon Materials, Zhejiang University, Hangzhou 310027, China;
E-Mail: xingfamazju@yahoo.com.cn (X.M.)

[3] School of Environmental and Material Engineering, Center of Advanced Functional Materials, Yantai University, Yantai 264005, China

[4] Computer Learning Research Centre, Royal Holloway, University of London, Egham, Surrey TW20 0EX, UK; E-Mail: Zhiyuan.Luo@cs.rhul.ac.uk (Z.L.)

[*] Author to whom correspondence should be addressed; E-Mail: zhangwm@zju.edu.cn; Tel.: +86-571-8795-2268 ext. 2230; Fax: +86-571-8795-2279.

ABSTRACT

Sensitive, selective and fast detection of chemical warfare agents is necessary for anti-terrorism purposes. In our search for functional materials sensitive to dimethyl methylphosphonate (DMMP), a simulant of sarin and other toxic organophosphorus compounds, we found that zinc oxide (ZnO) modification potentially enhances the absorption of DMMP on a manganese dioxide (MnO_2) surface. The adsorption behavior of DMMP was evaluated through the detection of tiny organophosphonate compounds with quartz crystal microbalance (QCM) sensors coated with ZnO-modified MnO_2 nanofibers and pure MnO_2 nanofibers.

Experimental results indicated that the QCM sensor coated with ZnO-modified nanostructured MnO_2 film exhibited much higher sensitivity and better selectivity in comparison with the one coated with pure MnO_2 nanofiber film. Therefore, the DMMP sensor developed with this composite nanostructured material should possess excellent selectivity and reasonable sensitivity towards the tiny gaseous DMMP species.

Keywords

Quartz crystal microbalance; gas sensor; volatile organic vapor; DMMP; nanowire; manganese dioxide; zinc oxide.

1. INTRODUCTION

Sensitive and selective detection of a wide variety of chemical species has become a necessity in many applications, including the quantification of chemical warfare agents (CWAs), explosives, environmental pollutants and many other toxic industrial compounds [1,2]. The threat of terrorism has greatly increased the need for fast detection of CWAs, so it is urgent to develop CWA sensors with fast response, high specificity, low detection limits and easy operation [3,4]. Dimethyl methylphosphonate [DMMP, $CH_3PO(OCH_3)_2$] due to its nontoxicity and organophosphorus compound elemental composition that mimics nerve agents, is commonly considered as a simulant for CWAs and insecticides, such as the G-series nerve agents tabun (GA), sarin (GB), soman (GD) and paraoxon [5]. DMMP has also become a significant environmental and food chain pollutant due to its large consumption as a common additive for anti-foaming agents, plasticizers, stabilizers, textile conditioners and antistatic agents [2]. Consequently, DMMP sensors with high sensitivity, rapid response, low energy consumption and good reversibility at room temperature are highly desirable, not only for neurotoxin detection for counter-terrorism purposes, but also for environmental protection and medical diagnoses for risk management [6-8].

Focusing on highly sensitive functional materials with advanced fabrication technology, metal oxide semiconductors [8-11], carbon nanotubes [12-14], conducting polymers [15] and organometallic compounds [16,17] have been studied. Of these, metal oxides are well-known for their industrial applications as adsorbents, catalysts and catalyst supports, especially manganese dioxide (MnO_2) and titanium dioxide (TiO_2) which are widely used as molecular sieves and electrode materials in batteries and sensors due to their unique electronic and surface properties [18-22]. Furthermore, they are often used as sensing materials for different gases now due to their large gases adsorbent capacity [8,23,24]. Several investigations have been carried out concerning the adsorption and reaction of DMMP or other CWAs on the surfaces of different metal oxides, including MgO [25,26], Al_2O_3 [26-28], TiO_2 [26,29-31], Fe_2O_3 [32,33], ZnO [26,34,35], SiO_2 [36], MnO_2 [37] and WO_3 [26,38]. Zinc oxide (ZnO) in particular shows very strong sensitivity toward toxic substances, such as halogens, sulfur, volatile organic compounds [35] and organophosphorus compounds [34,35]. Furthermore, the experiments on powdered TiO_2 have revealed three distinct modes of adsorption: DMMP condenses on the outer surface of TiO_2 below 160 K; molecularly diffuses into the TiO_2 interior and chemisorbs on TiO_2 from 160 to 200 K; and dissociatively chemisorbs above 214 K [30]. Enhancement to the absorption and reactivity of their nanoparticles and other nanostructures to sensitively detect various pollutants and harmful substances, including organophosphorus compounds, is anticipated due to their unique electronic properties, morphological features and high surface area [39]. These details indicate the nanostructured metal oxides like TiO_2, ZnO and MnO_2 may show sensitive and dissociative adsorbent of gas-phase DMMP at ambient temperature, promising sensing behavior for the mass detection of gas-phase organophosphorus compounds. However, only a few experimental studies have been reported on the adsorption of toxic chemicals or chemical warfare agent simulants on nano-structured MnO_2 at ambient temperature [40].

In addition, it is interesting that the behavior of alumina-supported iron oxide may be significantly different from that of pure alumina [32,33]. Furthermore, zinc oxide doped in SnO_2 may improve the reliability and sensitivity of the SnO_2 sensors for simulants of the CWAs at 250 to 400 °C [40]. It is suggested that the absorption capability of DMMP on MnO_2 surface is possibly improved by ZnO modification. In this study, we attempted to modify the MnO_2 nanostructure with ZnO to explore new sensing materials and furthermore to evaluate the DMMP absorption properties at ambient temperature by constructing a sensitive DMMP sensor based on the MnO_2 nanostructured film and the quartz crystal microbalance (QCM). Consequently, the adsorption behavior of DMMP on the composite material was characterized through the detection of tiny organophosphonate compounds with QCM sensors coated with ZnO-modified MnO_2 nanofibers and the comparison of these properties to those of the same sensors coated with pure MnO_2 nanofibers. We thus concluded that these features make the developed DMMP sensor possess excellent selectivity and reasonable sensitivity.

2. EXPERIMENTAL SECTION

2.1. Materials

Chemicals and regents for nanostructured metal oxides were manganese (II) sulfate ($MnSO_4$), potassium permanganate ($KMnO_4$), zinc nitrate [$Zn(NO_3)_2$], hexamethylenetetramine and ammonia. These chemicals and reagents were analytical grade and commercially available. The chemicals for volatile organic vapors (VOCs) were dimethyl methylphosphonate (DMMP), acetone, p-dichlorobenzene (p-DCB), p-dimethylbenzene (p-xylene), ethanol, n-hexane and trichloro-methane (chloroform) purchased from Sigma-Aldrich (Shanghai, China) and Wako Pure Chemicals (Osaka, Japan). All these chemicals and reagents were used directly as received without further purification. The AT-cut 6.0 MHz (HC-49/U) quartz crystals with aluminum electrodes on both sides were purchased from Hosonic International (Hangzhou) Ltd., China. The crystals were rinsed by ethanol and then deionized water prior to use. All experiments were carried out at room temperature (about 25 degrees Celsius in an air-conditioned room).

2.2. Preparation of Pure and ZnO-modified MnO_2 NW-structured Films

2.2.1. Preparation of NW-structured MnO_2

In the preparation process, 1.0 g $MnSO_4$ and 0.5 g of oxidizing reagent ($KMnO_4$) were dissolved in 20 mL of distilled water at room temperature to form a homogeneous solution. The solution was then transferred into a 100 mL Teflon-lined stainless steel autoclave, sealed and maintained at 120 °C for about 24 hours. After the resulting solid product was filtered and washed with distilled water to remove the possibly remnant ions in the final products and finally dried in air, MnO_2 NWs, the final product was obtained.

2.2.2. Modification of NW-structured MnO_2 with ZnO

About 0.5 g of home-made MnO_2 NWs was dispersed in 20 mL distilled water; sequentially, 0.5 g zinc nitrate and 0.5 g hexamethylenetetramine were added. The mixture was then transferred into a Teflon-lined stainless steel autoclave. The pH value of reaction solution was adjusted to around 10 with ammonia. The hydro-thermal treatments were carried out at 90-95 °C for 5 hours. After the resulting solid product was filtered, washed with distilled water repeatedly, and finally dried in air, the ZnO-modified MnO_2 nanowires (NWs) were obtained.

2.2.3. Preparation of QCM Sensors with ZnO-modified MnO_2 NW-Structured Film

About 1 mg of ZnO-modified NW-structured MnO_2 was weighed and dispersed in deionized water to form a dark brown colored stock solution with a concentration of 2 ug/uL. After standing for 24 hours, 2.5, 5, 7.5, 10, 12.5 or 15 microliters of the aqueous solution was dispensed onto the electrode surface of QCMs using a micropipette, forming a sensing film with an area of 0.2 cm^2 and a thickness index of 25, 50, 75, 100, 125 or 150 ug/cm^2, respectively. Then the device was dried in a dry cabinet at room temperature. After these steps, the QCM sensors coated with ZnO-modified NW-structured MnO_2 film were obtained.

2.2.4. Structural Characterization of the Pure and ZnO-modified MnO_2 Films on QCMs

The nanostructure of MnO_2 films were characterized by scanning electron microscopy (SEM). This observation was performed using a Field-Emission Scanning Electron Microscope with Energy Dispersive Spectrometer (FESEM-EDS, HITACHI S4800, Japan), operated at 25.0 kV. Both the pure and ZnO-modified MnO_2 films on the QCM sensors were deposited by platinum on the surface for SEM observation. The NW-structures of MnO_2 films were observed and recorded.

2.3. Experimental Procedure for Gas Sensing

The Sauerbrey equation was developed for oscillation in air and only applies to rigid mass attached to the crystal [41]. It gives the change in the oscillation frequency of piezoelectric quartz (Δf) as a function of the mass (Δm) added to the crystal:

$$\Delta f = -\frac{2 f_0^2}{A \rho_q \upsilon_q} \Delta m$$

Here, Δf is the observed frequency change (Hz), f_0 is the fundamental resonant frequency of crystal, A is the active area, the area where the crystal is coated with electrodes on both sides, υ_q is the density of quartz and ρ_q is the shear wave velocity in the quartz. A home-made experimental system was set up to evaluate the as fabricated QCM sensors. The gas sensors set in the 500 mL sealed chamber of experimental setup were thus characterized at around 25 °C with either analytic gases for measurements or high-purity nitrogen gas for cleaning. The sensing-film-coated QCM was used as sensing unit while an uncoated QCM was used in the experimental system as reference. The variation of the frequency difference between the reference and sensing QCMs was defined as the response of QCM gas sensors. When the QCM DMMP sensor was exposed to the analyte DMMP, the sensing film would absorb the analytic gas, therein inducing a decrease in the working resonant frequency of the QCM DMMP sensors and the frequency change increase from a value at the start of the experiment. The working frequency of the QCM DMMP sensor in experimental setup changed from the fundamental resonant frequency of crystal coated with sensing film to a lower steady frequency of crystal decided by the sensing film-absorbed target gas. In this way, as the adsorption process approached equilibrium between the adsorption/desorption at a given concentration of DMMP, the frequency change or the response of QCM sensor increased, and finally reached a plateau phase. According to Equation (1) and adsorption mechanism [30], the steady value of frequency change at the plateau phase would determine the amount of gas absorbed in the film at ambient temperature. The absorbed DMMP analyte could be desorbed by high-purity N_2, due to its dissociative adsorption at ambient temperature. The sensing QCM would thus be recovered; consequently the frequency change of the sensing QCM would be zeroed.

DMMP, acetone, p-DCB, p-xylene, ethanol, n-hexane and chloroform vapors were used as the analytical gases for this investigation. The experiments were performed as follows: first, a target analyte was injected into the testing chamber, the sensing film then absorbed the analyte, thus decreasing the output frequency of the sensing QCM. The response of the QCM gas sensor, or an increase of the frequency difference between the sensing and reference QCMs to the analyte, was measured continuously at a 1-second interval [42,43]. After the response reached a plateau phase, the measurement course in a cycle was finished. The chamber was then purged with high-purity nitrogen gas to expel the analyte and recover the sensing films of sensors; the cleaning course for next cycle of measurement was started. This process was repeated several times for each analyte to get reliable results.

3. RESULTS AND DISCUSSION

3.1. Morphology of the Sensing Films

Both the pure and ZnO-modified MnO_2 films on QCM sensors were investigated. The composite nanostructures in the ZnO-modified MnO_2 film can be seen in Figure 1.

Figure 1. SEM image of the ZnO-modified MnO_2 nanowire.

For comparison, the morphology and structure of the pure NW-structured MnO_2 film is shown in Figure 2. The observed data indicate that the ZnO-nanoparticles modified the surface of MnO_2 NWs.

These SEM images confirmed that the composite nanostructures were formed in the ZnO-modified MnO_2 film by the ZnO nanoparticles' joining to the MnO_2 NWs. Therefore, we attribute the distinct sensing properties of the ZnO-modified MnO_2 NW-structured QCM sensor to the composite nanostructures of the ZnO nanoparticle-modified surface of the MnO_2 NWs.

Figure 2. SEM image of the pure MnO_2 nanowires.

3.2. Sensitivity and Repeatability of the QCM DMMP Sensors

The QCM sensors based on ZnO-modified NW-structured MnO$_2$ films were repeatedly tested at predefined DMMP concentrations for assessing their sensitivity and repeatability. At first, the QCM sensors coated with ZnO-modified nanostructured MnO$_2$ films were exposed to a series of defined concentrations of DMMP vapors diluted in high-purity nitrogen gas to assess their sensitivity. This investigation was performed by alternatively exposing the sensors to 300-second DMMP vapors and 400-second nitrogen gas. The responses to the alternating inputs between the DMMP vapors of 0.35, 0.70 or 1.75 ppm and cleaning gas of high purity nitrogen gas at room temperature were recorded and displayed in Figure 3.

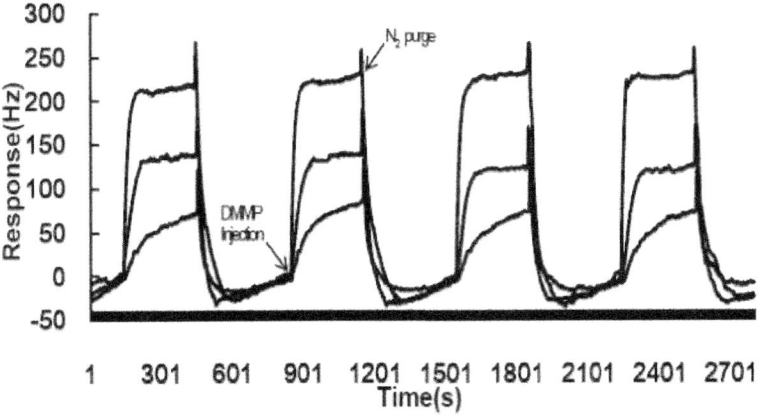

Figure 3. Response cycles of a QCM sensor coated with ZnO-modified NW-structured MnO$_2$ film towards 0.35, 0.70 and 1.75 ppms DMMP purged by high-purity nitrogen gas at room temperature.

The response curves in Figure 3 indicate that the sensors were sensitive to the vapor concentrations and quickly responded to the high-purity nitrogen gas purge, thus possibly possessing a high sensitivity and good repeatability towards gaseous DMMP.

3.3. Relationship Between the Responses (sensitivity) and the Thickness of ZnO-modified MnO$_2$ NW-structured Films

The Sauerbrey equation generally gives a good prediction of the linear responses of the QCMs working in air, but only applies to rigid films [41]. The real situation of our developed QCM gas sensors to DMMP vapor sorption is much more complex, so there might be many challenges in predicting the linear relationship between the QCM sensor responses and the concentrations of DMMP vapors. Clearly, besides the nature and thickness of crystals (such as fundamental resonant frequency and Q value), various factors related to the sensing films and target gases, such as the analyte-surface interaction and thickness of sensing films, can possibly contribute to the sensitivity and linearity of the sensor responses.

For the QCM DMMP sensor based on NW-structured MnO_2 film, the film thickness is an important factor for suitable response when the material nature and fabrication method are given. The thickness of the sensing films on electrodes thus effects on both the mass-sensing properties of QCMs and physicochemical adsorption of DMMP.

To address this problem, we designed test experiments to investigate the relationship between the sensitivity of the sensors and the thickness of the sensing films on the sensors. Through these experiments, we could maximize the sensitivity and optimize the linearity of the QCM sensor by appropriately selecting the thickness of sensing films. The electrode surface on crystals was deposited with 2.5, 5, 7.5, 10, 12.5 or 15 microliters of the 2 µg/µL aqueous solution of the ZnO-modified MnO_2 on 0.2 cm² working area forming sensing films having six different thicknesses. Although the abstract values of these thicknesses could be not figured out due to lack of the mass density of the film, the thickness indices could be easily obtained by supposing that the mass density of film is constant for all films produced with same material and fabrication method. Thus, the ZnO-modified MnO_2 composite film deposited on sensors had a thickness index of 25, 50, 75, 100, 125 or 150 µg/cm², respectively. This thickness index reflects the variation of film thickness, thus defined and used as thickness in this study. The test experiments were performed during the produced QCM sensor was exposed in 0.7 ppm, a predefined concentration, of gaseous DMMP. The experimental findings, displayed in Figure 4, reveal a significant nonlinear relationship between the sensitivity of the sensors and the thickness of the sensing films.

Here, we tried to explain the nonlinear properties of the QCM sensor through the behaviors of QCM oscillation and DMMP adsorption. Both of them are closely related to the thickness of sensing film deposed on QCM substrate. The nonlinear phenomena seems partly contributed to the interference of the sensing film with the vibration state of QCM substrate. Nevertheless, the Sauerbrey equation still works on the QCM gas sensor. In principle, the responses of QCM sensor generally arise from both gravimetric and viscoelastic changes in real sensing films, whereas the Sauerbrey equation predicts the outputs according to the gravimetric changes in ideal rigid films.

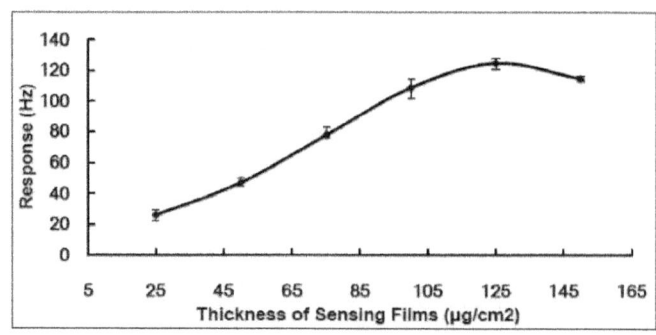

Figure 4. Responses of the QCM sensors with various thicknesses of sensing films to a predefined concentration of DMMP (0.7 ppm).

The sensing film moves synchronously with the underlying QCM substrate if the thickness of the coated sensing film is small enough relative to the acoustic wavelength in the film [45]. In this situation, the motion imparted by the QCM substrate displaces the sensing film parallel to the surface of the substrate just as the situation in a rigid film. The responses to the sensing film thus obey the Sauerbrey equation and reflect the gravimetric changes of the sensing film regardless of its shear modulus [46]. On the other hand, when the sensing thickness is big enough or acoustically thick (thicker than a few percents of the acoustic wavelength) the thickness stress gradients will become important. In this situation, the responses to the NW-structured MnO_2 sensing film will depend on both the acoustic thickness and the thickness shear modulus of the sensing film prior to and during DMMP vapor exposure [45-48], thus no longer accurately obeying the Sauerbrey equation. The experimental results, shown in Figure 4, reveal some of these effects. As is known, these nonlinear phenomena contribute to the variation of the QCM sensor's static working status consisting of the resonant frequency, Q value and other parameters prior to DMMP vapor exposure, as well as the dynamic sensing status of the film during DMMP vapor exposure, which will be described in the next paragraph. Despite working in a static working point different from that of theoretical crystal, the real QCM DMMP sensor can detect the mass changes of sensing film according to the Sauerbrey equation. That is why the QCM had been utilized in chemical and biological sensors in so many applications for so many years.

In our opinion, the nonlinear phenomena seem to contribute more significantly to the gas-surface interaction and adsorption of DMMP inside the ZnO-modified NW-structured MnO_2 sensing film. The uniform and porosity structures of the NWs in film lead to a huge surface-to-volume ratios; these ratios, in turn, make the interaction between the sensing film and DMMP vapor efficient, thus, leading to quick and thorough equilibrium analyte absorption/desorption state. When the sensing film on the QCM substrate is thin, the whole sensing film, as an effective surface adsorption area, effectively adsorbs the target analyte. Thus, while the QCM sensor's responses reflect the gas diffusion and adsorption/desorption process in sensing film exposed in a fixed concentration of DMMP vapor, the QCM sensor's plateau responses reflect equilibrium status of adsorption/desorption to a given DMMP concentration. The readouts of the plateau curve are consequently the responses of the QCM sensor to a given DMMP concentration. The effective surface adsorption area will increase together with the amount, or the thickness of sensing materials deposited on a given area of substrate. Therein, the total amount of the DMMP absorbed, furthermore, the responses of the sensor exposed in a given DMMP concentration will also increase together with the thickness of sensing materials, as the linear line between 0.03 and 125 $\mu g/cm^2$ shown in Figure 4. When the thickness is big, the DMMP molecule will penetrate a larger depth inside the sensing film to diffuse to the bottom of sensing film and reach equilibrium status of adsorption/desorption in whole sensing film. The responses of QCM sensor present more sensitive readouts to a defined DMMP concentration in spite of taking a longer time. However, when the amount is big enough, the sensing film is so thick that the DMMP molecules can only penetrate through the upper layer, and cannot reach to the bottom of the sensing film. Thus, the DMMP molecules will approach the equilibrium status of adsorption/ desorption in the somewhat steady upper layer of the sensing film. Consequently, the effective surface adsorption area will no longer increase with the thickness of sensing films and the responses of the sensor to a given DMMP concentration tend to be stead too, shown as the plateau phase of the curve. The findings also illustrate that a sensing film with a thickness near to 125 µg/cm^2 is most sensitive to 0.7 ppm of DMMP vapor. We recognized this thickness value as the optimal thickness of the ZnO-modified NW-structured MnO$_2$ film coated on QCM under the predefined experimental conditions. We consequently thought that the gas sensors set at this working situation would have a maximized sensitivity to 0.7 ppm, therein designing the sensors with this thickness of sensing films for the measurement range containing 0.7 ppm and all other experiments in this study.

Anyway, how to find out the optimal thickness for designing QCM sensors with the best linearity and sensitivity to organic vapors is a challenge. As analyzed in the paragraphs above, despite many efforts, the gas diffusion and gas-surface interaction in sensing films are still far from being well understood for QCM metal oxide gas sensors. Further theoretical, empirical, or experimental results would be expected to understand how the gas concentration profile develops inside a thin film of metal oxide NWs after its exposure to a target gas, thus enabling us to design the sensing film more rationally.

3.4. Relationship Between the Sensitivity and Concentration of DMMP Vapor

For each designed sensor, its sensitivity was assessed at various concentrations of DMMP vapors. In order to evaluate the repeatability, three or four consecutive measurements at each concentration were required. Thus, the response (R) to each given DMMP concentration (C) was repeatedly measured three or four times. The results over a range from 0.035 ppm to 2.8 ppm were shown in Figure 5, where the small plot indicated the QCM sensor based on ZnO-modified NW-structured MnO$_2$ film began to illustrate discernible nonlinearity at a DMMP concentration of near 3 ppm.

In contrast, as shown in the large plot in Figure 5, the responses of the sensor were almost linearly proportional to the lower DMMP concentrations, ranging from 0.035 to 1.05 ppm. The regression equation could be expressed as R = 176.04 C + 3.01 with a correlation coefficient of 0.9987, where C is the concentration of DMMP vapors and R is the response or sensitivity of the sensor, respectively. This relationship plots the calibration curve of the ZnO-modified NW-structured MnO$_2$ based QCM sensor. Accordingly, the limit of detection (calculated as three times the signal-to-noise ratio) could be estimated and given as 35 ppb.

This result represents a great improvement compared to those previously published by other groups. Brunol *et al.* reported their study to deal with the

DMMP detection, using tin dioxide-based gas sensors. They used a DMMP vapor concentration level of around 200 ppm [6]. Ying *et al.* studied a PVDF coated QCM as the DMMP sensor [7]. The sensitivity was 3.19 Hz/ppm over the range from 5 to 60 ppm DMMP in N_2 and the limit of detection was about 0.94 ppm. In comparison with their results, the sensor reported in this paper is highly sensitive, and thus suitable for low level DMMP detection.

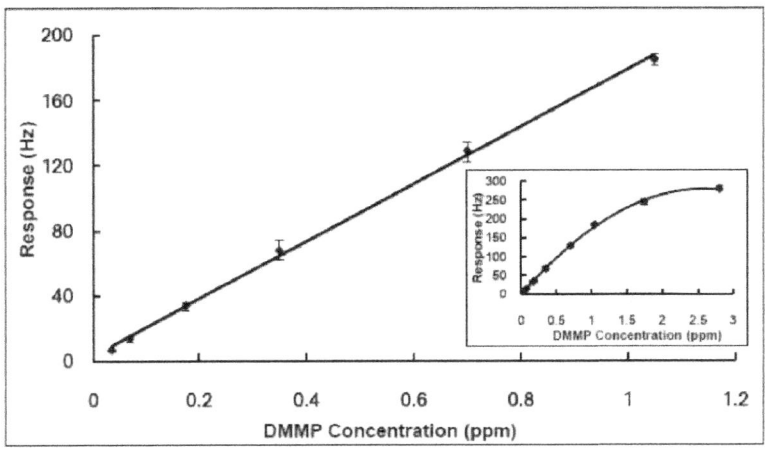

Figure 5. Linear plot of the reciprocal of a ZnO-modified nanowire MnO_2 coated QCM sensor's response against the concentrations of DMMP vapor purged in high-purity nitrogen at room temperature.

Although the frequency change of QCM responded linearly to the amount of gas absorbed and illustrated a high sensitivity to DMMP vapors, this sensor demonstrated nonlinearity at a large concentration range of DMMP, which was different from the description based on the Sauerbrey equation [41]. This nonlinearity might be contributed to by the thickness effects of the sensing film because the QCM works on a balanced status between the sensitivity and linearity. In principle, the thicker thickness of ZnO-modified NW-structured MnO_2 film possesses much more effective surface adsorption area to adsorb DMMP; this makes the QCM sensor much more sensitive. However, this thicker thickness also leads to the static working status of the QCM sensor being different from that of the theoretical QCM, thus producing slight nonlinear effects on the Sauerbrey equation. More importantly, as described in Section 3.3, a thicker sensing film also induces nonlinear DMMP adsorption. These facts indicate that the QCM DMMP sensor shows more observable nonlinearity. Thus the thickness of the sensing film is one of the key factors affecting not only the sensitivity, but also the linearity of a sensor.

3.5. Response of the Sensor Towards Various Organic Vapors (Selectivity)

To investigate the selectivity, the QCM sensors were tested against several VOCs according to the instructions described above. The responses of the sensor exposed to a defined concentration of 0.7 ppm DMMP and potential interfering

VOCs including acetone, chloroform, *p*-DCB, ethanol, *n*-hexane and *p*-xylene were measured. Each of these VOCs is usually used as solvents and may act as potential interferences. The amplitudes of the responses to the target gas––DMMP vapor as well as control vapors at same concentration of 0.7 ppm were shown as in Figure 6. As we can see from the Figure, the response to the target DMMP vapor was much larger than those to the acetone, chloroform, *p*-DCB, ethanol, *n*-hexane and *p*-xylene vapors. Therefore, these findings indicated that the developed ZnO-modified MnO$_2$ NW-structured sensor possesses a very high selectivity to gaseous DMMP from VOCs including acetone, chloroform, *p*-DCB, ethanol, *n*-hexane and *p*-xylene.

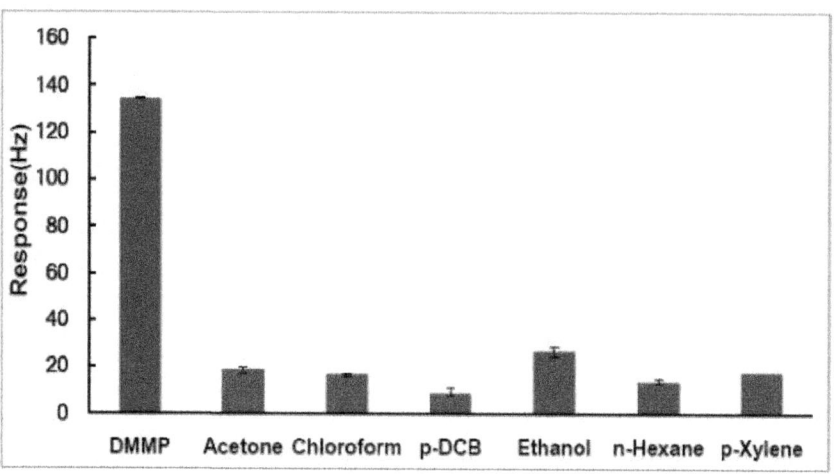

Figure 6. Comparison of the responses of the sensor to various organic vapors diluted to a predefined concentration of 0.7 ppm.

3.6. Comparison of the Sensitivity Between the Sensors Coated with ZnO-modified MnO$_2$ and Pure MnO$_2$

In order to assess the effect of the ZnO-modified MnO$_2$ NW-structured film on DMMP sensing, a reference QCM sensor with a same structure but pure MnO$_2$ NW film was fabricated for comparison. The responses of the reference QCM sensor to both the DMMP vapor and interfering VOCs including acetone, chloroform, *p*-DCB, ethanol, *n*-hexane and *p*-xylene at a same predefined concentration of 0.7 ppm were investigated according to the instructions described above. The response amplitudes of these paired QCM sensors were compared and are shown in Figure 7. The response of NW-structured pure MnO$_2$-based QCM sensor to DMMP vapor was not higher than those to the potentially interfering VOCs; one example is that the sensitivity amplitude to the DMMP vapor was even slightly smaller than the one to chloroform. Clearly, these facts indicate that the QCM sensor based on NW-structured pure MnO$_2$ did not present a usable selectivity to DMMP. In contrast, as shown in Figure 6, through the ZnO modification of MnO$_2$ NWs, the sensitivity of the MnO$_2$ NW film based QCM sensor to the DMMP vapor was greatly increased although that to the potentially interfering VOCs was almost unchanged. We thus contributed this improvement of the selectivity to DMMP to the ZnO modification on MnO_2 NWs, or the formed composite nanostructures shown as in Figure 1.

Understanding the interaction of phosphonate esters with the surfaces of ZnO-modified MnO_2 NWs at room temperature is a challenge; this problem is critical for the development of sensors to measure CWAs. However, very little is known about mechanism of DMMP specific binding to metal oxides although a few researchers have studied the details of DMMP absorbance. In this study, the MnO_2 nanowires-based gas sensor responded to the VOCs very well in comparison with most gas sensors, but as displayed in Figure 6, presenting an average to low selectivity. In this study, the nanocrystalline ZnO was utilized to modulate the strong catalytic activity of MnO_2, forming heterogeneous interfaces for VOCs testing. The heterogeneous interfaces formed possibly reduce the catalysis of MnO_2 to some VOC vapors, but increase that to other VOC vapors. These distinctive properties of this developed composite nanostructured material were evaluated through the QCM gas sensing. They displayed as the sensitivity and selectivity of QCM sensor to DMMP on device level.

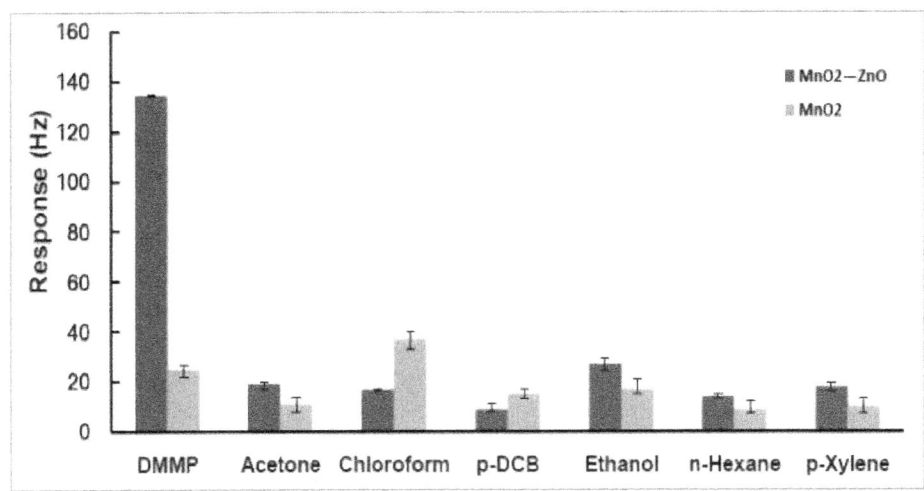

Figure 7. Comparison of the responses between the sensor based on pure MnO_2 nanowire film and ZnO-modified MnO_2 nanowire film to various organic vapor diluted to a predefined concentration of 0.7 ppm.

Interestingly, the experimental results indicated that the sensitivity and selectivity to the DMMP vapor was greatly increased whereas that to the VOC vapors was only slightly changed. We can thus contribute the improvement of the selectivity to the DMMP vapor to the modulation effects of the ZnO nanoparticles on the MnO_2 NW-structured film. Of course, further study would be helpful to understand the detailed mechanisms at molecular level, and furthermore to improve the sensitivity and selectivity. Mitchell *et al.* have explored the uptake mechanisms of DMMP to the effective sorbent and reaction material TiO_2 and considered that the DMMP molecule interacts through the electron-rich phosphoryl oxygen with surface-bound hydroxyl groups and with Lewis acid sites of the TiO_2 [31]. Similar studies will be valuable to understand the mechanisms of DMMP effective absorption to the surface of the ZnO-modified MnO_2 NWs at room temperature.

3.7. Long-term Stability

In order to assess the long-term stability, the response of the sensor to a predefined concentration of the DMMP in nitrogen gas, that is, 0.54 ppm, the average concentration of the linear range from 0.035 to 1.05 ppm, had been regularly examined for 10 days as the instructions described. After each testing, the sensor was immediately stored in a dry cabinet at room temperature. The response changes over ten-day period are shown in Figure 8. The results shown in Figure 8 indicated that the response was relative stable and remained above 90% of its original value after 10 days.

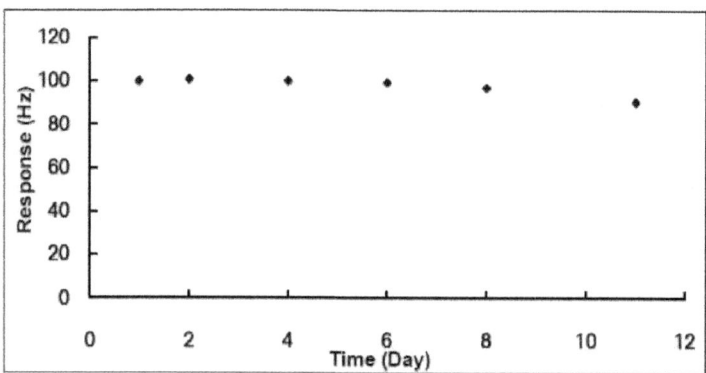

Figure 8. Responses of a nanowire ZnO_2/MnO_2 composite film based QCM sensor to a predefined concentration of 0.54 ppm DMMP in nitrogen gas during 10 days.

4. CONCLUSIONS

The preparation of highly active sensing film is believed to be a crucial step in obtaining sensitive and selective CWA sensors. For this purpose a composite nanostructured MnO_2 material was produced in this study by the ZnO modification of MnO_2 nanofibers.To evaluate its specific absorbing properties, both pure MnO_2 and ZnO-modified MnO_2 NW-structured films were prepared on QCMs for DMMP sensing. With the ZnO-modified device working at room temperature, a linear response to DMMP and a limit of detection of 35 ppb were obtained, even though the DMMP was diluted to a concentration lower than 1.05 ppm. The data presented in this study thus show that the ZnO modification induces the NW-structured MnO_2 film effectively absorb DMMP vapor, thus improved significantly the sensitivity and selectivity of sensors. Therefore, this work provides valuable data with CWA simulant supporting the development of new nanostructured material as a sensitive film for the nerve agent and insecticide detection. The composite nanostructured DMMP-sensing film combined with the simple and low-cost QCM detection provides a promising configuration to develop practical chemical warfare gas sensor.

ACKNOWLEDGEMENTS

This research is supported by the Natural Science Foundation of China (Grants No. 60874098 and 60911130129), the National High Technology Research and Development Program of China (863 Program 2007AA042103) and the National Creative Research Groups Science Foundation of China (NCRGSFC: 60421002).

REFERENCES

1. Munro, N.B.; Ambrose, K.R.; Watson, A.P. Toxicity of the organophosphate chemical warfare agents GA, GB, and VX: Implications for public protection. *Envir. Health Persp.* **1994**, *102*, 18-38.

2. Rudel, R.A.; Perovich, L.J. Endocrine disrupting chemicals in indoor and outdoor air. *Atmos. Envir.* **2009**, *43*, 170-181.

3. Wils, E.R.J.; Hulst, A.G. The use of thermospray-liquid chromatography/mass spectrometry for the verification of chemical warfare agents. *Fresenius J. Anal. Chem.* **1992**, *342*, 749-758.

4. Otake, T.; Yoshinaga, J.; Yanagisawa, Y. Analysis of organic esters of plasticizer in indoor air by GC-MS and GC-FPD. *Envir. Sci. Tech.* **2001**, *35*, 3099-3102.

5. Bartelt-Hunt, S.L.; Knappe, D.R.U.; Barlaz, M.A. A review of chemical warfare agent simulants for the study of environmental behavior. *Critical Rev. Envir. Sci. Tech.* **2008**, *38*, 112-136.

6. Brunol, E.; Berger, F.; Fromm, M.; Planade, R. Detection of dimethyl methylphosphonate (DMMP) by tin dioxide-based gas sensor: Response curve and understanding of the reactional mechanism. *Sens. Actuat. B* **2006**, *120*, 35-41.

7. Ying, Z.; Jiang, Y.; Du, X.; Xie, G.; Yu, J.; Wang, H. PVDF coated quartz crystal microbalance sensor for DMMP vapor detection. *Sens. Actuat. B* **2007**, *125*, 167-172.

8. Comini, E.; Faglia, G.; Sberveglieri, G.; Pan, Z.W.; Wang, Z.L. Stable and highly sensitive gas sensors based on semiconducting oxide nanobelts. *Appl. Phys. Lett.* **2002**, *81*, 1869-1871.

9. Zhang, J.; Hu, J.Q.; Zhu, F.R.; Gong, H.; O'Shea, S.J. ITO thin films coated quartz crystal microbalance as gas sensor for NO detection. *Sens. Actuat. B* **2002**, *87*, 159-167.

10. Tomchenko, A.A.; Harmer, G.P.; Marquis, B.T.; Allen, J.W. Semiconducting metal oxide sensor array for the selective detection of combustion gases. *Sens. Actuat. B* **2003**, *93*, 126-134.

11. Lee, D.; Kim, Y.; Huh, J.; Lee, D. Fabrication and characteristics of SnO_2 gas sensor array for volatile organic compounds recognition. *Thin Solid Films* **2002**, *416*, 271-278.

12. Chopra, S.; Pham, A. Carbon-nanotube-based resonant-circuit sensor for ammonia. *Appl. Phys. Lett.* **2002**, *80*, 4632-4634.

13. Modi, A.; Koratkar, N.; Lass, E.; Wei, B.; Ajayan, P. Miniaturized gas ionization sensors using carbon nanotubes. *Nature* **2003**, *424*, 171-174.

14. Kong, J.; Franklin, N.; Zhou, C.; Chapline, M.; Peng, S.; Cho, K.; Dai, H. Nanotube molecular wires as chemical sensors. *Science* **2000**, *287*, 622-625.

15. Matsuguchi, M.; Io, J.; Sugiyama, G.; Sakai, Y. Effect of NH_3 gas on the electrical conductivity of polyaniline blend films. *Syn. Metals* **2002**, *128*, 15-19.

16. Miyata, T.; Kawaguchi, S.; Ishii, M.; Minami, T. High sensitivity chlorine gas sensors using Cu-phthalocyanine thin films. *Thin Solid Films* **2003**, *425*, 255-259.

17. Rakow, N.; Suslick, K. A colorimetric sensor array for odour visualization. *Nature* **2000**, *406*, 710-712.

18. Huang, H.; Mao, S.; Feick, H.; Yan, H.; Wu, Y.; Kind, H.; Weber, E.; Russo, R.; Yang, P. Room-temperature ultraviolet nanowire nanolasers. *Science* **2001**, *292*, 1897-1899.

19. Ammundsen, B.; Paulsen, J. Novel lithium-ion cathode materials based on layered manganese oxides. *Adv. Mater.* **2001**, *13*, 943-956.

20. Giraldo, O.; Brock, S.; Willis, W.; Marquez, M.; Suib, S.; Ching, S. Manganese oxide thin films with fast ion-exchange properties. *J. Am. Chem. Soc.* **2000**, *122*, 9330-9331.

21. Wang, X.; Li, Y.; Synthesis and formation mechanism of manganese dioxide nanowires/ nanorods. *Chem. Eur. J.* **2003**, *9*, 300-306.

22. Perez-Lopez, O.W.; Farias, A.C.; Marcilio, N.R.; Bueno, J.M.C. The catalytic behavior of zinc oxide prepared from various precursors and by different methods. *Mater. Res. Bull.* **2005**, *40*, 2089-2099.

23. Berna, A. Metal oxide sensors for electronic noses and their application to food analysis. *Sensors* **2010**, *10*, 3882-3910.

24. Kanan, S.M.; El-Kadri, O.M.; Abu-Yousef, I.A.; Kanan, M.C. Semiconducting metal oxide based sensors for selective gas pollutant detection. *Sensors* **2009**, *9*, 8158-8196.

25. Michalkova, A.; Ilchenko, M.; Gorb, L.; Leszczynski, J. Theoretical study of the adsorption and decomposition of sarin on magnesium oxide. *J. Phys. Chem. B* **2004**, *108*, 5294-5303.

26. Aurian-Blajeni, B.; Boucher, M.M. Interaction of dimethyl methylphosphonate with metal oxides. *Langmuir* **1989**, *5*, 170-174.

27. Bermudez, V.M. Quantum-chemical study of the adsorption of DMMP and sarin on γ-Al_2O_3. *J. Phys. Chem. C* **2007**, *111*, 3719-3728.

28. Bermudez, V.M. Computational study of environmental effects in the adsorption of DMMP, sarin, and VX on γ-Al_2O_3: photolysis and surface hydroxylation. *J. Phys. Chem. C* **2009**, *113*, 1917-1930.

29. Trubitsyn, D.A.; Vorontsov, A.V. Experimental study of dimethyl methylphosphonate decomposition over anatase TiO_2. *J. Phys. Chem. B* **2005**, *109*, 21884-21892.

30. Rusu, C.N.; Yates, J.T., Jr. Adsorption and decomposition of dimethyl methylphosphonate on TiO_2. *J. Phys. Chem. B* **2000**, *104*, 12292-12298.

31. Mitchell, M.B.; Sheinker, V.N.; Mintz, E.A. Adsorption and decomposition of dimethyl methylphosphonate on metal oxides. *J. Phys. Chem. B* **1997**, *101*, 11192-11203.

32. Tesfai, T.M.; Sheinker, V.N.; Mitchell, M.B. Decomposition of dimethyl methylphosphonate (DMMP) on alumina-supported iron oxide. *J. Phy. Chem. B* **1998**, *102*, 7299-7302.

33. Mitchell, M.B.; Sheinker, V.N.; Cox, W.W., Jr. Room temperature reaction of ozone and dimethyl methylphosphonate (DMMP) on alumina-supported iron oxide. *J. Phys. Chem. C* **2007**, *111*, 9417-9426.

34. Paukku, Y.; Michalkova, A.; Leszczynski, J. Quantum-chemical comprehensive study of the organophosphorus compounds adsorption on zinc oxide surfaces. *J. Phys. Chem. C* **2009**, *113*, 1474-1485.

35. Oha, S.W.; Kima, Y.H.; Yoob, D.J.; Oha, S.M.; Parkb, S.J. Sensing behaviour of semconducting metal oxides for the detection of organophosphorus compounds. *Sens. Actuat. B* **1993**, *13*, 400-403.

36. Bermudez, V.M. Computational study of the adsorption of trichlorophosphate, dimethyl methylphosphonate, and sarin on amorphous SiO_2. *J Phys. Chem. C* **2007**, *111*, 9314-9323.

37. Segal, S.R.; Suib, S.L. Photoassisted decomposition of dimethyl methylphosphonate over amorphous manganese oxide, catalysts. *Chem. Mater.* **1999**, *11*, 1687-1695.

38. Kanan, S.M.; Lu, Z.X.; Tripp, C.P. A comparative study of the adsorption of chloro- and non-chloro-containing organophosphorus compounds on WO_3. *J. Phys. Chem. B* **2002**, *106*, 9576-9580.

39. Richards, R.; Li, W.; Decker, S.; Davidson, C.; Koper, O.; Zaikovski, V.; Volodin, A.; Rieker, T.; Klabunde, K.J. Consolidation of metal oxide nanocrystals: Reactive pellets with controllable pore structure that represent a new family of porous, inorganic materials. *J. Am. Chem. Soc.* **2000**, *122*, 4921-4925.

40. Yun, K-H, Yun, K-Y, Cha, G-Y, Lee, B. H.; Kim, J-C, Lee, D-D; Huh, J.S. Gas sensing characteristics of ZnO-doped SnO_2 sensors for simulants of the chemical agents. *Mater. Sci. Forum* **2005**, *486–487*, 9-12.

41. Sauerbrey, G. The use of quartz oscillators for weighing layers and for micro-weighing. *Z. Phys.* **1959**, *155*, 206-222.

42. Li, G.; Zheng, J.; Ma, X.; Sun, Y.; Fu, J.; Wu, G. Development of QCM trimethylamine sensor based on water soluble polyaniline. *Sensors* **2007**, *7*, 2378-2388.

43. Zheng, J.; Li, G.; Ma, X.; Wang, Y.; Wu, G.; Cheng, Y. Polyaniline-TiO_2 nano-composite-based trimethylamine QCM sensor and its thermal behavior studies. *Sens. Actuat. B* **2008**, *133*, 374-380.

44. Grate, J.W.; Kaganove, S.N.; Bhethanabotla, V.R. Comparisons of polymer/gas partition coefficients calculated from responses of thickness shear mode and surface acoustic wave vapor sensors. *Anal. Chem.* **1998**, *70*, 199-203.

45. Martin, S.J.; Frye, G.C.; Senturia, S.D. Dynamics and response of polymer-coated surface-acoustic-wave devices––Effect of viscoelastic properties and film resonance. *Anal. Chem.* **1994**, *66*, 2201-2219.

46. Lucklum, R.; Behling, C.; Hauptmann, P. Role of mass accumulation and viscoelastic film properties for the response of acoustic-wave based chemical sensors. *Anal. Chem.* **1999**, *71*, 2488-2496.

47. Bodenhofer, K.; Hierlemann, A.; Noetzel, G.; Weimar, U.; Gopel, W. Performances of mass-sensitive devices for gas sensing: Thickness shear mode and surface acoustic wave transducers. *Anal. Chem.* **1996**, *68*, 2210-2218.

This page left intentionally blank.